PRAISE FOR

Will Bonsall's Essential Guide to Radical, Self-Reliant Gardening

"If you wish to live well and eat well no matter what is going on in the rest of the world, this book is for you. Thresh your own grain and press your own oil. Can't buy seeds, no problem. Can't buy fertilizer, no problem. Will Bonsall will help you enjoy the good life under any and all conditions."

—ELIOT COLEMAN, author of *The New Organic Grower* and *The Winter Harvest Handbook*

"Every gardener and small farmer can benefit from Will Bonsall's decades of focused, quality experience. Will's book is *one of the key practical resources* you should read—as you reach for *full* sustainable soil fertility in your garden or farm!"

—JOHN JEAVONS, author, executive director of Ecology Action, and developer of sustainable, biointensive mini-farming

"Will Bonsall—Mr. Scatterseed himself—has done it all, and this book covers it all, from maintaining soil fertility with minimal external inputs to growing annual and perennial vegetables, fruit, nuts, grain, beans, and even oilseed crops. Will's methods are all vegan-based and garden scale, with little resort to tools beyond hand tools and a rototiller and shredder. His description of making oil-seed seed meals and cooking with them is particularly interesting.

Will's book is a great introduction to gardening for the beginner, and it also offers enough brand-new original material to delight even the most expert. Best of all, the interweaving of Will's coherent personal philosophy, decades of gardening experience, down-to-earth style, and touches of humor all make for an interesting, entertaining read."

—CAROL DEPPE, author of *The Tao of Vegetable Gardening* and *The Resilient Gardener*

"Here is a bright star in the constellation of voices for land-based sustainability. Not only is Will Bonsall incredibly learned—the result of decades of careful studies in the field and out—he is bawdy and brave and bold. His credibility is a PhD in homesteading, and his rambunctious wisdom is very worth reading. If you want to learn from a master, you need this book."

—JANISSE RAY, author of *The Seed Underground*

WILL BONSALL'S
ESSENTIAL GUIDE TO
Radical, Self-Reliant Gardening

WILL BONSALL'S
ESSENTIAL GUIDE TO
Radical, Self-Reliant Gardening

Innovative Techniques for Growing Vegetables, Grains, and
Perennial Food Crops with Minimal Fossil Fuel and Animal Inputs

— WILL BONSALL —

Chelsea Green Publishing
White River Junction, Vermont

Developmental Editor: Fern Bradley
Copy Editor: Laura Jorstad
Proofreader: Brianne Bardusch
Indexer: Shana Milkie
Designer: Melissa Jacobson

Printed in the United States of America.
First printing May, 2015.
10 9 8 7 6 5   22 23 24 25

Our Commitment to Green Publishing

Chelsea Green sees publishing as a tool for cultural change and ecological stewardship. We strive to align our book manufacturing practices with our editorial mission and to reduce the impact of our business enterprise in the environment. We print our books and catalogs on chlorine-free recycled paper, using vegetable-based inks whenever possible. This book may cost slightly more because it was printed on paper from responsibly managed forests, and we hope you'll agree that it's worth it. *Will Bonsall's Essential Guide to Radical, Self-Reliant Gardening* was printed on paper supplied by Versa Press that is certified by the Forest Stewardship Council.

Library of Congress Cataloging-in-Publication Data

Bonsall, Will, 1949– author.
 Will Bonsall's essential guide to radical, self-reliant gardening : innovative techniques for growing vegetables, grains, and perennial food crops with minimal fossil fuel and animal inputs / Will Bonsall.
    pages cm
 Other title: Essential guide to radical, self-reliant gardening
 Includes index.
 ISBN 978-1-60358-442-5 (pbk.) — ISBN 978-1-60358-443-2 (ebook)
1. Sustainable agriculture. 2. Self-reliant living. I. Title. II. Title: Essential guide to radical, self-reliant gardening.

S494.5.S86B68 2015
635—dc23
            2015002105

Chelsea Green Publishing
85 North Main Street, Suite 120
White River Junction, VT 05001
(802) 295-6300
www.chelseagreen.com

—CONTENTS—

PART II. The Seed

PART III. The Crops

PART IV. The Garden in Context

—ACKNOWLEDGMENTS—

The idea for this book had smouldered in my mind for many years, but it didn't ignite until 2011, when Chelsea Green editor Makenna Goodman approached me following a grain-growing conference at the University of Vermont and asked me for assistance with someone else's book ("maybe a sidebar or something"). In the course of the conversation she asked if I'd be willing to write a book myself. About what? She first suggested garden-scale grain growing, since that was the topic at hand, but then she also associated me with seed saving, so perhaps I might work that in, too? But then she also knew that I had a system of growing stuff without livestock manure, and maybe I could stick that in somewhere? I agreed and offered a few other ideas. She jumped at all of them, and I began to suspect that if I suggested a unit on marriage counseling or neurosurgery, it might also be accepted. We ended up with a plan for the very book I had always wanted to write: a far-reaching combination of broad philosophical discourse interspersed with specific details. Makenna was to be my editor, but then Theo got in the way. He's the newborn baby who made her go on maternity leave at just the wrong (for me) moment. My manuscript was passed along to Fern Marshall Bradley, with Ben Watson serving as project manager and shepherding things along. With towering patience Fern walked me through the process, gladly taking on tasks that should have been the author's responsibility. Her suggestions for developing certain thoughts and expanding some topics were a great improvement on the original and would have caused the demise of many more spruce trees were it not for Chelsea Green's emphasis on using paper with recycled fibre. Her gentle restraint in deleting some of my rowdier humour will probably keep the book from being banned in Boston. In the final push to meet the deadline, she was always focused and good-humoured during many long phone conversations with me to resolve innumerable details; there's no thanks that covers that.

At Chelsea Green Pati Stone was exceedingly generous with her expertise, going beyond what's reasonable in helping to organize all the graphic images, improve my mediocre shots, and put it all into place. Thanks also to Margo Baldwin and the others at CGP for having confidence in me and providing the opportunity to make this book a reality. Without them it would have continued as an unfulfilled fantasy.

Other than CGP I am beholden to a long line of apprentices and hired workers who have enabled me to experiment with a wide array of crops and practices.

I cannot but be mindful of three long-gone friends who contributed to my vision in a different way: my neighbour Lucian Keniston and his good friends (and later mine) Orlando Small and Edgar Keyes. These old-time Yankee gardeners loved working the land in their traditional manner, yet each understood that there were some flaws in the old ways. They understood sustainability better than many movement people today, but couldn't quite see how to improve on those traditional methods, and they had difficulty grasping the concept of eco-efficiency (which I myself was only beginning to formulate). Their knowledge and experience and our many discussions helped me to develop the ideas contained in this book.

To all these and many others I am profoundly grateful.

The Vision: Beyond the Boundary

When you look at the borders of your garden, I assure you that you're seeing a mirage. There *are* no borders around your garden. Your garden is more like a pebble dropped in the water—what you perceive as a boundary is just the innermost ripple; countless other ripples spread out from there and bounce back from afar. This must be so, else how can we call a garden organic? I mean *organic* as in *organism*: a unified living system whose parts function to the benefit of one another and the whole. Sounds grand, eh? Well, it *is* grand, and minuscule at the same time. For me—and, I believe, for everyone else—that's the challenge: to focus on the minute details while constantly remaining aware of the big picture. It's what motivates me to farm, indeed to live, as I do. I'm always peering beyond the apparent edge of my garden, to see how I fit in with the cosmos. And it is what motivates me to write this.

Some of what I do in my garden is radical, but much of it is ordinary. The radical essence of my garden is the visionary context underlying it, a set of assumptions that inform what I do and don't do there. Our assumptions always arise from the scenarios we envision (consciously or not), how we perceive our present, and what we expect in the future. Here are some of my scenarios.

Coping with the present. Times are tough, at least for most of us, always have been for me. However, I've usually been able to mitigate the lack of income by a lack of outgo, producing a substantial part of my own food and fuel needs and doing without those things I considered superfluous (in more recent years that strategy has been challenged by a growing family with their *own* ideas of what is superfluous). Hence my emphasis on personal self-reliance; I can't afford to do otherwise. I must often make do with what I already have at hand. This approach is self-serving; it doesn't do much to address the larger problems of the world, but it's fine as far as it goes.

Surviving a future catastrophe. (Remember Y2K?) I call this the *Titanic* Scenario: Most of us passengers can see the iceberg ahead, but no one can seem to change the ship's course in time. A self-reliant lifestyle might seem like a viable lifeboat, but it's probably useful only as a short-term strategy. Reliance on self-reliance presumes that you're so fully detached from the ship that you will not be sucked down with it (but it's a *very* big ship, and the suction will be incredible). It also presumes that you have an ample supply of every necessity and that other survivors (who will all be in the same boat) will let you keep it to yourself. (Should you even want to, considering that you will all need one another's help.) I want to note that the *Titanic*'s lifeboats never even tried to row to Newfoundland— that was out of the question. Rather, they sat tight until another ship could save them. What if there had been no other ship?

No, by far the only realistic approach is to change course *before* it's too late. Mind, we're already scraping the berg, but we're not sure whether the hull is broken through and beyond repair. The captain and crew may not be open to our suggestions (don't waste time blaming them, we insisted they get us to New York in record time!), but we must convince them (and one another). It's time for lots of dialogue, and fast! It couldn't hurt if we knew of an alternative course, a path through the pack ice that we had already paddled ourselves, something that might give the helmsman some ideas. That of course brings us to a third scenario.

Transforming civilization while we still can. We can vote at the ballot box, we can vote at the checkout counter, we can teach a new vision. Although it is terribly frustrating to try to influence the juggernaut that is modern civilization, we have many allies. And while we must tend to our own little patch (*il faut cultiver mon jardin*), we can keep noticing that our garden has no borderlines.

Since we can't be sure exactly which scenario is most likely, perhaps we should act on all of them; they are by no means mutually exclusive. As someone has said: "What if all this global warming stuff is a lot of bunk, and all we do is make the world a nicer place?"

Distortions of the Big Picture

As a teenager I began my working career in the mining business, prospecting for copper, lead, zinc, silver, and the like. This was before recycling was such a watchword, but even then it occurred to me that aside from all the environmental havoc wreaked by the extraction and smelting of all that metal, much of it would wind up in dumps in the form of discarded motors, batteries, and appliances, where it would continue to poison our world. If all that stuff could be salvaged and reused over and over, how much mining would humanity need anyway?

Glad to be out of that business I eventually found my way into organic farming. I met Robert Rodale, who lived just over the hill from the Pennsylvania farm where I visited on school breaks; his wife showed me around the farm, including the Sir Albert Howard plots. It made lots of sense, as far as it went, but it still seemed to lack something: a wider perspective. Removing the chemicals was great, and turning "wastes" into food was a no-brainer. But no one in the organic movement was talking about the more fundamental questions: Where does this organic stuff come from, why isn't it needed there, and where does it wind up? And all this mining of lime and greensand and Sul-Po-Mag, to be shipped hundreds or thousands of miles—hadn't I left that business behind?

When I finished college and went into homesteading (it was 1971 at the time), I had the great good fortune of not getting into market gardening. I tried, but I wasn't very good at it—I don't mean growing the food, I mean selling it. I quit rather easily when I discovered how much more profitable it was to eat my crops rather than sell them and do something else to earn money. I've never been too good at that, either, but my focus on spending less rather than earning more had some unforeseen advantages: It cleared my vision. I gained a better understanding of the economy of the land than does someone who buys in soil amendments and ships them off in the form of farm products. I'm not suggesting it's wrong to grow food to sell; I'm warning that in the marketplace our calculations are skewed by externalities—hidden costs that no one recognizes or deals with. For example, even though meat and other animal-derived food products are a very inefficient use of land, the public craves them enough to pay a price in dollars that obscures the inefficiency. Things that make sense economically may not make sense ecologically and vice versa.

Fortunately, I couldn't afford the country lifestyle magazines, which portrayed a vision that was more fringe culture than actually counterculture, even before it became gentrified. I quickly gave up on livestock after discovering that they were an obstacle, not an avenue, to self-reliance. The effort to grow their feed, instead of merely buying it in, was a real eye-opener.

Perhaps the greatest distortion of our thinking about sustainability is due to the elephant in the room, which almost no one mentions though everyone stumbles around it. That elephant is the sea of petroleum that washes over our entire civilization, including agriculture. If the price of petroleum included the externalities, or hidden costs, the flaws in our global food system would become starkly clear.

These two distortions—the marketplace and the sea of petroleum—are barriers to sustainable agriculture, indeed to sustainable civilization. The marketplace, in addition to distorting values, is the second, if not the first, greatest form of soil erosion. Even if the farmer avoids any nutrient loss through runoff or leaching, a vast amount of tilth-building substance is lost when a crop is sent to market. In bygone days that tilth (in the form of night soil) would have found its way back to the farmer's fields, creating a fairly tight loop. Today of course that fertility (waste) ends up as sewage in landfills or the ocean, or as turf builder for golf courses, or in places other than the land that produced it and still needs it. It robs the ecosystem whence it came and poisons the place where it is dumped.

Petroleum use is obviously unsustainable because the source is finite and non-renewable. New oil-field discoveries and extraction technologies merely postpone the day of reckoning, while releasing yet more dinosaur farts into the atmosphere. Even most "recycling" of plastic isn't really recycling; rather, it is salvage. Reprocessing used milk jugs into parka filling and park benches is a onetime linear extension of the use chain—it's a *chain*, not a cycle. A use chain ends with a waste product; a cycle, such as aluminum recycling, can go around and around endlessly.

One way I avoid the perils of the marketplace is to be as self-sufficient as possible for food. The few things I grow to sell are niche crops, such as seeds and nursery stock (hardy kiwis and hazelnuts). These seem to return a lot in terms of cash income in proportion to what they cost the soil—that's because rather than selling biomass, I'm selling information in the form of DNA. All the same I do have to replenish what they remove from the soil. Their

production is not a sustainable cycle, so my priority is the subsistence crops.

My region is hardly a banana belt—even with global warming, I can rely on barely 120 frost-free days and not very hot days at that. Nevertheless I can grow and eat a wide assortment of plant foods. Though "exotic" foods like peanuts, bananas, and citrus are occasional treats for me, I find a vegan locavore diet to be not very constraining at all. There are so many food plants that people could grow here in the North, but don't, that most animal-based store-bought diets look rather monotonous by comparison. Aside from the fact that I enjoy a varied diet, an added benefit is that I'm not ruled by the marketplace: So many market gardeners don't make room in their plantings for unusual food crops because "it won't sell," and therefore they won't even bother to grow any for their own use. My goal is not to feed the world, but to feed myself and let others feed themselves. If we all did that it might be a good beginning.

I am sometimes asked whether I am a gardener or a farmer, and I have to ask myself, what are the criteria for those? Farmers have many acres; I own many acres, but only farm a couple of them. Farmers sell their crops for income; I sell very little of my crop, and it is not a very significant part of my income, although it certainly slashes our grocery bill. Indeed over the years my greatest farm income by far has been from talking and writing about it; therefore my main cash crop has been information. Of course, gardeners grow vegetables (mainly tomatoes), not staple crops like wheat and soybeans and oilseeds and sugar beets. Farmers do grow those things, but by the ton, not in single-digit bushels. Ah, but farmers usually keep livestock; yes, but gardeners usually buy in their fertilizer and seed. I give up; you'll have to decide what I am.

Since I'm trying to minimize food purchases in general, I don't limit my gardening to vegetables. There may be more profit in mesclun mixes or salsa ingredients, but they do not mean self-reliance. I'm more concerned with staple foods—grains and pulses and oilseeds, as well as the usual (and often unusual) greens and root crops. Although my preferred diet consists of those plant foods that I can grow to maturity here in western Maine, no one should feel sorry for me. When people hear that my family and I do not eat meat, milk, or eggs, they often ask: "So what *do* you eat?" Silly geese, little do they know that the diversity of tastes and textures found in plant foods dwarfs the paltry range of flavours found in animal foods (which even as a meat eater I found rather boring). I'm aware that because I eat so largely from my own soil (although it comprises a great assortment of glacier-borne igneous rocks), I run some risk of deficiencies in obscure nutrients found in the global foodshed. Therefore I keep my diet as varied as possible (plus I like it that way). By the way I feel extremely healthy for my 65 years, thank you; my doctor confirms that.

I try to keep my use of petroleum products, especially fuel, to a minimum. My tractor hasn't run for years, and anyway it's meant for work in the woods. Most of my cropland has not been ploughed for two or three decades. Some areas see occasional rototiller use, by no means every year, usually for incorporating grain stubble and heavy green manures, and not always then. My walking tractor has a mower attachment and a chipper/shredder. The latter sees frequent action, as I'll discuss later, yet its consumption of gas per unit of work done (cost–benefit ratio) is very moderate. I use no plastic mulches. Add in my chain saw and weed whacker (which others use; I prefer my scythe for tight places) and that pretty well accounts for the petroleum use in our farming system. Although it is proportionately small I am always mindful that *no* amount of petroleum is sustainable, since to my knowledge none is currently being created. My ultimate goal is to have done with it altogether, about which I'll have more to say later.

Of course, no amount of gnat straining and camel swallowing will change the fact that our biggest use of petroleum is in moving stuff (including people) around, a huge argument for locavorism and anti-consumerism. In general I prefer to avoid going anywhere or buying anything, and when I must I'm careful to factor those costs into my decisions.

Whence Fertility?

For a very long time the US Department of Agriculture (USDA) couldn't have cared less about "organic." According to Earl Butz (secretary of agriculture during the 1970s), it was a recipe for global starvation (which his system was preventing?). Only when the marketplace began clamouring for organic commodity crops (an oxymoron?) and when US farmers began to recognize the enormous potential for organic exports (yet another one?) did the USDA decide it needed to weigh in. Not to promote organic, mind you—the marketplace was way ahead of them there—but to concoct some watered-down standards (including GMOs!) that would pave the way for imperialistic agribiz. A useful side effect of those USDA attacks on the integrity of organic principles has been to make the movement define itself more precisely and to engender a lot of soul searching about our values. A less useful side effect has been a tendency to separate "local" from "organic" and to pit them against each other. I want to address that.

Lately, I've noticed some bumper stickers that say: TO HELL WITH ORGANIC—BUY LOCAL! It saddens me because it shows an erroneous assumption that the two concepts can be somehow separated. They cannot. When someone in my community brings to market "locally grown" food that has been grown with applications of chemical fertilizer shipped in from hundreds or thousands of miles away, how local is that food? Conversely, when winter lettuce is shipped in from certified-organic factory farms located in California or Chile, how organic is that lettuce? Truly local and genuine organic are inseparable— you can't sell out one for the other.

Fortunately for me personally, defining organic is barely relevant, since I have no need to certify what I myself eat. I hew to a line that is more strictly organic than the standards set by USDA or any other certifier. What I *do* call my garden is eco-efficient, because that reflects what is important to me—not to the USDA.

There is a movement called the "veganic" (vegan-organic) movement, composed mainly of animal rights advocates, that eschews the use of manure and other animal products, largely because of the cruelty associated with livestock raising and slaughter. I don't use that stuff, either, but for somewhat different reasons, which go beyond compassion. I'm into the compassion thing, too, though it doesn't stop me from swatting blackflies and squishing potato bugs. My big problem with moo-poo and the like is that it is inefficient, unbusinesslike really. It is unbusinesslike to view manure (cow, horse, chicken, yours) as a *source* of fertility, because it isn't. Remember way back in high school biology, that textbook chart that showed the Food Energy Pyramid (soil fertility is another form of food energy)?

In the bottom layer—Producers—the only organisms shown were plants; everybody above the base layer (including earthworms, gerbils, and us) was given the undignified label of Consumer. Oh, there were different "Orders," or hierarchic labels to make you feel better or worse—bigger numbers carry some sort of stigma—but the idea conveyed was that we critters were not doing anything useful (at least not compared with, say, ragweed). In fact we are serving a good purpose, but we are *not* producing food energy and we're *not* producing fertility. No one denies that a cow flop contains lots of fertility, but the *cow* did not produce that fertility, the *grass* she ate created it out of dirt and rain and air and sunshine. (How cool is that?) The cow, bless her, merely collected that energy and concentrated it and

moved it around, destroying a great deal of it in the processes of living. We mustn't blame the cow any more than ourselves for living, but neither should we look upon the cow as a *net* source of fertility; that's just kidding ourselves.

If you really feel a need to keep animals for their meat, milk, eggs, et cetera, I say go for it. And of course you will then have a supply of manure—kinda hard to avoid—and of course you will use it on the land, foolish not to. But if you find, as I do, that plant foods are more easily produced and satisfying, you will have no further use for livestock and thus no manure. No loss, because what is more businesslike is to use that grass for compost or mulch, which will put the nutrients back in the soil directly.

Although the veganic model looks mainly at grass for its fertility, there is another model, or vision, that takes an even deeper look into the "business" of soil maintenance. *"Eco-efficiency"* (a word I coined, only to discover that somebody else already had, fortunately with a similar meaning) describes the ratio between an organism's intrinsic food energy and the food energy (or soil fertility) required to produce it in the first place. As the Food Energy Pyramid showed us, only the plants with their photosynthetic chlorophyll have a positive ratio; the rest of us—a bunch of losers really—have a negative ratio. But just as different animals vary in their eco-efficiency (pigs have a better feed-to-weight-gain ratio than horses), plants

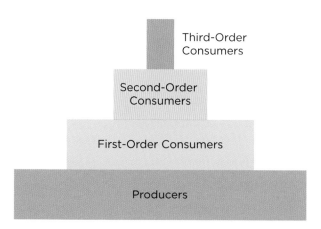

Figure I.1. This is the traditional textbook graphic for explaining the Food Energy Pyramid.

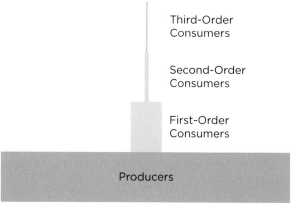

Figure I.2. Here's a more accurate image, based on known eco-efficiencies.

likewise vary hugely in their ability to transform little into much. Let's look at some examples, starting with the typical vegetable garden.

Most of the vegetables we like to grow and eat are basket cases when it comes to soil building. For example, if you sow a plot to leeks or pumpkins but harvest nothing and instead turn under the crop at maturity—the whole works: roots, leaves, fruits, seeds, eating nothing—would the soil be enriched thereby? No doubt, though not in good proportion to the fertility it took to produce the crop. That's probably why seed catalogues never have leeks or pumpkins listed in their green-manure sections. Rather they suggest more vigourous grasses like rye and oats or pasture legumes such as clover, vetch, or alfalfa. Those are far more eco-efficient.

The problem with most veggie plants as soil-builders is that they are "Band-Aid species," by which I mean that they are nature's quick-fix short-term remedy for disturbed soil. They don't need to be particularly eco-efficient; they just have to sprout and spread quickly to cover and protect the soil. In a natural system they will soon be replaced by perennial weeds and grasses, then in turn by woody shrubs, eventually succeeding to forest. The problem is that humans cannot eat grasses or woody shrubs or forest trees. Our tummies have adapted to those Band-Aid species—the succulent, easily digested weeds from which our vegetables have evolved. They're all well and good in our pampered gardens and on our plates, but for powering a dynamic soil community, we had better look elsewhere.

What about those perennial weeds and grasses that make up our hay fields? They are certainly no wimps at creating much from little: Consider that they generate a huge surplus of biomass that is removed in the form of stuff from cows, with very little returned for the plants themselves. Yes indeed, grassland is a very eco-efficient ecosystem; one British veganic writer described it as the ultimate. Understandable error, I suppose, since they have lots of pastures over there and very little of something my fellow Mainers take for granted: forests. You see, forests (especially old-growth hardwoods) are even more eco-efficient than grasslands. How do I know? For one thing, if you look around the planet, most climax (self-perpetuating) ecosystems are forests. The exceptions, places like Kenya, Mongolia, Patagonia, and Wyoming, are climax grasslands—prairies—because they lack one essential: water. Add rainfall to those regions and they, too, will begin the succession to forest.

It isn't hard to see why forests are so eco-efficient: Grassland sod is only inches deep, and the top growth is rarely over 6 to 8 feet (1.8–2.4 m). Every year it dies back to the sod, which must reinvest itself the following spring. The main component, grass, has most of its "solar panels"—long narrow blades—arrayed in a rather awkward manner for sun catching, which is why the broadleaf dicots can live so well among the grasses, grabbing the escaped light. In contrast, forest trees don't die back. They merely shed their leaves and go dormant; their incremental gain (as trunks and boughs) is retained year after year. Although the larger supporting tree roots are within the top few feet, I have found feeder roots at the bottom of a 15-foot (4.6 m) well, pumping up minerals that would otherwise never be part of the biosphere. Most hardwood trees grow to 80 feet (24 m) or more and have many layers of chlorophyll-laden leaves, all arranged to trap a maximum of sunlight. Clearly forests are nature's most eco-efficient ecosystem—if only we could eat directly from them! Well, perhaps we can a little; some archaeologist calculated that in all human history more folks have been nourished by acorns and chestnuts than all other foods combined. I've been experimenting with those and other tree crops for years, and I hope to make them a much bigger part of my diet. However, I do not see them supplanting wheat, corn, potatoes, et cetera, and therefore I ask myself: Is there also some way I can eat from the forest *indirectly*, by using the residues—leaves, brush, and the like—as soil-building materials for my tilled cropland? Can I parlay stuff that's deemed of little fertility value (yet there's *so much* of it) into the foods I love to eat? Can I spin straw into gold? I can, and so can you; in chapter 12 we'll discuss how.

How to Use This Book

I apologize for dragging you through so much abstract explanation, but I feared that otherwise you might try out some of those ideas that serve my purpose but perhaps not your own. For example, if your goal is simply to make a good living growing and selling stuff from your land, then much of what I'm saying will seem altogether pointless. Society does not generally expect its farmers to be visionaries; indeed, society really needs farmers to keep their sights within a finite horizon and not question the dictums of policy shapers. I cannot show you the borders of your garden; I can only try to point out that there aren't any. And so, I now propose to share with you a whole slew of ideas from which you can pick and choose those you feel will serve your interests, regardless of whether you share my ideology.

Most of what I write about draws from my own direct experience over 40 years of growing and eating my own food. But I also share a lot of hearsay, which is important but risky. What do I mean by hearsay? It's things I've heard said, or have read about, but never directly experienced with my own eyes or hands. For example, when I talk about nitrogen, understand that I've never actually *seen* any nitrogen, but I've "heard say" that there is such stuff and that the air we breathe is laced with it, so who am I to quibble? Just please don't go telling others that Will Bonsall saw some nitrogen somewhere—that would be misleading. It's hearsay and I share it for what it's worth. Also because I've noticed that if you only talk about your own personal experience without sticking in a bit of hearsay, then folks just don't take you seriously.

My farm and the gardens therein are not linear, and it's impossible to describe them and my way of cultivating them in a linear fashion. And so this book tends to cycle back around to key ideas and principles, with specific ideas and practices running sometimes haphazardly from one part of the book to the other. I begin with what seems most fundamental to my way of gardening, and that is building soil fertility (part 1 of the book). From there, in part 2 I spend time telling you how I grow and save seeds and propagate plants, because without employing those techniques, gardening cannot be a truly sustainable cycle. Part 3 is my thoughts about growing particular crops—not just veggies but also grains, pulses, oilseeds, and permacrops.

In part 4 I take those first three parts and explain how the garden is integrated into the context of land, rocks, and water; how plants are intermingled in gardens to make the most efficient use of space and fertility; and how I cope with the other living creatures—the insects and bugs, the animals, and the disease organisms—that may want to feed on those crops. And finally, in part 5 I share a hodgepodge of techniques for processing crops into flour, meal, juice, puree, sauce, kraut, and much more. Better still, you'll find a few of my favourite ways of preparing foods for the table, including some wholly new food items.

A Place Called Esperia

It is difficult to discuss truly sustainable agriculture outside the context of a sustainable civilization. The ideas behind my or any other food system make certain assumptions regarding the society it feeds. For example: Is there a marketplace, and what is it like? How does it work? And what industrial and technological inputs are available? Are those all sustainable? While these questions demand far more answers than I have to offer, they may at least clarify our vision beyond the immediate borders of our garden, and perhaps serve as a scorecard for judging our progress.

Let me digress and tell you about a place called Esperia. It is neither here nor now, but rather a fictional place described in a novel I wrote a few years ago called *Through the Eyes of a Stranger*. I mention it here only because it offers a paradigm, a framework for discussion. Here and there in this book, I pose the question, *How do you suppose they do that in Esperia?* Understanding the principles embodied in my mythic Esperia can help you understand the question.

Esperia (several centuries in our future) is a sustainable society, in that it is based on endless cycles and ever-renewable resources. Fossil fuels and petrochemicals have no place there. There is a strong focus on eco-efficiency, for many reasons, but largely to maintain a viable population level without excessive impact on the land base; hence there's a preference for permacrops, especially trees, and a lack of livestock.

Out of a desire for stability and security, Esperia is governed by the Vine Laws, which mandate a high level of self-reliance at various levels of society; for example, each household is responsible for producing its own staple foods and domestic fuel. Even tool and machinery production is as locally based as possible.

If this sounds totally alien, perhaps it is because the civilization we have today is so unsustainable and so insecure that we have difficulty envisioning anything else. The fact that there is no such place as Esperia (and never has been) is only because we haven't built it yet. I believe, though, that we can and must do so if we are going to continue this experiment we call civilization. The marketplace could actually be a useful vehicle for making this happen, but only as long as we learn to control it instead of letting it control us. In the meantime Esperia might serve as a paradigm, or a frame of reference, giving us a canvas on which to paint our vision, if not the vision itself.

If I were to suggest that we are headed for some disastrous times when our access to the essentials of life will be disrupted by unforeseeable events, most people will not challenge that, beyond disagreeing over details of when, how, and so on. From religious fanatics to money market managers, there is a general sense of malaise that's hard to ignore, though most of us tend to avoid taking any action beyond praying, buying gold, and stocking up on peanut butter and ammo. Some such Great Crisis (Esperians refer to it as the Calamitous Times) must be factored into our vision of a garden-without-borders, if we are to come up with solutions that are anything more than personal stopgaps. But how to prepare for a crisis when we cannot predict its time or the scale of it? Perhaps by at least preparing for those scenarios, we *can* predict. For example, we can assume that supplies of petroleum-based fuels and materials may become scarce or prohibitively expensive. We can assume a much more challenging climate, with generally higher temperatures and erratic weather extremes. Even basic tools and seeds may become hard to obtain as our manufacturing and transport infrastructure are impaired or decay.

Of course, any strictly personal solutions only guarantee that you will become a target for those who have not developed any of their own. Thus the most stable and reliable security measures will be those involving cooperative, collective, community action. Those solutions also give some efficiency of scale and specialization. In short, while it is imperative that we all develop some degree of self-reliance, we must also look beyond any particular strategy for self-reliance. We must consider all possible solutions and more: We must be adaptable and resilient. That's a lot to ask.

Of course, the Esperia in my novel is my own particular Esperia, just one vision of a Land of Hope, and it includes details that are uniquely mine and may not be essential to a sustainable civilization. That Esperia, like my garden-without-borders, is not predicated on utopian perfection; its members are not saintlike and the children are *not* all above average. It has flaws like any other society, but it has one asset that compensates for every other wart and pimple, and ultimately trumps every achievement of our own civilization: It is sustainable. Without sustainability all other accomplishments—including democracy, arts and culture, technology, and civil rights—are superficial and transient. However important those may be it is not enough merely to rearrange the deck furniture on the *Titanic*; we have to change course.

PART I
Soil Fertility

Not all soils are naturally fertile, and even those that are must be *kept* fertile. As soon as we break the natural vegetative cover of the soil by clearing and ploughing, the soil ceases to be able to maintain its own fertility, and we must assume responsibility for maintaining it henceforth. In fact we do it very poorly; our best efforts can only hope to slow the rate of humus burnout caused by our excessive aeration of the soil. We must bring in tilth-building stuff from elsewhere, but from where and at what cost? The bottom line is that some imports make more sense than others, and how we balance those efficiencies says much about the long-term sustainability of our agriculture and our civilization.

CHAPTER ONE

Composting as if It Mattered

I make compost and lots of it, and not just because it's something hippie homesteaders are expected to do, but because I get a kick out of doing it. No, really. In my lifetime (so far, that is) I've made easily 200 tons (181 tonnes) of compost. (Well, not easily; it was a bodacious amount of work.) I still think it's more fun than a barrel of monkeys to take a mess of useless stuff and turn it into a valuable product.

Compost is not my exclusive source of fertility; I also use green manures on some areas, and last year's layer of decayed mulch is a significant source of nutrients; often it is enough by itself. However, the compost is a biggie, especially on certain heavy-feeding crops. It is crucial that I make enough compost and that the quality be good. And making good-quality compost requires more thought and effort than simply making a pile and letting compost happen.

I've known very few gardeners whose compost making furnishes most of their crops' nutrient needs. (How far can grapefruit rinds and coffee grounds go toward building up the soil?) Typically, gardeners say that their compost is a source of humus, a source of trace minerals, or a bioactivator. They rely on hauled-in animal manure or purchased lime and other mined minerals to do the heavy lifting. This is where my garden-without-borders is different: My compost, in conjunction with green manure rotations and mulch, is intended to supply *all* the needs of my crops and the soil in which they live—the humus, the

Figure 1.1. Truly sustainable fertility arises from the very land it nourishes. This compost was made from grass, leaves, crop residues, and kitchen garbage, all sourced from the farm itself.

Figure 1.2. For easier access to my compost bins while I fill them, I remove the poles or planks from the front side.

NPK, the good cooties, and so forth. And the ingredients in my compost all come from my immediate neighbourhood. I do *not* bring in significant amounts of other stuff from afar. The main exception is leaves from the nearby town of Farmington, which I could as well collect from my own forest (and do), but theirs go to the landfill anyway, and I like to prevent that when I can.

Of course, many gardeners go beyond that and add their yard waste—leaves and grass clippings—which is a huge improvement. (You can see how nicely these things fit in with that eco-efficiency business I was talking about earlier.) Even so, typical compost systems are often wasteful and counterproductive. The ingredients consist largely of weeds, crop residues, and kitchen wastes whose nutrients arise from the garden or the marketplace. They get piled in a nondescript heap in a corner of the yard. There are no precise boundaries around the heap, so the stuff at the edges kind of moulders into the ground (a net *loss* to the garden system). Since stuff is added in dribs and drabs, the pile never really heats up. The weed seeds, the pest bug eggs, the disease spores are all concentrated there where they can ripen, hatch, or fester in rich luxuriance. Meanwhile the rains leach

much of the goodness into the soil beneath the pile, the very place it is least needed. It puts me in mind of *Julius Caesar*: "The evil that men do lives after them, the good is oft interred with their bones." What's really aggravating about this is that most of that lost fertility *originated* in the garden.

A Multiple-Bin System

To avoid loss of nutrients from compost, I take great pains to keep all the materials well contained within a series of bins consisting of upright posts and parallel planks and poles. There are five bins in my system, which requires 12 posts, each 10 feet (3.0 m) long, to construct. The posts are set apart 8 feet (2.4 m) on centres for the length of the bins, and 5 feet 4 inches (1.6 m) between the near surfaces of opposite pairs. Each pair is connected at the top by a spiked 2 × 4, which prevents the posts' tendency to spread when the bins are filled.

The posts are sunk 3 feet (0.9 m) into the ground, so the frost doesn't heave them about (if you live on Oahu, that last line might be lost on you). Now, digging a straight 3-foot vertical hole in Industry, Maine, is apt to involve as much quarrying as digging, so

Figure 1.3. The scorched part of this post will endure a very long time, even when buried.

once I place those posts I'd appreciate it if I didn't have to replace them for a while. I use cedar, which I hew with either a broadaxe or a chain saw to 5 × 5 inches (12.7 × 12.7 cm) by 10 feet (3 m) long. Cedar is rot-resistant, but not enough for me, so I scorch the bottom few feet in my sap furnace to a depth of ⅛ inch (3.2 mm). (Guess; you can't measure it while aflame.) I do this because I know that bits of charcoal have been dug out of ruins thousands of years old, the rings just as clear as the day they were cut. It will not rot. Moreover, if the posts are green-cut or wet when I scorch them, the heated resins will form creosote, which is boiled into the interior—something like pressure-treated telephone poles.

When I set the posts I don't fill around them with dirt. Instead I use small stones (if you have a shortage, bring over your pickup) tamped in firmly. The frost will not shift them in stone as much as it will shift them in dirt, and the risk of decay is further reduced if the underground portion is not in direct contact with moist soil. By the way I make sure the scorched part comes up at least to the surface, but not too much above, lest I blacken my clothes every time I rub past it. A reason why the posts are 10 feet (3.0 m) long is so they end up 7 feet (2.1 m)

Figure 1.4. I dig holes 3 feet (0.9 m) deep to ensure that posts don't frost heave. (Don't worry, after we finished using son Fairfield as a yardstick, we took him out of the hole before we set the post.)

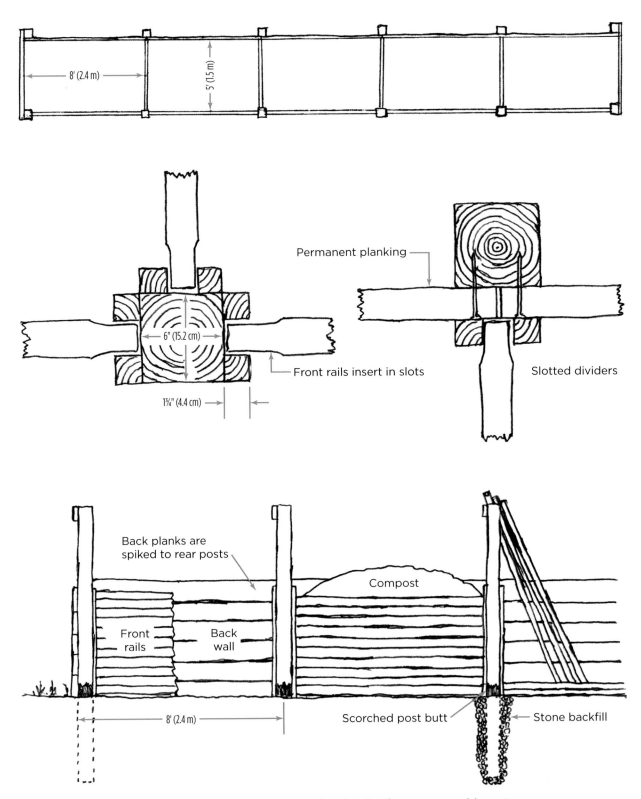

Figure 1.5. These drawings show some of the construction details of my compost-bin system.

aboveground once they're set. Thus the attached top crosspieces don't get in my way, especially when I'm transferring stuff from one bin to the next.

The planks that form the back wall of each bin are spiked to the inside of the back-wall posts. When the bins are full the compost pushes the planks outward against the posts. The front-wall planking slides into slots, so I can remove any part of them for easier filling and emptying. I made the slots by nailing a 1½ × 1½-inch (3.8 × 3.8 cm) square onto the face side of the 5-inch (12.7 cm) wide beam, leaving a 2-inch (5.1 cm) gap between them. The planks that form the interior walls between bins also slide into slots. I use 2-inch cedar planks, which I mill with my chain saw. Add up the two layers of cedar planking and subtract them from the 5-foot, 4-inch (1.6 m) gap between posts, and that leaves an interior space 5 feet (1.5 m) wide.

By the way I don't use plank siding on the front of the bins for the entire height of the pile. Above 3 feet (0.9 m) poles that are 2 to 4 inches (5.1–10.2 cm) in diameter serve just as well. I have acres of them and they're easily prepared. When they become too decayed to serve the purpose, I simply put them through the chipper/shredder and they join the next bin (don't get sentimental, we all return to the soil eventually). Why don't I use poles for the whole thing? Kinsman Tom Vigue opines that their loose fit allows too much drying of the outer several inches of the heap and impedes its thorough heating, and I have come to agree with him in part (more on this topic later). The posts at each end of the system are also planked across to form the end walls, which helps prevent spreading, too.

Why did I choose 5 × 8 feet (1.5 × 2.4 m) as the inside dimensions of my bins? Mostly because of hearsay. I heard it said that a pile less than 4 feet (1.2 m) in any dimension would tend to dry out and not have enough critical mass to generate enough heat (I explain later why heat is so important). On the other hand at least one dimension should be less than 5 feet or else air would fail to penetrate the heap and it might become anaerobic (that's Latin for "air failing to penetrate the heap," and it's not good). Okay, so that explains 5 feet wide, but why 8 feet long? I

discovered, by measuring many batches of finished compost, that they all averaged around 50 pounds (22.7 kg) per cubic foot (0.03 cu m). That meant that a 5 × 8-foot bin—an area of 40 square feet (3.7 sq m)—would hold 1 ton (907 kg) for every foot (0.3 m) of depth. How convenient to calculate my supply at a glance! Moreover, there's the matter of turning the pile with a compost fork: How far did I wish to fling that stuff? Experience suggested 8 feet for a maximum. If you cannot assemble enough material to fill such a large bin in a reasonable time, you can build it smaller, but no less than 4 × 4 × 4 feet (1.2 × 1.2 × 1.2 m) to ensure critical mass for proper heating.

My original system included a crude roof to shed excess rain, but it seemed those huge piles were always cooking themselves dry, at least in the early two stages—they needed more water, never less. When I eventually rebuilt the system, I didn't bother to include roofs. Having said that, there might be a real advantage to some kind of protection in the finished stage, when there are more soluble nutrients and no heat is being generated. Someone suggested I use plastic tarps, which would also prevent blown dandelion or bird-dropped seeds from landing on the pile. Perhaps, but I'm trying to use less plastic (I call it ticky-tacky, from the classic Malvina Reynolds song), not more. Anyway, when my compost is finished I'm usually quick to spread it—organics aren't meant to sit around.

Compost Ingredients

So once it's built what do I put into this system? The simple answer is: whatever I have and in whatever proportion I have it. Obviously not animal manure. Well, there will be plenty of worm poop and some droppings from the wild birds that eat them, but no domestic livestock manure. Since I'm not burdening myself or my land with domestic critters, why should I require it of others or their land? Anyway, I'm not big on moving lots of stuff around—eco-efficiency implies economy of energy as well as economical land use. Of course we are animals ourselves, and I certainly make use of our privy cleanings (*night*

Figure 1.6. Shredding leaves before composting allows me to spread thicker layers without their wadding up.

soil is the euphemism used by folks who don't like to call something what it is), but I don't use these in the general compost system I'm describing here. The topic is discussed later in this chapter.

I use all the usual crop residues and weeds from the garden; actually I'm embarrassed when weeds constitute a significant part of the pile, as it reflects poorly on my weed management ability. What I *do* appreciate about the weeds is the soil clinging to their roots, although I try to shake most of it off in the garden where it lives. A certain amount of soil in the pile is very desirable, much like the salt in a casserole, and since none of the other ingredients contain much of it, when I don't have enough weeds I go dig up a few bucketfuls of dirt from the most organically sterile area I can find (it's the minerals in the dirt I'm after; the rest of the pile is all about organic matter).

A visitor once commented with some hubris that she no longer wasted energy making compost; she just let everything rot where it lay in the garden. I didn't venture to comment, as she clearly knew all she needed to know, but I continue to make compost nevertheless, thank you. Most of my compost material comes from outside the garden, and thus the compost I make provides a net gain in the garden's fertility, not merely shuffling around what's already there. But I should note, even that shuffling process is extremely important: Those old cabbage leaves and cornstalks, tomato vines and carrot tops, are often full of pest eggs, borers, blight spores, septoria, anthracnose, and so on, all of which may fester there on the soil surface, waiting to attack next year's crop. Keep in mind that nearly all of our crop species are not native, and many of their pests are not found in the surrounding fields and forests, but only in our gardens. In a properly made compost heap, nearly all those pests and diseases will be destroyed by the heat generated by via aerobic fermentation (which does not occur in crop residues simply left lying in the garden), plus the myriad biologies that accompany accelerated decay.

The early stages of composting are totally dominated by thermophilic (heat-loving) bacteria, which cause most other bacteria, plus fungi and earthworms, to either leave the pile, go dormant, or die. Only after the pile cooks do the other decomposing agents go to work. Adding them to the pile before the fever has passed would be futile.

That being said, I don't consider garden residues a mainstay of the heap—after all these plants are not very eco-efficient and their fertility arose from the garden in the first place. The heavy lifters in my pile, the stuff which brings fertility into the garden from outside, are the grass and tree leaves and ramial chips. Ramial chips is another name for small brush (under 3 inches, or 7.6 cm) which has been put through a chipper/shredder. It is distinct from trunk wood, which is mainly cellulose. For more details about how to make and use ramial, see the section *Ramial Research* in chapter 3.

As I explained before these are the ingredients that make the overall system eco-efficient—make it work—since they build up long-term humus with a much smaller input of "earth-blessing" (space, nutrients, water, air, sunshine, et cetera). They do require something, of course—in particular, nitrogen.

Although I put *all* my garden residues and weeds through the compost system, that's not always true of the grass and especially of the leaves and chips. As I said, they're the mainstay of the heap—but I also use major quantities as mulch and in other ways (and that's described later in the book).

Someone asked me once, in a theoretical vein I assume, whether I would use deer droppings in my compost piles if I chanced upon some in my woods. For one thing, since I've never known deer to poop in piles of any consequence, the time spent chasing around trying to gather it up would be much more profitably used mowing grass and shredding leaves. Moreover, I'm not sure how valuable a manure derived mainly from spruce and fir browse would be for cultivated crops. In fact the deer do occasionally loiter around my bins, nibbling on cider pomace that gets spilled there. If a deer were to inadvertently drop a few doe-berries in the pile, would I discard the whole batch as non-veganic? No, I think not.

Ideally I build a new compost pile every three or four weeks during the season (April through November), and one challenge is to have enough material at hand to build a complete 5 × 8 × 4-foot (1.5 × 2.4 × 1.2 m) pile within a few days so that it will heat up thoroughly and uniformly. Again, I'm counting on the heat generated by the pile to: kill pest eggs, destroy weed seeds, and cook any disease spores, plus commence the breakdown of fibrous materials. The materials in a compost heap built up over time do also break down—slow fungal activity would be the driving force—but without the initial bacterial fever, there would be too little heat to wipe out the baddies.

A second challenge is to have a reasonably consistent balance of ingredients. There's plenty of leeway here, but I would prefer that the summer piles not be only grass, autumn only leaves, and spring only leftover kitchen garbage. Therefore I stockpile certain materials: hay, leaves, and kitchen garbage.

Figure 1.7. Apprentice Arica is mowing this ragweed for the compost pile! We're counting on the intense heat and bioactivity to destroy the ripening seed.

I put lots of fresh-cut grass directly into summer compost piles, but I make plenty into hay for autumn and spring piles. By "hay" I do not mean that I carefully cure and bale it for long storage as a dairyman does. Rather, I let it dry enough before raking so that it will not be fire-fanged and half rotten *before* I compost it. I would mention that my concept of "grass" or "hay" is much broader than a dairyman's. When fed to big ungulates "grass" should be mainly grass with some succulent legumes, as in a timothy-clover mix. My bacteria and earthworms are not so finicky, so if my pasture contains oodles of buttercups and milkweed and goldenrod, so much the better, as long as it's oodles. I want a dense yield of not-too-woody biomass; eco-efficiency is more sought after than digestibility.

Tree leaves especially tend to come all at once—October—so I try to collect them soon after they drop, shred them while crispy dry, and store them, mostly in a deep 8 × 10-foot (2.4 × 3.0 m) rain-proof bin, where I can access them at any time of year.

Likewise during the winter we generate many buckets of kitchen garbage, with no place to put it. The tight-lidded buckets accumulate outside until snow-go, at which time I can start the season's first pile using this stored garbage, stored leaf shreds, and stored hay.

Building a Compost Pile

Now that we have an ingredients list let's consider how to build a pile. It's important to keep in mind that the main direction fertility moves in a compost pile is down. There may be upward loss through volatilization, especially of ammonia compounds, but most nutrients are held in a water-based solution that is susceptible to gravity. This is even more true if we add enough water to keep the pile cooking—the biological fires are stoked, not quenched, by additional water. Therefore there is some risk that soluble goodies in the bottom layers may be leached into the soil beneath, and that's about as helpful as a screen door in a submarine. I minimize this by starting a new pile with a few layers of highly absorbent, high-carbon materials, stuff that will actually benefit from the nitrogen-laden leachate from above. A typical sequence is ramial chips followed by spoil-hay (that's the trashy stuff that was mown too late or let lie too long), then ramial chips again or shredded leaves, and again spoil-hay, repeated three times for a total of six bedding layers.

By the way, when I say a "layer" of hay, I'm thinking of a jumbo-sized wheelbarrow as full as I can pack it and pile it. When I say a "layer" of shredded leaves, I'm talking about that wheelbarrow filled to the brim and mounded, or about 25 or 30 gallons (94.6–113.6 l). The hay will pile higher than the leaves, but it will also settle more in the heap, so they're probably roughly comparable. When I say a "layer" of kitchen garbage, I picture from two to four 5-gallon (18.9 l) bucketfuls.

Lacking spoil-hay I might simply alternate ramial with leaves, but I really prefer to alternate "forest stuff" with "pasture stuff." In fact I like to follow that general pattern throughout the heap-building, as it better reflects the mutual role of those two ecosystems in feeding me. As a rule the forest stuff is high-carbon and will be acted upon slowly by fungal agents; the pasture stuff is higher in nitrogen, which fosters feverish bacterial growth. I'd hate to live in a world without either one; apparently my compost heap and the crops it nourishes feel the same way.

After the initial bedding layers I switch to something richer—maybe clover or comfrey or kitchen garbage—but henceforth I add whatever I have, trying to alternate wet/dry, nitrogen/carbon, mineral/organic. I want to end up with a pile as diverse and self-complementary as possible.

I should emphasize the importance of using a good proportion of dry trashy stuff to rich succulent matter, like at least three to one. The usual concern is that too much high-carbon material might moulder along without enough nitrogen (remember, that's hearsay) to spark the bacterial heat. A reasonable concern, although a little nitrogen goes a long way, but people tend to ignore the opposite extreme: a high nitrogen-to-carbon ratio will make a pile heat up fine and in fact will drive off the surplus nitrogen in the form of ammonia and methane. That wastes fertility while filling up the atmosphere with ozone-destroying gases. People fret altogether too much about nitrogen in the soil and not enough about humus, just as they overrate protein in the diet while ignoring fibre. Either produces a toxemia of the system.

As the pile grows, usually over a period of several days, I insert poles into the slotted front so I can heap it ever higher. If I sense that my ingredients are all on the dry side, I add a few bucketfuls of water as I go along, rather than relying on a massive soaking at the end, which may not penetrate evenly. I have usually not paid much attention to aerating the piles; indeed, in the first stage there is sometimes a concern about them being too fluffy with all that coarse dry stuff. But after watering the air pockets settle out and the heap becomes much denser. Too dense? Cousin Tom practises a variant of the traditional Indore method of composting. (The Indore method incorporates poles laid sideways across one or more layers. As a pile is built the poles are pulled out, leaving passageways for air.) Tom creates two or three "chimneys" into the core of a pile by bundling several sticks together (a bit of crookedness creates more air passageway; that's good) and placing them upright as the pile builds. I'm uncertain whether it's worth it: That new internal exposure allows moisture and heat to escape, whereas adequate oxygen has never seemed to be a problem for me. Still, Tom knows a good thing.

Eventually my heap reaches a height of 6 or 7 feet (1.8–2.1 m), and I can barely reach up to pile more stuff on. But the heap will soon shrink down to half that height, mainly due to air spaces settling out.

One last step before I call it complete: I fork a last extra-big layer of hay on top and dump on 50 gallons

Skip the Lime

Despite the need for minerals in a compost pile, I never add lime (which I don't have in my system anyway) or wood ash (though I have plenty of that). In the course of all its chemical permutations, the pile goes through both acidic and alkaline extremes, and I don't wish to have such a strong base (alkali) interfere with any of that. Moreover I intend to use my compost on any and all crops, including some—like potatoes—that might take offence at the alkali. Not to worry, though: I have other ways of using wood ash, as described in chapter 4.

Figure 1.8. This 4-ton (3.6 tonne) pile is completely built and awaits the final watering and cap layer of soil.

(189.2 l) of water. I used to haul that in buckets from a nearby pond, but now I have a hose from an irrigation tank on the hill, lazy me. Lastly, I cap it all off with a few bucketfuls of sifted (pebble-free) soil for weight and moisture retention and, of course, minerals.

The Heap in Action

Soon after I've finished the heap, then watered and capped it, it begins to heat up in a very big way. You probably know that putting damp hay bales into a hayloft can set the barn afire. The same process is at work in a compost pile, but since the heap is not tightly enclosed (as in a stuffy hayloft), the heat can dissipate without combusting. If a pile doesn't heat up, the probable cause is lack of moisture. Without moisture the bacteria that generate the heat just can't get started.

One year we had an apprentice who made a scientific study of our process: He bored a hole in the end of a 4-foot (1.2 m) 2 × 2-inch (5.1 × 5.1 cm) cedar square and inserted a cooking thermometer. Every morning he took an iron pry bar and bored a hole into the heap, inserting the thermometer into the centre. Using graph paper he kept a "fever chart" of every pile we made that summer. It was remarkable how similar the lines were, given the varied content and timing of each pile. The temperature of every pile spiked within the first three days, from ambient to 162°F (72.2°C), then slowly it would slide back to around 140°F (60.0°C) over a week or so, then it would plummet to something under 100°F (37.8°C). At that point (always on Day 19) it would nearly level off and presumably might take weeks to reach ambience, so great is the insulating power of compost.

From the data we concluded that Day 19 days was a good time to turn a heap into the next bin (usually we waited more like three weeks, to make sure). Turning the heap is important for several reasons, and I explain that below. It's okay to wait longer than three weeks, but from the time a pile cools time's a-wastin', and we need that bin free to start another pile. For this reason I rarely start the first pile of the season in Bin #1 of my five bins. As soon as it is capped, I must wait 19 days for that bin to be freed up to start a new heap, and meanwhile grass and other ingredients are pouring in. If I start by building a pile at Bin #1, then for the next 19 days I am at a standstill, whereas if I start in Bin #2 or even #3 (out of five, remember), there is still room for everything to be turned at least twice. More important, as soon as the first pile (in #2, say) is built, I can immediately commence a new pile (in #1), delaying the holdup until perhaps there is a lull in the inflow of ingredients. This all may be of little concern to the backyard gardener who may be building only one heap for the whole season, but since I am typically generating between 12 and 20 tons (10.9 and 18.1 tonnes) of finished compost per year, I need to keep things moving. This is not a salad garden. Of course you can be much more casual about these matters when your food (and the fertility to produce it) comes from somewhere else.

To turn a pile I use a five-tined compost fork (I'm shocked to discover that my hardware store actually labels them thus, no longer as "manure forks"—a sign of who's purchasing them?). It used to take me about 15 minutes to throw 1 ton (0.9 tonne) from one bin to the next, a forkful at a time. Now that I'm grown older it usually takes me a good 20 minutes or an hour for the whole heap. It's not that I'm feebler; I just like to stop more often to contemplate how much fun I'm having.

There are several reasons for this turning, the main one being that the material on the top and sides of the heap was not wholly involved in the first round of heating to 162°F (72.2°C). The turning will incorporate all this stuff and there will be a second fever, though not as long or intense. However, that second heating is important to assure that everything is properly cooked. But is it indeed properly cooked? Cousin Tom has an interesting take on this: He thinks that covering the top and sides more tightly, to retain moisture and heat, would obviate the need for turning. Furthermore, he makes the excellent point that if the turning is supposed to leave the remaining ingredients cooked, it is of dubious value, since the secondary heating is definitely tamer and may be

Figure 1.9. Here the system is all filled up; we must wait three weeks before a bin will open up for starting a new pile.

inadequate to kill some of the things we need killed. Whether or not that is true I have another compelling reason for turning: to create a more uniform product. Turning mixes all those layers while breaking up clumps of coarse junk so the agents of decay can better access it. That alone justifies the effort in my eyes.

If I turn a bin promptly every three weeks for a maximum of three times (or four stages), then the whole process takes twelve weeks, or less than three months, from beginning to end. Thus, a pile built in early June will be ready to apply in early September.

Three months from beginning to end does not of course mean that a pile commenced in late November will be finished in late February (except perhaps on Oahu). Obviously composting activity does not continue unabated through freezing weather, but neither does it come to a complete halt. I have commenced heaps in mid-December and had them seethe on into the New Year, melting the snow that fell on them to further water their interiors. This happens because even a bit of thawed stuff at the core will generate its mite of heat, which would be snuffed out by the surrounding chill were it not for the high insulating value of compost. Instead that little locus of living warmth thaws its surroundings a bit, generating yet a little more heat, and so forth until the whole heap is a steaming inferno. At some point, however, this process must succumb to the cold, at least where I live. Depending on the timing, this halt may come before the living heat can kindle; if a bunch of frozen materials are assembled in mid-January, it may be impossible for any microbial spark to catch, as even the bacteria drift into the suspended animation that is a northern winter. If the initial fermentation *is* able to begin and sustain itself through the first stage, once that cycle has cooled off, *then* it must fall into the icy grip of winter and be locked at *that* stage, even though there is lots more decaying to do.

Just as that insulating inertia keeps the pile cooking despite its bleak environs, that same inertia grasps the pile in its new permutation: a 3-ton (2.7 tonne) organic Popsicle. Nor will it hasten to relent, even when the sun is climbing and the robins are back. The ground has thawed enough so that plenty of earthworms are there for those robins, yet in the compost bins is a great gelid mass; it has made a contract with nature from which it is not easily released. If I'm in no rush, there's no problem. But if I'm needing that bin free to build a new heap (yes, I've started piles in mid-April), or if that batch is finished and I want to apply it to early-spring beds, I speed things up by busting the stuff out with a maul or dull axe and spreading the chunks where I want them, where the sun and rain will thaw them much more quickly. Or I may scrape off the thawed layer on the top and sides, exposing the core to much quicker thawing. Otherwise I just wait.

Using Compost

Ultimately, whether sooner or later, despite your clever ploys and your stupid blunders, you will end up with compost. It's really quite hard to prevent; it's just a question of how long it takes and how valuable is the finished product.

So then, what do you do with it? The living community within it will continue feeding upon itself, mould becoming earthworms, centipedes becoming bacteria (and ultimately vice versa); life and death continue, decay marches on. Not, however, without a net loss of energy at each stage. If you added no fresh matter, I suppose the decomposition theoretically would result in ash, the mineral residue from which all the organic energy has been wrung out. For that reason if no other, compost should be put to work in the soil as soon as it reaches a finished condition (obviously it *never* reaches a finished condition—nor do we—but when the necessary biology has rendered it useful for cultivated plants). I call it finished when it is broken down into uniformly small particles that are unidentifiable.

An obvious way to use compost is to incorporate it into the soil shortly before sowing the crop that is to benefit from it. Further decomposition should occur in the soil, where the crop plant rootlets are prepared to take up the nutrients as they become available.

For example, if I start compost in June and it's ready in early September, there is usually an open space where I've removed an early crop and another late crop is about to replace it. That's a fine time and place to spread the new compost. But if that late-succession crop is to be a green manure, then the compost might be more prudently used elsewhere, or it might be spread in late fall for a spring crop. An advantage of this is that if the compost does contain some viable weed seeds, say due to inadequate heating, then they may sprout and get done in by late tilling or by winter. Again, I don't like to have finished compost sitting around when it could be in or on the soil, powering a community of microbes whose services will be wanted anon.

My favourite way to incorporate compost is with a wheel hoe, using the crossbar or stirrup attachment, although the tines could also do a good job. This *stirring* in—as opposed to *turning* in (as with a plough)—has the advantage of keeping the organic matter in the top few inches, rather than burying it away from the life-giving air.

You can also use finished compost without incorporating it, but rather spreading it on the surface and letting earthworms and the rain incorporate it. An advantage of this is that it also serves as a mulch, protecting the soil surface from the elements. Of course, what then protects the compost from the elements? I believe it largely protects itself by trapping the nutrients until they can be moved downward (worms, rainwater, et cetera) and appropriated by crop plant roots. A possible disadvantage is that the more volatile nutrients (like ammonia) may evaporate, especially in direct sunlight. I usually avoid this by immediately adding a thin mulch of shredded leaves. Thus the first mulch is more for fertility; the second more for protection.

This double mulching is especially useful when I want to use unfinished compost: stuff that has been thoroughly heated by two cycles in the bins, yet still contains plenty of recognizable material. Of course many crops resent growing in such crude stuff—wouldn't you?—but at least one crop loves it: tomatoes. Those semi-weeds revel at having their feet in near-garbage and will be disgustingly healthy, provided the compost contains no diseased tomato plants

from last year. Another crop that seems to appreciate this rude treatment is squash (see the *Cucurbita pepo* section in chapter 8), but not cukes or melons, which take themselves far too seriously. Ever notice which plants grow happily right in raw compost heaps? Certainly none that has any sense of decorum.

Take care if you decide to use finished compost to topdress crops that are already up and growing. At best compost may spatter all over the plants to their detriment (lettuce and broccoli?—forget it); at worst, young plants may get buried or broken over.

I never make enough compost to adequately fertilize *all* my crops, but fortunately I don't need

Figure 1.10. Three loads of compost per bed is ample even for the heavy feeders. That's about 900 pounds (408.2 kg) per 180 square feet (16.7 sq m). Photograph courtesy of Scott Perry

to. Areas that were last in a green-manure rotation should need none, nor crops following a crop companion that included a living mulch (see chapter 2 for more on living mulches). Most of mine goes on the intensively spaced veggie crops. When a legume is one of the companions, I use much less compost, relying on it more for humus than for specific nutrients. I'm far more generous with members of the cabbage family, which are overdomesticated shallow-rooted crops requiring spoon-feeding—if only I didn't love them so much.

Although I make all the compost I can, and of the best quality I can manage, I do not rely on that alone for my garden's fertility. I also use green manures, mulches, and living mulches, and that's the subject of the next chapter. But there's one more material to discuss before leaving the subject of composting, and that's humanure.

Composting Humanure

One can hardly discuss long-term sustainable soil fertility without considering the waste products of our own bodies. If we consume the products of our soil and fail to return our urine and feces to that soil, then *we* become a form of erosion, a vehicle for the unsustainable removal of vital nutrients from our food system. Many people are repelled at the thought of personally recycling their own bodily wastes; we are so used to dropping it in 3 gallons (11.4 l) of potable water and flushing that water off to Neverland where someone else deals with it. We neither know nor care what *they* do with it—out of sight, out of mind. Does it get treated only to take up less space in the landfill or to fertilize a distant golf course? What is certain is that it is lost forever from the land that produced it.

The loss might seem less significant because the volume is very small compared with the huge amounts of other stuff (grass, leaves, and so on) used to build the soil; however, we mustn't overlook that its fertility is very dense, like any other manure. Humans eat a rich and varied diet, far more so than any livestock, and our digestive systems are comparatively inefficient at absorbing and utilizing all that goodness.

This is reflected in the intense richness and smell of our wastes, even our sweat. The more our offal smells awful, the more fertility it contains. If we aspire to a sustainable food system, we must recognize that humanure is not a four-letter word, that our filth is a resource. We cannot have sustainability without responsibility; it all has to come home and stay there.

Having said all that, we must acknowledge that there is a very good reason for feeling disgust at our bodily wastes: Not only are they nasty, they are hazardous. You do not simply spread them on the land and plant crops there; that is a sure recipe for disease. They must first be properly composted and proper hygiene practised by the compost maker.

Now I must distinguish between feces and urine. Urine is generally not pathogenic; its main issue is the awful smell of ammonia formed by bacterial action on urea—not particularly harmful but certainly obnoxious. Urine can be spread on the soil without health risks. If it is spread just before or during a rain and not too much in one place, the odour will immediately disappear. Of course collecting pee separately from poo is easier for people who have a Y chromosome and thus are equipped to direct it more accurately. The main advantage of segregating urine is to keep the feces drier and needing less absorbent bedding. Also, segregated urine can be used directly without composting. Otherwise there is no problem with them remaining together. In dry weather urine can be stockpiled in tight-lidded plastic buckets.

I should mention that urine must *not* be applied directly to growing plants (even if they're not for food), else it may burn the foliage. It is best to spread pee on the ground near growing plants. Although there are some species—like corn, sunflowers, squash, leeks, and garlic—that delight in such insult, others (potatoes, most root crops, and legumes) may react badly to it. Potatoes may grow knobby and develop scab, carrots will be less sweet and not store as well, legumes will find their ability to fix nitrogen compromised.

For crops like zucchini and chard where the goal is prolonged production, a midseason boost of urine can be very useful; for crops like winter squash a late boost of nitrogen (from urine or any source) may be

Figure 1.11. This well-maintained privy is comfortable, weather-proof, and odour-free.

Figure 1.12. The hatch is flipped open to empty the catchment chamber twice each year.

counterproductive, favouring more vine and foliage growth when we want the plants to focus on ripening their fruit.

Feces requires an altogether different treatment; it is full of coliform bacteria (*E. coli*) whose very purpose in life (though perhaps they don't see it that way) is breaking down our yuckiness and converting it to soil-building humus. It is really very thoughtful of them. Still, through no malice on their part, they pose a grave health risk to us. Although they live in our colons, our bodies have developed a system to ensure that they only move downward, never backwards into our upper digestive tracts. If a stray *E. coli* bacterium finds its way into the top of the system (say, if we eat contaminated food), it can really wreak havoc, even to the point of death. We shouldn't resent *E. coli* being what it is, but if we wish to avoid the collywobbles or worse, we must keep it in its place.

I must point out that everything I am suggesting is only relevant if you live in the country, with a bit of distance between yourself and neighbours. Whether you use an outdoor privy or a composting toilet, you will raise some eyebrows, especially if at certain moments your place has a certain ambience that wafts downbreeze. However lovable and environmentally aware your neighbours may be, they may come to find you a bit tedious, even if your system is odourless all but a few hours of the year. I always empty the privy (twice a year) when everyone is away and the breeze is right, followed by a bath in the farm pond. By the time folks return one would never suspect there was ever an unseemly whiff on the place. Keep in mind that however sanitary your operation, some neighbours will be grossed out by the mere knowledge that doo-doo is being processed next door (of course they don't think twice about whoever is experiencing their own).

Our humanure composting system is completely separate from our regular compost system, and that's critical. Do not add human manure to a compost pile—always compost it separately. Our own humanure system begins with our privy.

It is a 5 × 7-foot (1.5 × 2.1 m) two-room building about 150 feet (45.7 m) from the house. It sits atop a fly-proof catchment chamber, basically a mini cellar. I began by pouring a 4-inch (10.2 cm) thick concrete slab (with rebar and wire embedded for crack prevention) on a 1-foot (0.3 m) deep layer of small stones (the stone layer prevents frost heave). On three sides of the pad I laid a 2-foot (0.6 m) high wall of 8 × 8 × 16-inch (20.3 × 20.3 × 40.6 cm) cement blocks, with one long side (downhill) left open. Upon the wall I laid 6 × 6-inch (15.2 × 15.2 cm) cedar sills and the building itself. Inside the building it's a typical two-compartment privy, and materials simply fall into the chamber below.

On the open downhill side of the 2-foot (0.6 m) deep chamber, I installed a heavy, hinged hatch (see figure 1.12). I prop boulders against it to hold it tightly closed against vermin. Likewise, both seat covers are hinged and close tightly to exclude pests.

Each compartment has a bucket of shredded leaves for throwing down the hole after each use. An overhead space holds several feed bags of packed leaf shreds—I don't intend to run out. Even fastidious guests have commented on the lack of odour, especially compared with public privies (campgrounds, say) with their ammonia-generating chemicals and no bedding at all. We never throw ashes onto our humanure, which only generates more ammonia. For years we used hardwood sawdust bedding, but shredded leaves are more effective odour quenchers and produce higher-quality fertilizer (if you never viewed your droppings in terms of quality, you need to rethink your self-image). With the leaves, unlike sawdust, I'm never concerned about using too much, as even straight leaves would be a valuable soil amendment. This ensures not only that no odours will be emitted, but also that there will be no seepage out the back side of the chamber. That is also an advantage to collecting as much as possible of the urine separately, even though it may all end up in the same place.

Twice a year, spring and fall, I shovel out the catchment chamber and wheelbarrow the contents, typically six to eight large wheelbarrow loads each time, to a nearby double-bin compost system used exclusively for humanure compost (see figure 1.13), where I empty it onto a base layer of yet more shredded leaves. When the pile is finished I cover it thoroughly with another layer of leaf shreds, so that

Figure 1.13. The compost on the left will soon be emptied into the right side and replaced by the latest privy cleanings. Shredded leaves keep odour in and flies out. It helps neighbourly relations that this humanure heap is as odour-free in reality as it is in the picture.

Figure 1.14. The fertility for this plot, which will soon be planted to corn, didn't have to travel far.

none of the humanure is exposed to view or, more important, to flies. If the final layer of shredded leaves is not enough to thwart flies, a light sprinkling of screened soil will be very effective. After about six months that pile is turned into the adjoining bin to make space for the next privy cleaning; again it is covered with shredded leaves. At the end of the process—12 months after emptying the privy—this twice-turned material is hardly recognizable as having any fecal origin. Notwithstanding all the leaves it is a very rich material, richer than the best livestock manure and, thanks partly to the leaves, having a highly diverse mineral content. It is ready to use.

While properly composted humanure is safe to use on anything, I do not assume that I've done everything perfectly, and therefore I apply it only to appropriate crops. This makes sense anyway; I end up with four or five barrows full for the year, which doesn't go very far in the general garden. Many crops don't need it and indeed would be better without anything so rich. However, two crops—corn and squash—do not understand the concept of "too rich"; those gluttons will welcome all they can get and make good use of it. Furthermore the amount of humanure compost we regenerate yearly is just about right to satisfy the needs of those particular crops. How elegant. It helps that they are both large-seeded crops, for which I can make a wide, deep furrow, fill it with the compost, sprinkle ashes on top, and drop on the seed on that before covering with soil.

It may leave a slight ridge, which is ideal, but the humanure is well buried, so even if any parts are less than perfectly cured, it will not pose a health hazard. When the crop is hilled up in early July, it will be further buried, and when the land is next tilled—next spring at the very earliest, maybe the end of the next season—it will be totally part of the soil, and any remaining *E. coli* will have been completely neutralized by the soil community.

Someone once suggested that the manure produced by one human would be adequate to grow all the food for that one person. That is ludicrous; there are no perpetual-energy systems, certainly not with human animals as part of them. In fact I always point out that humanure, like any animal manure, is a relatively insignificant piece of the overall fertility picture, a drop in the bucket, if you'll pardon the expression. On the other hand it is far too valuable, far too rich a resource to waste, especially since wasting involves robbing one part of the world while polluting another.

CHAPTER TWO

Green Manures

While composting is a strategy for bringing nutrients to cropland in a form suitable for immediate soil building, growing green-manure crops is a means of building up the soil from within itself. Green manuring is an effective way to create organic matter (including nitrogen), though it does not "create" any minerals. It can, however, make minerals already present in the soil more accessible to plants.

I'm always a bit concerned that the words *green manure* imply an inferior substitute for the real thing—brown manure—when indeed it is quite the opposite. Livestock manure is a waste product left over after the animal has extracted its own life force from it. Green-manure crops are a more direct source of life energy. Furthermore, manure's fertility is derived from the tops of pasture plants only, whereas with green manure the entire plant is converted into fertility, including the root systems, which are a good source of phosphorus and other neat stuff.

As I mentioned earlier an important advantage of composting is that the generated heat destroys a lot of disease spores and other problems. However, the plant species I use for green manures generally do not contain the same species I will be planting there later for food, and so I don't have the same concern about disease. Of course a green-manure crop left to go to seed (such as buckwheat) would create a weed nuisance in the ensuing crop, but that's just a matter of timing. Typically green-manure crops

are cut and tilled in at their peak of lush maturity, but before setting seed. A big advantage of green manuring over composting is that I do not have to gather large amounts of biomass, move it to a central location, and process it with all the labourious turning, watering, and so on. Yet green manure is not without its drawbacks: A great deal of energy is required to incorporate all that stuff. I avoid ploughing in organic matter, but even with rototilling I'm still using petroleum (which doesn't grow here). Hand-tilling, as with an Italian grape hoe, does a superior job, but is labour-intensive; even though it's my own homegrown sweat-energy, it still must be factored into my considerations.

Planting a green manure crop requires taking the planted area out of direct food production, which is okay if you have enough room for both. I do, so I rely on both systems of soil building: I make compost plus I use green manures in rotations. Actually my green-manure areas often function both ways: as in situ soil building *and* as a source of compost material.

Here's how it works. Some green-manure crops such as clover offer maximum benefit when left in place to grow for a full season or parts of two seasons. That's partly because legumes like clover need time to allow rhizobacterial root colonies to become fully developed (that's not hearsay; you can pull them up and see clearly for yourself). Turning the crop under prematurely greatly reduces the benefit gained therefrom. However, a clover crop allowed to grow all season long develops a tangled mass of viny tops that are difficult to incorporate. It is best to mow the piece—as many as two or three times. But the huge mass of mowed tops will tend to smother the plants and impede further growth (assuming I'm not yet ready to turn the whole crop under). To prevent this I rake up that tangled mass and haul it to compost (where it is very welcome) every time it needs cutting.

Wait a minute, I hear you saying. That crop was meant to build up the soil, but you export great masses of the high-nitrogen stuff to enrich the soil elsewhere! Absolutely; such is the enormous soil-building power of clover and the other legumes that it can do both without twitching an eyelash, well, a nodule then. The nitrate-laden root mass, plus the last fulsome growth of tops, will enrich the piece enormously, and the raked-up tops are a copious surplus for compost (which may or may not return to that particular piece). No such thing as a free lunch?—hah! And how those clover tops heat that compost! A relatively small amount will turn that heap into a raging furnace. Keep your moo-poo and give me clover. Furthermore, while manure (especially bovine) sends polluting ammonia (NH_3) into the atmosphere, clover and other legumes pull nitrogen *from* the air, sequestering it in the soil where it does much good.

Buckwheat (*Fagopyrum esculentum* and *F. tataricum*)

I may use green manures as part of a longer-term rotation—as much as two full seasons—or as short fill-ins of perhaps three weeks. I learned long ago that there are right and wrong choices of which plants to use for green manures in different situations. For example, if I have a weed-infested patch with low fertility (including nitrogen and humus), then buckwheat may be the best candidate. I don't try to combine any other species with the buckwheat, even though I generally prefer polycultures

for green manure. That's because buckwheat does not play well with others—buckwheat is all about buckwheat and does not make a good partner with anything that I know of. Its aggressive root system is extremely competitive, hogging whatever nutrients are available (including water) and starving the neighbours. In addition to its assertive root system, buckwheat has oodles of horizontal leaf surface, enabling it to shade out whatever it cannot starve out. If those neighbours are noxious weeds, then hurrah for buckwheat.

Other than being a terrific smother crop, what is buckwheat's value as a soil builder? The hearsay on this is somewhat inconsistent. I'm told (and observation confirms) that buckwheat does not contribute as much nitrogen to the soil as other options might. It consists mostly of hollow watery stems, but not so much solids. Yet even that is a huge asset, especially since that mass is easily digested in the soil. And let us not ignore, as so many do, the bigger issue of eco-efficiency: Compare the output of the crop with whatever it took to make it. In this respect buckwheat shines. Its faculty for concentrating phosphorus is especially important in closed systems like mine. (I don't bring in phosphate, not even mined rock phosphate, yet phosphorus is essential for crop health.) Buckwheat doesn't merely grab the available phosphorus from its competitors, but also mines its own by dissolving the calcium phosphate or apatite in the surrounding dirt (I use "dirt" here as opposed to "soil" to distinguish the non-organic mineral component of the soil). It is one of those amazing plants that actually eat the rocks—as horsetails do with silica and lichens with feldspar—building soil quite literally from the ground up.

Buckwheat is not without its limitations, however. It is not at all cold-hardy, so its use is restricted to the frost-free months, which for me is between Memorial Day and the autumn equinox. Buckwheat's value as bee forage is widely touted, but this comes with a hitch: It must be mown and/or tilled in before it goes to seed, ideally when it is in full bloom. Now, there are those among the nectar harvesters who might take umbrage at this untimely intrusion into

Figure 2.1. Buckwheat biomass fertilizes the ground from which it grew. This buckwheat will lie and wilt right where I felled it until it's ready to be tilled in. Photograph courtesy of John Paul Rietz

Figure 2.2. Wild bees help me by pollinating the buckwheat while helping themselves to the nectar.

their labours, though I must say I have never been stung while tilling buckwheat. While I assume that unflattering remarks are being made about me back at the hive, out at the patch the nectar crew just keeps bunching up on the ever-dwindling crop until it's all gone and they simply move on to the clover or something else, with a grumble of disappointment. They seem to realize that the nectar resource is just too vast to fight over, especially when fighting involves tearing their little guts out. I'm a personal advocate of bee philosophy.

A strategy that avoids the bees is to till in the buckwheat *before* it flowers. That gives a less valuable plough-down, but it leaves time to plant a successive crop of the same thing, which, if timed right, will be killed down by the first frost before it can make seed. Personally I'm inclined to give the bees (which I do not keep) a crack at the nectar flow, especially since they're so gracious when the time comes for me to take it from them. I have enormous affection and admiration for the little folks and appreciate our alliance.

Rye (*Secale cereale*)

Buckwheat is particularly suitable for planting in weed-choked areas and for areas where you want to get a lot for your investment but have very little (in soil fertility) to invest. Now let's change that scenario slightly. Let's still assume the soil is wrung out, but weeds less of an issue; what's needed is lots of massive nutrient-laden biomass, but again without much input. Let's also assume it's too late in the season for any frost-tender crop to make much headway; perhaps the piece was home to an earlier food crop that has just been removed. Then let's take a good look at rye. In the section *Rye* in chapter 9 I discuss it as a food crop, but for now it's all about green manures.

Like buckwheat, rye does comparatively well with little input. But compared with buckwheat rye returns a greater quantity of valuable biomass for the initial investment—at least if weed suppression is not the primary motive, although rye itself is far from helpless in the face of competition. "Rye" usually means winter rye (in the United States, that is), and therein lies its key advantage (or not). Planted in late summer winter rye makes a modest amount of growth before winter stops it in its tracks. It doesn't kill it, but makes it go dormant until snow-go, when the rapidly waxing warmth of April wakes it up. There are still plenty of frosty nights, but rye thumbs its nose at those and makes growth spurts during the warmer hours. By the way the same factors that make rye suitable for certain situations—low fertility and timing—also apply to hairy vetch, so intercropping them makes for an especially strong team. Winter wheat works much the same as rye, though more demanding. Personally, I've never been able to resist letting it go for a food crop.

By midspring rye will have bulked up considerably, and that is the time to turn it in. It is in fact a very rank mass of stuff to turn in, and many people are daunted by that, which is perhaps why many people procrastinate until the intended green-manure crop has become a de facto grain crop. All well and good if it does, but it's best to know what your intention is before you get to that point.

One way of dealing with all that rank growth is by first mowing it, with a tractor sickle bar—not an option for me—or rotary mower (mine is an attachment on my walking tractor), or by hand with a scythe. While the scythe is clearly the most sustainable method, cut rye is easier to incorporate if it is shredded up, as by my rotary mower. If the rye is left to lie there a couple of weeks, you could chop it in with an Italian hoe/

mattock, but it is pretty tough stuff (see chapter 13 for more on Italian hoes). A cautionary word here: I've heard say that rye residues form some kind of allelopathic (that's Greek for "un-neighbourly") chemicals that can inhibit the growth of succeeding food crops, but this is not a problem if the rye has a few weeks to commence decay before the new seed goes in.

Winter wheat works much the same as rye, though it's more demanding. Personally, I've never been able to resist letting it go for a food crop.

An advantage of winter grains is that they overwinter and make that quick early-spring flush of growth when it would be difficult to get a new crop in—I'm talking mud season here. That advantage can turn into a problem if you're not prepared to deal with it and it gets away from you, and that's why I tend to favour oats.

Oats (*Avena sativa*)

Many years ago we discovered that oats were a perfect catch-crop (a short-season filler crop) for us: something we could easily sow in a single row where a crop had been removed, or in a whole bed or block of beds. For quick lush growth during cool weather, oats are hard to beat. When we planted an oat crop in late summer, it would grow rampant, often reaching knee-high before freezing out. You see, as perennials, winter wheat and rye recognize the coming of winter and stoically hang it up until spring. But oats—an annual—knows it can only stand so much cold before it succumbs, and so it makes a desperate attempt to reach maturity, futile but useful to us.

For me the fact that oats eventually winter-kills is a big plus. It does what I need: makes lots of soil-building biomass right up to the coldest weather and then conveniently dies, covering the soil with its residues. Come spring it's barely there, having significantly decayed during the winter. Few people appreciate how much life happens under that blanket of snow, more fungal than bacterial. Pity the land that lies bare to winter's breath!

Every year Molly would suggest sowing oats where some late crop had stood, and I would demur, thinking it too late, but then agreeing to try it any-way.... Every year we sowed it later and later. At some point, of course, she would be wrong—the skimpy growth would barely justify the cost of the seed—but even then there was a benefit: Sown thickly oats as short as 4 inches (10.2 cm) would lay over and give protection against the drying winds and late freezes of early winter, before a proper snow cover arrived.

Most folks give no thought to what *kind* of oats (or other crops) they buy for green manure. They simply ask for oats—if a varietal name is mentioned, it is meaningless. It might be the variety that the seed company, or perhaps USDA, has recommended as the best for that specific purpose. Most likely not. What is offered is what is available on the general market. Oat varieties, for example, are chosen according to which yields best for the seed grower. A seed grower wants to maximize the yield of *seeds*. But what you and I need is yield of *biomass*: leaves, stems, roots.

The oats readily available on the market are good for horse feed, which is fine if that's what you want to grow oats for. There are varieties of "hay oats," which are relatively scant (or very late) seeders but make tons of foliage, and that's what I want. I have trialed dozens of them and have a few favourites. One very robust cropper is called Sir Douglas Haig—he was that World War I general who invented the brilliant strategy of hurling endless waves of British soldiers across no-man's-land, giving trench warfare a bad name. Anyway they named an oat after the guy, and it is a first-rate green-manure variety. The problem is that it is not commercially available—I got it from the National Germplasm System (that's not as gross as it sounds; *germplasm* is all about seeds), but I haven't yet built up a sufficient stock that I can offer it to others. I do offer some green-manure varieties through my Scatterseed Project, but only as propagation-sized samples, just like NGS sent me. If you wish to grow these special green manures, as well as other uniquely adapted varieties, you absolutely *must* learn to grow and save your *own* seed. (Don't worry if you missed that, I'll be saying it again.)

Even in midsummer oats may be a better choice than buckwheat et al.—*if* you have reasonable fertility to begin with and want to ratchet it up still higher ("it takes money to make money" comes to mind). In that case I always mix field peas with the oats. A grass-legume mix is always a strong suit. In this case the tall oats hold up the peas, which would otherwise sprawl, while the shade of the pea foliage makes a living mulch between the oat plants. The two combined capture a maximum of solar energy for photosynthetic efficiency. If I'm planting only for a very short lay (early or late), I omit the peas. Although they are very cold-hardy, I sometimes feel their growth rate is too slow to justify the expense of the seed, even if the seed is homegrown.

Sweet Clover (*Melilotus indicus*)

I've found sweet clover to be a very potent ally under certain conditions. I take my cue from the fact that it grows wild along roadsides, where it gets lots of salt. Most legumes have a high demand for the alka-line minerals calcium, potassium, and magnesium, but sweet clover is notable among clovers in its tolerance of (and need for?) sodium. Perhaps that's because it is in fact not a "true" clover. Although it is a powerful fixer of atmospheric nitrogen, perhaps its greatest asset is its ability to send down taproots *many feet* deep, where it extracts minerals—notably phosphorus—from the subsoil. These roots are supposedly nodule-bearing throughout the system, though I cannot imagine how they find air at those depths unless the plant itself carries the oxygen down to that subterranean workplace.

Life is so incredible. Sweet clover is a true biennial: It uses most of the first season developing that great mass of root tissue, and the next spring it goes into high-gear nitrate production. It can grow to 3 feet (0.9 m) the first year, and all that mass may threaten to smother the subsequent spring regrowth. There-fore I have mown and removed that first-year crop for compost, but here's the thing: Regrowth arises from buds on the base of the stem, so I'm careful not to mow too short.

A good application of wood ash will get sweet clover off to a great start (for more about applying wood ash, refer to *Supplying Minerals Through Wood Ash* in chapter 4). You'll get that back and much more from those deep-delving taproots, which are also quite adept at breaking up compacted subsoil. My land suffered under the mouldboard plough from 1803 to 1962, and a bit beyond that under me, so there's lots of forgiving to be done. Now alfalfa is often used for this very same purpose and rightly so, but sweet clover is much more cold-hardy, whereas alfalfa abides heat. Alfalfa (an Arabic word) was introduced into Europe by the Moors from North Africa; our modern sweet clover varieties originated in northern Eurasia.

To use sweet clover most effectively requires proper timing. I seed it in spring, mow it in late summer (which stops the oats), and give it a light mulch of ramial chips before winter. In the second spring it makes rapid growth as it goes to seed. It would reach 5 feet (1.5 m) if I let it, but I cut it down and turn it in when it is at its lush peak, not yet seedy. We are cautioned about feeding sweet clover to livestock, especially horses, as it can sicken them if handled improperly. For my "livestock," however, which includes earthworms, the rotting root systems serve as convenient passageways to the underworld, where they do their finest work.

Japanese Millet
(*Echinochloa esculenta*)

A few times in the long-ago past I have used Japanese millet for green manure. Its main asset is that on reasonably tilthful soils it can make bodacious growth in a long hot summer. A drawback is that the seed is very late to ripen, so I cannot reliably grow my own, unlike other species we've discussed. Of course most folks don't even try to grow their own green-manure seed anyway, but in my garden-without-borders I try to account for all the inputs, and the surest way to achieve that is by doing it myself. Not to say that I grow all my own green-manure seed; I buy plenty, but I am always searching for routes that lead away

To Scythe or Not?

I always planted sweet clover with a light nurse crop of oats, and that makes it more difficult to mow high. I no longer have the option of mowing with a raised sickle bar, so a scythe is really the appropriate tool for me to use. For most scythe amateurs mowing high is no problem; it's the cutting short that bothers.

By the way, if it seems that hand-scything is an awfully tedious way of dealing with acres of cropland, it is. I'm not dealing with acres of any one thing, but small fractions of acres that add up to the whole farm. If I were keeping livestock, it would be an entirely different game, but simply trying to control my own destiny is ever so much easier.

from the marketplace. Of course there's an upside to Japanese millet's failure to mature seed: It won't get by me to become a weed nuisance next season. Just because I can't mature Japanese millet seed reliably doesn't mean others can't—I mean, I don't exactly live in a banana belt. Japanese millet is very frost-tender, so wherever it's grown, it should be sown after the last spring frost.

Incidentally, Japanese millet is totally different from the millet we use for human food. For a description of millet for food, see chapter 9.

Creative Combinations

I want to share an exciting idea with you, as it combines features of both green manuring and companion cropping; moreover it builds up the soil *at the same time* that a food crop is growing there. Start by considering squash or pumpkin: a sprawling heavy feeder that cannot be planted much before Memorial Day in my area. When the soil is warm and frost is but a memory. Between snow-go and squash planting time, we have seven or eight weeks during which the ground is bare and doing little or nothing, except maybe sprouting some early weeds. Why not put it to better use? say I. Here's how I do it. As soon as I see some snow-rid ground, albeit half-frozen mud, I seed it down thickly to oats and field peas (you know, the sort you use as pea soup, only not split). Then I rake or tread the seed in as conditions allow. Seed depth is not important; I merely need to hide the seed from newly arriving birds until it sprouts. As April slides into May the growth rate accelerates to form an ankle-deep carpet. A week or so before planting squash, I need to chop in the vegetation, but only where the actual squash will be planted; that's because too much decaying matter can actually cause a nitrogen deficiency, as the decay process pulls nitrogen *from* the soil. Once the decomposition has commenced apace, it will of course *release* lots of usable nitrogen *for* the squash plants.

Since I mean to plant the squash in hills, I only chop 2-foot (0.6 m) circles every 6 to 8 inches (15.2–20.3 cm), leaving the oats/peas in between to continue growing; they're not in my way yet. I generally use an Italian hoe—the poor man's rototiller—to chop stuff in (see chapter 13). I may or may not fertilize those circles for the squash; I typically plant it on lean land just to utilize the benefits of this method. Although the oat/pea mix will soon become a *source* of fertility for the squash, it may not be soon enough. I use either fully composted humanure or regular compost that is not yet thoroughly decayed (I have so much need for finished compost elsewhere, whereas squash is not

Figure 2.3. While the young squash plants are getting established, the oat/pea green manure keeps right on growing.

so picky about the state of the compost). I don't turn in the compost; I merely topdress each ring, using as much as I can spare.

I used to plant squash seed directly until I wearied of feeding the voles; I now start seedlings indoors in 4-inch (10.2 cm) peat pots (thinning to three plants each), at least two weeks earlier. That head start allows me to be in less of a rush to set them out: I can afford to let the chopped oat areas break down for a few days longer than I had previously. The rest of the area—oats and peas—continues to put on lush growth until the squash plants are sitting in wells of light with walls of oats/peas surrounding them.

Up to now the compact squash plants have been demurely occupied with sending roots down and out into the decaying stuff, but by the Fourth of July they are really ready to sprawl. The squash vines would easily scramble up and over the knee-high oats/peas, but I do not make them do that.

So far both crops have shared the space, but now the squash needs it all to itself, so I help it out. I could mow the oats/peas or till them in and that would be the end of it, but I demand further service. I take a 4 × 4-foot (1.2 × 1.2 m) sheet of plywood and stand it on edge next to the row of squash hills, where I flop it down and tread on it, flattening out the green manure while careful not to damage the vines (should we say I don't want to crop the squash or squash the crop?). Timing matters here. I continue flip-flopping the panel and treading it until the whole piece is flattened and the squash looks to be totally in control. This lays the oat/pea mulch in a neat tightly thatched layer that few weeds can penetrate and that will now decay to nourish the out-reaching squash roots.

It looks quite elegant, but for one problem. The oat/pea mulch isn't really dead; the stems are merely crimped, and a few days' sunshine will make it all try to straighten up again. The trick is to exclude sunshine. Black plastic sheeting would probably work well enough, but it doesn't grow on my land, and besides, it's made of ticky-tacky (yes, I know many people consider it organic; some people think Elvis is still alive). I do have tree leaves, lots of them; not the nice dry ones I shredded up back in October, but

old piles of matted leaves that sat under the snow all winter. They'll do just fine. I spread those in a layer over the mulch; it needn't be particularly thick, just solid enough to exclude any sunlight. The only problem now is that those whole leaves will dry out and tend to blow away, so I add a thin third layer of trashy old hay or chipped twigs—not thick, just enough to pin down the leaves, sort of like a hairnet.

The results of this treatment are several: Counting all three layers, I've added a huge amount of soil-building matter to the area while dooming any weeds to suffocation. Squash is a water hog, but the heavy mulch takes care of that, allowing rain to percolate through it but not evaporate out again.

Perhaps the greatest benefit is one I never would have expected: The squash plants have been

Figure 2.4. These oats, at full height, are about to be flattened before the squash sprawls.

Figure 2.5. Squash vines overspread the mulch that once grew here, now totally weed-free.

Figure 2.6. Post-harvest view: Except for the quack grass invading from the edges, this mulch could be left in place for next year's crop, perhaps cabbage.

completely freed of striped cucumber beetles (though not the squash bug). Don't ask me why, but for several years we've seen few if any of the flittery stripers, though they do still pester the cukes and melons, which are not mulched that way. Someone suggested that the stripers can't find the plants in their oat-walled canyons. Perhaps, but even after the mulch is flattened the beetles are conspicuous by their absence. Maybe it's something about the oats, because one year I used red clover instead and did not notice the phenomenon.

Complex Mixtures

I avoid using most of the perennial pasture grasses as green manure, even though they are potentially more eco-efficient than the annual grasses. To realize their potential requires a longer lay, or rotation cycle, usually two years and ideally more. Given a longer time period they establish a dense root system that sequesters phosphorus and lots of other goodies, which are then released for the ensuing crop. For shorter season lays the succulent annuals like oats and peas seem more practical to me.

A few times I have experimented with more complex mixtures and observed them over a full two-year lay. I used a combination of oats, barley, peas, herd's-grass or timothy, red clover, white or ladino clover, alsike clover, hairy vetch, and alfalfa. I broadcast each separately, one after the other, because if I mixed them beforehand they tended to settle out slightly in the mixing bowl, grasses floating to the top, and I wanted a consistent stand overall. I also spread wood ash on the piece. As expected the annual oats, barley, and peas leaped ahead, and for several weeks the entire patch looked like a nearly pure stand of those three alone. When it grew too rank I scythed it and carried away all the top growth to compost—even though the main object was to build up *this* piece. The oats, barley, and peas were nearly finished off by the low mowing, and that was fine with me. I had intended them to serve as a nurse crop until the other species got off and running.

Within a few weeks the whole area was dominated by the red clover. The other species were all there, scattered throughout the area, each biding its time until conditions might favour it. A second mowing sometime in late July or August (if memory serves) gave the ladino and alfalfa their chance to shine. Both love hot weather and tolerate drought better than, say, red clover, which now kept a low-key presence.

The vetch was always a bit player in this show; it thrives in situations where lack of humus, minerals, and water keeps everyone else off the stage, and it then gets the leading role, which it plays admirably. In this mixed environment it pushes no one aside, but does a creditable job of filling in wherever others are weak. Appreciate vetch for what it does well, not for what others can do better.

The second year started off without the annuals of course—they had died at the end of last season, and were by now food for the long-termers. The red clover resumed the lead in the cool, wet spring weather. Again alsike lurked among it, thriving only where aeration was an issue. (Mind you, all the legumes are highly aerobic—for them air is food; however, alsike seems to manage with less than the others.) After mowing, ladino had an expanded presence, but as summer brought hotter and drier conditions, it was the alfalfa that really took charge. The timothy that had in the first year showed just a scattering of fine spears now became better established, which helped keep the various legumes from sprawling.

I believe if I had continued the experiment a third year, timothy would have been the main actor, with everyone else continuing in a supporting role. Instead I turned it all in and followed it a couple of weeks later with a glorious crop of russet potatoes, even though four or more cuttings had been exported from that area for use elsewhere. I've used this procedure on a few occasions, with comparable results each time.

As this experiment with a complex mixture of crops clearly showed, an important part of eco-efficiency is adaptability. A crop that is highly eco-efficient in one context may be surpassed by other species when some factor is altered. Temperature, moisture, humus, shade, day length, pH, compaction—any number of subtle factors could favour one over another. Maybe it's some obscure trace element like pandemonium. (You've never heard of it? It sounds no sillier than californium, and no one challenges that one.) The point is, the world is full of niches, and every niche has some organism or group of organisms that is ideally adapted for that place (and time). Well, no, not ideally—nature is not an ideal; it is a dynamic living reality—but rather something that is optimally adapted for those particular circumstances. Of course an organism cannot adapt unless it is present, which is a point in favour of complex seeding mixtures—I could have added many more. This is the principle behind fallowing, an ancient practice whereby a plot is left for a season to grow up to whatever it will (as in Leviticus 25:2–7). It assumes that whatever plants volunteer there—come up on their own—will automatically be the best adapted, better than whatever we might have decided to plant there. In fact what comes up on its own depends mainly on what seed happens to be already dormant in the soil and only secondarily on the conditions in the soil. Therefore the fallow population has been self-selected not for soil-building potential so much as for seed dormancy. That may be useful for a particular species, but less so for the soil community.

Sources of Green Manures

As eco-efficient gardeners-without-borders we must ask ourselves about every input: Where is it coming from? What are the underlying costs? It's only fair that we should ask this about the seed for all these green manures. For oats, peas, buckwheat, rye, and so on, it's rather simple: We can grow our own seed on a portion of the plot itself or on a small separate plot dedicated to that use. But what about those teeny-seeded pasture crops?

I have found timothy and most of the pasture grasses very easy to harvest and clean, simply by mowing when they are dead ripe and threshing them more or less like grain. If there is any trick to it, it is in maintaining a pure stand so you can harvest pure seed. In my case that is rarely if ever important.

Although I have not done it I assume sweet clover could be handled the same way: by cutting it, threshing it, and winnowing it like grain. I have difficulty with "true" clovers because when their seed is ripe most of the plant is still green and succulent. I suppose you could mow the crop at that stage, dry it as you do for hay, then beat the seed out of it. I've never done that with clovers, though I have accidentally seeded red clover by using old hay mulch, so presumably that could have been threshed and winnowed out before using it as mulch. I have hand-picked ripe clover blossoms, then further dried the flower heads before hand-rubbing the seeds out of them. This works fine if you only need tablespoons of seed, but it's clearly not a practical way of harvesting seed for green-manure plantings. I have had better results with mowing and drying vetch vines before threshing. In fact the problem is that they shatter *too* easily, so the timing is critical to avoid losing all the seed. This funky method of gleaning seed is probably not very *space*-efficient, but is that such a problem if the main object is the biomass (for compost or green manure) and the seed is merely a bonus?

I have also collected large amounts of free grass seed by simply sweeping out the floor of a hay barn after the hay has been removed or a truck bed that has been used to haul hay. I can winnow away most of the chaff to get a cleaner sample, which is worth doing because the seed will be easier to spread evenly. What I cannot control is what kind of seed it is. Although the hay may be mostly timothy with a bit of clover, the residual seed may be mostly clover, or vice versa.

Let me make it clear that I do *not* grow all the green-manure seed I use—I grow very little of it in fact. Nor is it a top priority when I can buy all I need in a year for less than $20. But that's not wholly to the point, is it? Like my fuel, my fertilizer, and my food crop seed, these are all inputs whose future availability is not assured, and therefore I wish to assert more autonomy over them. Regardless of how much green-manure seed I grow, it behooves me to get experience with growing it and learning what is and isn't practical.

CHAPTER THREE

Mulch

We know much about the beneficial effects of mulch: maintaining steady soil moisture, averting erosion, preventing evaporation of ammonia, nourishing a beneficial community of microbes and earthworms, and more. However, there seems to be some profound benefit that goes beyond even those. This was well pointed out by Caleb Harlan in his seminal book *Green Manures*. He notes that even mulches that are inert and contain no intrinsic fertilizing elements—such as slate tiles, glass panes, wooden doors, even standing water—seem to bestow a tilth far beyond what they contribute materially. How can this be? Harlan admits he doesn't know, but he suggests we'd be foolish not to exploit the phenomenon, which I call the "cover-the-earth effect."

Actually, I have a theory that may help to explain this effect. To paraphrase the cheeky bumper sticker: FERTILITY HAPPENS. In this case I am largely equating fertility with nitrogen (although there's ever so much more to it than that). We know that much nitrogen is fixed, or converted to nitrate, by biological processes, such as by rhizobacteria associated with legume roots. Some is also a result of inorganic chemical processes going back to the primordial pre-life era, such as nitrogen fixation during a lightning storm. I guess that's why snow contains a small amount of ammonia.

I submit that something akin to that is happening in the soil, even the most sterile dirt. The industrial process that produces nitrate for synthetic fertilizer requires lots of electricity, which is a main reason why chemical fertilizer plants tend to be located in places like Sweden, Switzerland, and the Columbia River Valley, where hydropower is abundant and cheap. Well, every soil particle—especially clay particles—carries a tiny ionic charge. If that minuscule reaction (of mineral ions with atmospheric nitrogen) precipitates an iota of nitrate, every moment of every day and night year-round, throughout the breadth of the soil, might not a significant amount of fertility (that is, nitrate) accumulate? However, nitrate can't accumulate if it is wicked up and volatilized back into the sky as fast as or faster than it can form. Therefore, *any* material that covers the earth closely and traps those teeny mites of nitrate as they form would, over weeks and months, accumulate a small but significant amount of life-promoting substance, quite apart from whatever biochemical processes may also be going on. This is just a hypothesis, but whether it's valid or not, the phenomenon of abiotic nitrogen fixation is real, and we should exploit it, in several ways, as we shall soon see.

I don't feel any overpowering prudish modesty about my own person, but Mother Earth is altogether different. The life-begetting processes in the soil are sacred (yes, some atheists use words like *sacred*), and to leave the soil bare to the elements is sinful. We should shelter this miracle under a respectful covering of mulch (and I don't consider black plastic very respectful).

Living Mulch

In chapter 2 I described a companion green manure—oats/peas/squash—that might also be described as a *living* mulch. In addition to providing tilth-building matter while the primary, or food, crop is still growing there, the oats/peas also serve as a cover crop, repressing weeds and protecting the soil from destructive drying wind and sun and erosion. However, in that example, the green manure is knocked down and made into a "dead" mulch while the primary crop is still rather young. There are other strategies in which the companion/green-manure crop is allowed to continue living right alongside the food crop. The only living mulches I have had

Figure 3.1. By the time it is turned under this clover crop, originally sown and grown in the shade of a corn crop (note the cut corn stems still sticking up out of the soil), will have replaced most of the nutrients the previous corn crop took out.

any direct experience with are legumes, especially low-growing clovers.

One year quite by accident I planted hull-less oats in an area that, unbeknownst to me, was rife with red clover seed from the previous year's potato mulch. Sometimes I undersow clover on a crop that is half grown, but this clover got an early start right along with the oats. I tried to hoe some of it out—ridiculous, really—but my apprentice suggested letting half of it go. I did, and the difference was undetectable: glad I didn't waste time trying to hoe it all. The clover didn't harm the crop, which was no surprise, but I anticipated a problem at harvest. Indeed the clover grew tall enough to be a nuisance, but the oats grew taller still. Hand-harvesting with a grain sickle I was able to grasp the oats near the top and cut the straw near ground level. Of course much of the clover was cut, too, but as I lifted off the oats most of the clover shook out. This was important because the clover was very succulent and would have interfered with the curing of the oats in the stook; the oats all would have rotted.

This was an extreme example, a bad example, yet even so it worked more or less. A more intentional method I have used is to undersow the red clover (or ladino, but not crimson) after the oat crop is half grown. The clover skulks in the partial shade of the dominant oats until that crop is harvested in mid-September, at which time the clover bursts into a luxuriant ground cover, looking as if it had been the only one there all along.

This trick works especially well with crops that are widely spaced and exceptionally tall, so there's no risk of their being overwhelmed by the ground cover. Corn, sunflowers, and amaranth all fit these criteria nicely. I plant each of these with another companion food crop, as I describe in part 3, but that's only the beginning. When each crop is about knee-high, I hill it up and broadcast clover seed on the area between hills or rows. I prefer ladino for this, as I think it makes better growth in a shorter season, especially when it is hotter and drier. The clover is not delighted with this situation—it has hardly sprouted when it is dominated by its overbearing heavy-feeding neighbours. Nevertheless it bides its time, sinking roots and lurking there with thwarted expectations.

In sweet corn the ladino has a particularly bad time of it. Almost daily it is trampled by someone going in to pick a couple dozen ears. It surely resents that, yet it soldiers on until the day the corn crop is finished and the stalks are cut and hauled to the shredder. And eventually the time comes when ladino looks up and sees the sky. The world is its oyster, but it knows it hasn't much time (yes, plants "know" many things more surely and precisely than you or I; nothing spiritual about it, they merely pay attention), so it doesn't waste a moment. It uses those waning autumn days to bulk up, a modest growth by most standards, but enough to carry it dormant through the winter with a positive attitude. Come spring that struggling beginning it made last fall will send it out of the gates running, in time to make solid new growth before it gets replaced by a later-planted crop like melons. Thus I get the sum of two short seasons' growth where I otherwise would have gotten none. (I describe this in more detail in the *Companion Planting* section of chapter 14.)

I have "heard say" that red and ladino clovers can also be used under shorter but wide-spaced crops like cabbage and also under (or around) peas. The latter sounds especially unsuitable to me, as the two species are too closely related—they would both be producing surplus nitrate while competing for the same calcium, et cetera. Moreover the lush clover might produce an environment conducive to powdery mildew (PM). The cabbage combination seems to make more sense, if the clover doesn't start too early and smother the young cabbage plants or harbour slugs.

I have had great success with white Dutch clover, which, unlike ladino, is a low creeper. It is too low to smother anything, plus the plants form spreading mats, so the clover roots can be kept at a safe distance from the crop (by thinning/weeding) while retaining the mulch effect everywhere. I now mulch virtually everything, including cabbage, with shredded leaves; however, that doesn't mean I can't use both leaf and living mulch as long as I give the clover a slight head start before spreading shreds. In this particular situation I consider the clover more a mulch than a green manure, since it doesn't produce a lot of biomass, but then I am probably underestimating all those hard-working nodules.

I continue this discussion about the combination of green manure and shredded leaf mulch in chapter 9, *Grains*.

Stuff from the Forest

The reason that green manure/living mulch strategies work so well is because the component plants give so much compared with what is required to make them (this is a great example of what I mean by "eco-efficient"). Therefore they are not very competitive with the crop, yet that is not to say that they require nothing. Everything living makes some demands on the earth. The dead make no demands, and therefore "dead mulches" are sometimes more appropriate. They do not create any new tilth-building substance (beyond what they created when they were alive, elsewhere), but they provide other precious services, some of which are well understood while some are not.

The conventional wisdom about tree leaves and other dead stuff from the forest is that they are not very valuable for soil building because their carbon-to-nitrogen ratio is not favourable. That is comparing them with livestock manure, which is undeniably higher in soluble nitrate. In fact the only thing higher in soluble nitrate is synthetic fertilizer, which most of us recognize as unsustainable. We all agree that animal manure is superior to synthetic fertilizer because manure also adds humus, which absorbs some of that surplus nitrate and releases it more slowly over time.

Tree leaves also contain plenty of nitrogen, albeit in a lower concentration than manure. In order to end up with the same amount of nitrogen in my soil, I have to add a much greater volume of leaves than I would manure. And the problem with that is? I mean, the concern is not about adding too much carbon to the soil, but rather adding too little nitrogen to go with it, since a certain amount of nitrogen is required to fuel the microbial activity that will digest all that carbonaceous matter into humus. I should be at least as concerned about the opposite imbalance: more nitrogen than the carbon-rich stuff can absorb, the result of which is excess nitrogen (as ammonia or methane) escaping into the atmosphere or leaching into groundwater. Manure does that; leaves don't.

As gardeners-without-borders we must ask ourselves bigger questions like: Where did these materials arise and at what cost to the place that begot them? Of course the synthetic fertilizer loses on every score; it's not even in the running. It gives us no answers; it ignores the questions. As for manure the fact remains that its fertility is quite expensive (in terms of eco-efficiency) compared with more direct sources (and that is after making allowance for composting them all). Now, an acre of forest produces vastly more biomass than an acre of cows on pasture. The lower concentration of nitrogen in that biomass is much more than compensated for by the sheer volume of it. The manure only looks better because its nitrogen is more concentrated, but by that same reasoning synthetic fertilizer is the best of all—no messy organic matter to dilute its strength. If our goal is adequate nitrogen contained in lots of humus, tree leaves are hard to beat. Moreover, there is so much more to tilth than nitrogen. For example, if we compare a mineral analysis of ash from leaves with one of ash from manure, leaves contain twice the levels of minerals, especially those trace minerals that must be pumped up from the depths of the earth where the glacier deposited them. If we wish to take our cue from natural systems, notice that trees first, then grasses, are what run the show; animal droppings are quite incidental.

Working with Tree Leaves

For those who question the fertilizing power of tree leaves, I would share a couple of anecdotes. Although I have many acres of hardwood forest where I could collect my own leaves, I usually prefer to haul them from the town of Farmington, 8 miles (12.9 km)

away. I'm doing the people of Farmington a favour by hauling theirs away (the town used to do it but no more), plus they're already raked into piles, relatively free of twigs and branches. Of course I always ask the owner if I may take them, even though the response is predictably something like "What! Is the pope Catholic?" At one home on a shady side street, I asked the owner, an elderly woman with a charming Austrian accent, and she answered graciously, "No, I want them myself." Intrigued, I asked wherefore, and she replied, "I need them for my garden." Delighted, I probed further. She, having no known ideological views on the subject, relied exclusively on the maple leaves because they were what she had—the shade-free portion of her yard was wholly occupied by her garden, so there was little lawn to supply grass clippings, and she had no access to manure. She had no formal system for shredding or composting; she merely said: "I mull them over from time to time." That had been her "system" for years, and the heavy-feeding cabbages and leeks I saw testified to her success. The huge pile of leaves in her yard was waiting to be converted into next year's sauerkraut!

Second story: I once persuaded the Town of Farmington street commissioner that they should truck their curbside leaves up to my place rather than to the town dump. Yes, he hastened to agree, it made good sense for everyone. The first two truckloads landed in my leaf dump, and I congratulated myself that henceforth I would merely wait for the annual windfall. Next year they didn't show up, so I asked the street commissioner, what gives? Well, it turns out that they need them at the landfill. Need them? I repeated dumbly. Ayuh, it seems they need them to add to the cleanings of the fairground horse stables. Ah, says I knowingly, it's to balance the excess nitrogen in the manure. Weeell, not quite, he corrects me; it's rather that all that sawdust bedding makes the piles cold, despite the pony doo-doo. What! They're using "my" leaves to heat up horse-hockey so it will be well composted? And does it? Yep, assured the commissioner with no irony intended, works slick as shit. Now understand, this good fellow was not a vegan, not into organic or sustainability—for all I know he could be a Republican—yet don't try to tell him that leaves lack nitrogen.

Shredding Leaves

Most of the ways I use leaves require shredding, and that involves some kind of shredder. In my case I have an Amerind-MacKissick chipper/shredder designed to work as an attachment with my Gravely 12-horsepower walking tractor, but the chipper/shredder can be powered by any other PTO source or can be purchased with its own self-contained engine. For chipping brush the 12 hp is really needed to accomplish anything, but for shredding leaves a lighter power source might be quite adequate. An advantage of mounting this shredder on something

Figure 3.2. Some landowner's dilemma is my mulching bonanza.

Figure 3.3. Shredding leaves is worth the effort, because the shreds will not blow around the garden or wad up in the compost. Photograph courtesy of Yaicha Cowell-Sarofeen

Figure 3.4. Shredded leaf mulch is efficient to apply and highly effective. Photograph courtesy of Yaicha Cowell-Sarofeen

is that it is easier to move about—on its own it's cumbersome to move any distance.

My Gravely, by the way, is a magnificent machine, which is probably why Gravelys are no longer made. Because such walking tractors have several interchangeable attachments (mowers, snow blowers, tillers, and more), you don't need to own several engines, all of which require maintenance. On the other hand that means that if the one tractor is out of order, you cannot do any of those jobs. Changing attachments is a rather minor nuisance and should not weigh very heavily against that option.

The safety grating on my chipper/shredder won't allow anything less than ¾ inch (1.9 cm) in size to pass through, so the product has a very uniform consistency, perfect for applying. I simply walk down the paths with a large rubber barrel full of shreds held on one shoulder and with my free hand create a blizzard of leaf confetti over the beds.

The crop plants are usually a few inches tall, and the shreds drift down and settle down nicely around the plants, making a tight and effective mulch with minimal effort. That's why it is important that they be consistently fine—confetti-like: so that no clumps of whole leaves will smother the crop. Also, shredded leaves stay put in a breeze much better than whole leaves, especially once they've been wet and settle into a tight "felt."

Another reason for shredding leaves is the twigs. Although I remove the larger branches for chipping separately, most leaves have a lot of fine twigs that are a nuisance to pick out, yet I don't want them left whole. This is especially an issue with leaves collected in the forest or from lawns with big old trees (the ones with deepest roots and thus better trace-mineral content). When fed through the shredder they are reduced to harmless little chips that bother nothing, although you may need to pre-snap longer twigs so they don't block the chute.

If I'm using partly wet leaves they will plug up the grate, but if I remove the grate they'll come out too fast, only half shredded. Half shredded is okay as long as I can put them right back through again, giving me a product somewhat like the confetti, though not as uniform. For mulching paths and wider-spaced crops, and for making compost, this is quite satisfactory. (Remember, once the grate is removed, there's nothing between you and the spinning hammer mills, so stand clear.)

I need to stockpile as much confetti as possible when they're crispy and dry, and so I race to shred those as fast as they come in, and pack them into a large (8 × 10 × 4-foot, or 2.4 × 3.0 × 1.2 m) covered plank bin that holds roughly 2 tons (1.8 tonnes; see figure 3.5). If I have a real lot of confetti, I can pack it into feed bags and cover them with a tarp against rain and snow. I use lots of those banked up against the exterior door of my cellar for insulation during the winter.

A well-powered chipper/shredder is the best way to shred leaves, but it is not the only way. My father used to pile his leaves up by the garage door and run through the pile repeatedly with his push power mower, stirring it up between passes. Cousin Tom makes his into a long, very deep pile (3 feet, or 0.9 m) and wades through it repeatedly with his rototiller, regathering it as it spreads out. And of course, we all know the hands-down most efficient way of shredding leaves: You gather them all into a huge pile in the middle of the lawn and tell your kids and all the neighbour kids: "Stay out of this." Within hours the pile will be reduced to molecules. (To accelerate the process you could hang a used car tire from an overhead limb.)

Figure 3.5. This rain-proof bin is about to receive 2 tons (1.8 tonnes) of shredded dry leaves.

The problem with all these alternatives is the lack of consistency; some will be crushed to powder while some large pieces will remain. This isn't a big problem for some uses (say, compost), but for mulching closely spaced crops, including grain, the crude shreds are more difficult to apply. For mulching more wide-spaced crops, like cabbage and tomatoes, and for paths, the cruder shreds produced by a lawn mower work well.

I often end up with more leaves than I can shred immediately. When I have run out of time I have also stored whole leaves loose in a hex-wire (aka chicken wire) enclosed leaf dump, which I cover with a well-weighted tarp. The leaves stay pretty dry until I can shred them the following spring or summer. During this time the bottom layer will get somewhat damp from wicking ground moisture. If I am short on confetti the following spring, sometimes I just spread the dampish leaves out on a tarp to redry for some hours and then shred them, although usually I just use the crude method, double-shredding without the grate in place.

Some degree of shredding is necessary for composting leaves, because whole leaves tend to form a soggy mat that will take forever to break down. In fact that was what got me into shredding in the first place: Because I could use only thin layers of whole leaves in the compost pile, I was always using much less than I wanted, in proportion to the hay and other stuff. Once shredded, leaves can be the major ingredient in my pile, quite compatible with my eco-efficiency obsession. Indeed, aside from leaves and brushwood, my composting of corn, sunflower, and amaranth stalks would be very difficult without first shredding them. By reducing the volume and exposing more surface of those materials, I greatly increase the rate of decay, plus the biological heat that renders a superior end product.

Of course, this is all driven by a gas-powered engine, which raises the question: "How would they do this in Esperia?" (See the introduction to learn about Esperia.) I mean, there's nothing inherently

unsustainable about these machines, *if* they could run on something other than dinosaur farts. And as it turns out they could be, which I will discuss in appendix A.

Working with Ramial Chips

If you're shredding tree leaves, it's safe to assume you have trees, whether a single yard specimen or a whole forest. And while a tree gives us a crop of leaves every year, sooner or later the tree itself dies—maybe not suddenly like an animal, but a limb here or there. Or perhaps the tree is harvested as it reaches maturity, giving us saw logs, pulp, firewood, and more. Inevitably there is a lot of residual slash—small branches and twigs that are of little use to most folks. Moreover, between the big forest trees are many other

Figure 3.6. The leaf dump holds incoming loads until I find time to shred them.

trees, small saplings growing too thickly. Most will be stunted and eventually die from competition unless we, as good forest stewards, weed them out. All that residual biomass, that woody stuff, has real potential as an agricultural resource. No, it doesn't contain much nitrogen compared with manure, or even with leaves, but it has cellulose, the basic building block of humus. And what is most important in our gardens-without-borders, it is highly eco-efficient. Trees are such highly evolved beings that their product—ultimately including their own bodies—is much greater than the amount of earth-blessing (yes, atheists can use words like *blessing*) it took to produce themselves. By "earth-blessing" I'm referring to any or all of the inputs, such as soil, air, water, sunlight, plus other less obvious stuff. (Call it "natural resources" or something else if you like; I just hate to sound too left-brained.)

The term *ramial chips* (or *ramial wood* or just *ramial*) refers to brush or branch wood less than 3 inches (7.6 cm) in diameter that has been chipped up. The chipping is important to expose a maximum of surface area, which allows for faster decomposition. The 3-inch diameter is from the "official" definition of ramial; I can't imagine using anything over 1-inch (2.5 cm) diameter, unless it were too punky for use as firewood. That's because the part of the tree we value most for agriculture is the bark and buds. Their nutrient value is much greater than that of trunk wood, which is basically cellulose. In terms of nitrogen-to-carbon ratio, ramial is typically 5 to 10 times richer than trunk wood (according to hearsay). Having said that, I note that any type of wood has real value as an energy substrate (cellulose is a complex sugar) feeding a micro-universe of wee beasties who, in the course of doing what they do to live, somehow manage to contain nitrogen, though it's not in the wood itself. Bigger wood is also older and more lignified, which makes it more resistant to breakdown. If the wood is big enough for lumber, pulp, or fuel, then it should be used as such, but I say anything smaller than a broomstick has its highest calling as a builder of soil tilth.

Generally, ramial excludes softwood (pine, spruce, cedar, fir, hemlock, larch), because softwood contains resins that interfere with proper decay in the soil. I do use lots of chipped softwood branches, though, as mulch on my blueberries, raspberries, and blackberries.

Once a well-meaning fellow was consulting with me on strategies for building fertility veganically in tropical soils. He wished to enable Jamaican peasants to produce niche products for the export market. (You know, the same market that makes them poor peasants.) As I explained the huge potential for tree-based stuff, he jumped ahead of me with what he saw as a brilliant inspiration: Instead of clearing rain forests to produce stuff like oranges, soybeans, or beef for export, why not export ramial wood chips for American and European veganic farms? I'm not

Figure 3.7. Converting useless brush to ramial chips.
Photograph courtesy of Scott Perry

Forest Bounty

Although my forest supplies a large part of the fertilizing material for my croplands and hay fields, its munificence doesn't stop there. Of course my heating fuel costs only my labour. And the forest has also supplied me with an endless stream of raw materials for buildings, equipment, tools, and furnishings. Instead of buying tool handles, I used to make and sell them. Years later I learned that local kids knew our road as Axe-Handle Road. The various carts and wheelbarrows that I've trundled over the landscape have all come from that land; only the screws and iron tire bands originated in the marketplace. Our farm-begotten cherry trestle table is set with bowls, spoons, and ladles all from wood sourced within a few hundred yards of where they now serve. A spinning wheel, flax rippler, bench vise, shaving horse, and cider press were all harvested and crafted from the luxurious assortment of wood species I have at hand.

As in the garden diversity is crucial. The naves and felloe-blocks of cartwheels are best made from crack-resistant elm, but don't look for it at your lumberyard. Lightweight shoulder yokes for toting maple sap require basswood, but that's not to be found on the shelf. Old-growth hornbeam makes the very toughest axles and axe handles; check out Walmart, tee-hee.

Even in the house itself the lumber is not generic: There is spruce for the weight-carrying summer beams and rafters; larch and cedar where rot is a concern. Poplar, which will dent but not splinter, covers all floors and walls; tight-grained black cherry countertops will resist spills and stains; butternut with its rich warm elegance invites us to relax on the flip-top benches.

Mind you, these are all things that if they *were* to be found in the marketplace, I could never afford to buy. Yet they are truly elegant, not only in the sense of beauty but also in their appropriateness—the right material for the right purpose.

Sometimes my frugal improvisation is someone else's extravagance. The sills under my house are 10 × 10 inches (25.4 × 25.4 cm), 24 feet (7.3 m) long, cut from enormous spruces in a steep ravine on a remote corner of my land. Hiring a skidder to yard them, then truck them to a sawmill, along with the milling bill would have bankrupted me. Instead I taught myself to hew them into the final shape using only a chain saw and chalk line. Then, very slowly, we used a farm tractor to drag them nearly ¾ mile (1.2 km) to the building site, where the final fancy joinery notches were cut, also by chain saw.

As a result of that learned technique, I was able to shape all the beams for the entire house, some with elaborate joints that shift and tighten over time, rather than loosening. The massive posts and soaring rafters are more evocative of a Loire château than a hippie homestead. Yet if I accepted the assumptions of off-the-shelf living, I would have had a skidder come in and haul those trees to market, and I would have bought a pile of 2 × 6s, Sheetrock, PVC, and asphalt roofing, and built a house out of ticky-tacky, like rich people live in. I don't mean to knock it; it's probably all they can afford.

In my impoverished lifestyle I regularly enjoy certain foods and other commodities that are unavailable to the very wealthy. On the other hand I suppose they may value them less than I do, so I suppose we're both happy.

sure I ever got through to him about the folly of his ideas, but I'm hoping you see it. How much sense can it make to move vast volumes of biomass thousands of miles? And how organic is that really? Humus-building stuff should be produced very near where it is needed; isn't it bad enough that I haul leaves 8 miles (12.9 km) from town? Moreover tropical rain forests are the worst possible place to be extracting soil-building material. People often assume that under all that dense canopy of vegetation must lie a very deep, rich soil, but in fact it's quite the opposite. The high heat and humidity of those regions causes stuff to decay or oxidize so rapidly that humus can hardly accumulate. Therefore most tropical rain forest soils are thin and sterile, since most of their organic capital is invested in the living top growth. Remove those forests, and the tilth is burned out of those soils in a very few years. No, we don't need to import humus-making stuff from the tropics; they need it right there. Like nitrate and most of the minerals, we can get our own from much closer to home with vastly less expenditure of fossil energy.

Ramial Research

Sometimes the most convincing arguments for something come from the most improbable sources, if only because an outsider's perspective implies an absence of bias. The main impetus for using tree/woody residues in agriculture came from a group of forestry researchers at Laval University in Québec. In much of the eastern townships of Québec, large areas of commercial agriculture (including feed grains) are found in close juxtaposition with hardwood forests (including maple syrup operations). The logging operations generate huge amounts of slash, or woody debris, which is generally regarded as worthless—indeed a potential fire hazard. (For now we will leave aside the question of its value to the long-term health of the forest.) It occurred to these foresters that such "waste material" might be useful for maintaining soil tilth in the nearby agricultural lands (I would emphasize "nearby"—it's important). I had long pondered the same possibility: To what extent could my woodlot contribute to my croplands? I was fully aware that chipped brush wasn't a significant source of nitrogen, but I was also aware that soil tilth is about a lot more than nitrogen (which, after all, I've never actually seen with my own eyes, whereas tilth is hard to miss). It always seemed to me that adding nitrogen didn't guarantee that you get humus, but create humus and nitrogen will "happen."

Based on decades of research the Québecois foresters came up with the following recommendation: Ramial wood should be chipped after the green leaves have fallen. That seems counterintuitive, at first. If ramial chips have such a nitrogen deficit, wouldn't the green foliage help balance that, so bacterial decomposition could be accelerated? It turns out the researchers aren't looking for *bacterial* decomposition at all, at least not in the beginning. They are counting on *fungal* activity to break down the cellulose and lignin in the wood. Nitrogen-fed bacterial action would only get in the way of that. In particular they are counting on *white* moulds (Basidiomycota) to do the work; brown moulds (which are more associated with leaf decomposition) are not as suitable. The obvious implication of this is that ramial should be treated separately, rather than incorporated directly into high-bacterial compost piles. Although I have done both the ramial that I add to compost has usually been largely acted upon by fungal action before or after I chip it.

Even if I accept all that (and I see no reason not to), I still don't see why I can't cut and pile brush whenever it's convenient, leaves or no. If there are leaves attached, they would usually dry and shake off in the pile and work their way to the bottom, where I can recover them later for compost.

One of the important points made by the Québec researchers is that most of what we organic types do to "build up" the soil is really just an effort to slow down our destruction of it by livestock and cultivation. Woody residues—ramial chips—actually *build* the soil tilth, relatively quickly and from materials that the ecosystem can most spare.

The Québecois researchers also insist that fresh woody residues, of any kind, should *not* be ploughed in. This would bury them in an anaerobic environment where all sorts of nasty polyphenols will form and resist further breakdown. Now, I've never seen an actual polyphenol to recognize one, but somehow it sounds crazy enough that I completely believe it (just dare tell me you never bought into something every bit as weird). Furthermore, these polyphenols are not good for the soil, and plants hate to grow in them—can you blame them? The remedy is to not bury this woody trash at all, but to spread it on the surface, an inch (2.5 cm) or so thick. Wouldn't you

Figure 3.8. A mountain of home-sourced soil-building material. Photograph courtesy of Scott Perry

expect it to cause a dearth of nitrogen *in* the soil as it decomposes? Not on the surface, it seems. Moreover we are told to spread the stuff in late autumn and leave it lying there over winter, to be shallowly tilled in come springtime. So what can happen to decay this stuff lying on the cold ground under 2 feet (0.6 m) of snow? Plenty, as it turns out. Again the key is *fungal*, as opposed to bacterial, decay. At those temperatures the bacteria spores either die or encyst (that's nerdy jargon, meaning "going to bed until things warm up"). But soil fungi, it seems, can remain active at much lower temperatures, especially under a heavy blanket of snow and chips.

Come spring the chips are still there but somewhat transformed. They can now be worked into the top 2 to 3 inches (5.1–7.6 cm) without causing soil imbalances, or so I was told. I was also told I could follow this treatment with crops—any crops—and they would do well. I was skeptical—well, that's just me—but our apprentice Michael was really gung ho— well, that's just Michael—so we gave it a try. I seeded hull-less barley on a strip that, aside from the chip treatment, was not exceptionally fertile. We watched, and our reactions say more about our expectations than the actual results. The barley never showed any yellowing or chlorosis. It grew only moderately tall and gave us a so-so yield. Mike was crestfallen; the chips seemed to have worked no miracle on the piece. I was very impressed; it seemed to have caused no harm at all and actually gave us a modest crop

while adding a major humus boost for succeeding crops. I called it a success, but not a panacea. Nitrogen must be there, but it must come from other sources. The chips provided an energy substrate, a carbohydrate base for the fungi and eaters of fungi, and eaters of eaters, and somewhere in there nitrogen showed up. And is it so surprising? Every woody flake spawns a microcosm, a living universe in miniature, much of it invisible: moulds and mites and millipedes and earthworms and . . . oh yes, some high-protein barley for me. Can I spin gold from straw? So it would appear.

That was but one of a number of experiments I did, parlaying leaves or chips into something of higher value to me. I conducted another on a terraced area (see chapter 13 for a full description of the terraces). When I first created the terraces with a tractor and mouldboard plough, I did it in such as way that the topsoil of the uphill side ended up piled on top of the topsoil on the lower side. That created a deep, rich rock-free soil on the outer half of the terrace, and a scalped area of glacial drift on the inner half. The outer half I immediately planted to all the usual crops; the inner half would need years of improvement just to become usable. Where to begin? Nitrogen, of course, the key building block for all life. Where to get it? My garden has no borders, and they say the sky is 79 percent nitrogen. Legumes (or rather their mycorrhizae) fix sky nitrogen (N_2) into soil nitrogen (NO_3, nitrate), so I planted clover seed. Disappointing results: stunted, 3- to 6-inch (7.6–15.2 cm) plants, even though I had spread some wood ash there, knowing how legumes love alkaline minerals. Then it occurred to me that although the clover bacteria could make nitrate from the air, in that compacted subsoil there *was* no air. What they needed was some material to create interstices in the soil, regardless of its own fertility value; they needed humus. So I tried again, this time with the chip treatment. Not as much as I'd have liked, but tilling shallowly; I didn't want to glut the soil with more organic matter than it could digest, and I certainly didn't want any of those nasty Polly Fenols. This time the clover made a modest showing, though still nothing amazing.

After a two-year lay to allow the nodules to do their full work, I tilled in the clover, noting that the soil was noticeably darker to the depth that I worked in the chips. This time I covered the ground with a heavier chip mulch, confident that the high-nitrogen clover residue would enable the soil to digest more woody material. In those days I failed to appreciate the importance of fungal as opposed to bacterial action, but it still worked well, supposedly because the white mould did its work over the winter and prepared the chips for the further decay by aerobic bacteria come warmer weather. Come snow-go I again seeded the ground to red clover (in retrospect I believe alsike clover would've worked better, as it seems to abide compaction more graciously). This time I actually had to mow the clover and remove it, lest it smother itself. Some paradox: I'm forced to *export* tilth-building material from the very piece I'm most needing to fertilize! I suppose I could've composted those cuttings and returned the compost to that piece, but in fact I did not, which just shows the high eco-efficiency of red clover.

At the end of the second two-year cycle the soil was "decent" to a depth of 4 to 6 inches (10.2–15.2 cm), which is remarkable considering that it's supposed to take centuries to rebuild 1 inch (2.5 cm) of topsoil. I will *not* say that this piece was "rebuilt" so quickly, but that it could now support moderate crops like beans. Even now when I plant those terraces I spread twice as much compost on the inner half as on the outer, and crops like corn will still be noticeably shorter on the inside edge, but the gap is closing fast.

How I Use Ramial

I have come up with my own ways of utilizing ramial. The most straightforward is to use it as mulch that *never* gets turned in, such as on woody perennials. A thick layer under apple and pear trees does them a great benefit, as well as shrubby crops like hazelnuts, quince, elders, and medlars. Cane fruits are a no-brainer for anyone who has ever seen raspberries and blackberries luxuriating in the slash of a logging clear-cut. I took my cue from that when I planted my blackberry patch: I just spaded a slice in the old sod and stuck in the plants, with no effort to remove or chop up the sod. Nor did I add a molecule of compost or other fertilizer. I *did* apply a powerful herbicide in the form of 3 inches (7.6 cm) of chips around the plants. Actually these were not even ramial chips with their bark and cambium, but sawdust and scraps from my wood-splitting yard. Of course the sod provided onetime fertility (especially phosphorus) as it decayed, but all that we added year after year was sawdust and chips, and every spring the blackberries popped up through that trash with exuberance, each year growing new canes even taller than the previous year's. They are now well over 10 feet (3.0 m) high and loaded for bear every August.

While I stuff my face and stain my hands purple, I gloat inwardly that I am really eating sawdust. When used on permacrops the chips need never be turned in. The earth itself, using moulds and earthworms and rain and gravity, eats up all that woody rubbish from the bottom, while I just replenish it with a new layer on top.

I am experimenting with use of ramial chips on terrasols (sunchokes to some) and groundnuts (*Apios americana*). After they're harvested in late autumn I cover the area with chips and let it sit over winter. These permacrops have the advantage that I needn't turn in the chips in springtime but can leave them on the surface all the following summer as a mulch, provided the new crop can poke up through them. When I dig up the next crop the old chips will automatically get turned in, hopefully not too deeply.

Aside from using ramial chips directly to build up garden soil, I can also use them *in*directly to maintain and enhance the productivity of my grassland, which in turn provides grass for mulch or compost. I describe this in much greater detail in chapter 5.

I must emphasize the profoundly radical importance of forest-based fertility. Indeed it takes our very paradigm of forests and turns it on its head. For countless centuries forests have been viewed as a lower level of use compared with agriculture or urbanization. Initially we saw them as a dark and dangerous no-man's-land of wild beasts, an impenetrable obstacle to civilization. Now we have

Figure 3.9. This tiny patch of blackberries thrives with a mulch of ramial chips. It will produce many gallons of sweet berries for us, friends, and the birds, with a lot going to waste.

"progressed": We recognize their value for recreation, climate services, forest products—all very good, but we still fail to recognize their critical role in *promoting* civilization, that is sustainable civilization. We still say and think: Ah, yes, forests are lovely, but when push comes to shove, what are we going to choose, trees or people? At some point you can't have them both. Now we can see it all differently: We won't preserve forests just because we're nice people who care (in the long run, I never trust "nice people who care"); we'll preserve them because we are selfish and don't want to starve. We won't choose trees over people; we'll keep trees *in order to* feed people. "Of course I'm a tree hugger, I like to eat, don't you?" The main "ideal" we'll need is that of self-preservation; business, you see, nothing more. Count on it.

CHAPTER FOUR

Minerals: Whence and Whither?

Living tissue consists of materials that bulk rather large in appearance, but are actually unsubstantial. The major components of living tissue—nitrogen, hydrogen, and oxygen—are gases in their pure form, and so is oxidized carbon (carbon dioxide). Burn them or evaporate them, and they pretty much disappear, going back into the atmosphere until they are called into use again. One part remains: the minerals. Phosphorus, calcium, potassium, magnesium, and a host of other minerals make up a minuscule part of biomass, but their presence is necessary for all the myriad organic structures to form. Some minerals, such as cobalt and nickel, may not themselves be part of any structure, but must be present to serve as catalysts for biochemical reactions. They do not actively participate in the reaction, but they enable it to happen by their presence, sort of like a facilitator at a meeting.

Where do all these components come from? Nitrogen, hydrogen, oxygen, and carbon ultimately come from the atmosphere, so they circulate around the world freely and are rather easily appropriated into the biosphere. As long as air moves and breezes blow, those materials are cycled around the planet; no shortage anywhere. It's just a matter of incorporating them into the soil in durable forms, and the soil itself does that, if we let it.

The minerals, on the other hand, arise from the lithosphere, the solid bit under our feet. Although rocks and dust are present almost everywhere, some particular minerals, such as boron, fluorine, and iodine, are not so ubiquitous, and there's an interesting explanation why. Many of the components of rock well up from the earth's depths to be deposited in a process called mineralization (the geologist means something rather different from the biologist). This is not always uniform, and certain minerals may occur so sparsely in some places as to be inadequate for the plants that grow there. In addition, and more important, there are secondary geological processes that cause local deficiencies of certain elements.

From Rock to Dirt to Soil to Crop

The geological processes that act on minerals are always re-sorting and reassembling them. When primary igneous rocks surge up from the bowels of the earth to either extrude (as lava/basalt) or intrude (as granitic formations), they contain a rich assortment of all the elements. Granitic rocks, for example, contain a dazzling array of major and trace elements.

The processes by which primary rocks are pulverized, such as glaciation and chemical erosion, yield a medium—dirt—that is all-sufficient and ready for living things to convert into soil. The quartz in the rocks becomes sand, the feldspar becomes clay, and so on. And as primary rocks break down and re-form as sedimentary rocks like shale, sandstone, and limestone, they travel around, mostly via the movement of water and ice. Their deposition is selective—according to their physical properties like density or chemical properties like solubility—so that certain minerals may be either concentrated or removed from the original mix. Therefore some soils lack the whole panoply of elements that the original igneous rocks provided.

I consider myself blessed to live on unsorted glacial debris, a hodgepodge of largely igneous and metamorphic rocks that the last ice sheet dragged down from northern Maine and Quebec, while grinding it up in varying degrees. If certain elements (boron, for example) have been leached out of my soil by nature or human abuse, then possibly the remaining stones still contain those elements, waiting to be released by further grinding.

But what about soils that are deficient in a particular mineral (especially trace minerals)? In such a situation an obvious solution is to import that mineral from someplace where it is superabundant, such as a phosphate deposit or a lime quarry.

In particular there may be certain trace elements, like boron, iodine, and fluorine, that are just not there. Perhaps they were absent in the original igneous rocks; it happens. In such cases it only makes sense to import these minerals from wherever they are found: borax from dried-up lake beds in California, say, and seaweed from the nearest coast—places where these elements have accumulated and from which they would not be missed. The needed amounts of these trace elements are not so huge as to create a large carbon footprint, even if carried from afar (unlike calcium, potassium, and phosphorus).

Members of the remineralization movement (founded by John Hamaker) are obsessed with the notion of grinding up rocks and spreading the dust to revitalize our depleted soils (depleted by the two forms of erosion: water and the marketplace). I am in complete sympathy with their ideas, so long as this grinding is not powered by any fuel, which would be simply stupid, since they attribute much of the underlying problem to deforestation and fossil fuel use. I see no reason why such grinders might not be direct-driven by small windmills or waterwheels and operate only when there is enough breeze or flow (that's how they do it in Esperia).

Smart Mineral Management

Although mineral amendments may have their place, especially in soils that have been depleted by generations of crop removal, I prefer to avoid them, mainly by finding ways of accessing the minerals that are already there but unavailable and by closing the circle of life so that they are not constantly being siphoned off to the marketplace. I got out of the mining business many years ago in favour of the recycling business, and I don't wish to turn back.

A major problem with moving some minerals, such as calcium and potassium, is that they are heavy, and their processing and transportation typically involve a huge carbon footprint, something "sustainable" agriculture has yet to fully include in its calculations. The ecosystem and our own wallets dictate that we avoid this as much as possible.

The key question, then, is how can we *keep* those minerals on the land, once gathered and applied? If our farm and food system is such that those once exotic materials get cycled over and over, right on the same land, then a onetime application seems pretty sustainable; but if our soil management practices allow them to be carried away, again in a no-return broken cycle (like the marketplace), then how sustainable, indeed how organic, is that?

Let me tell you the story of how I keep minerals on my land. First, a little history. Back in 1973 I bought two bags of lime from the local farm store; years later I applied about 20 pounds (9.1 kg) of borax (20 Mule Team, like Ronald Reagan used to peddle), mostly on the hay fields. One year I bought in a bag of phosphate, though that was for the plots for the Scatterseed Project (see chapter 6), which are generally separate from our family gardens. That's all the minerals I can recall importing over 42 years. Of course, the tons of leaves I've hauled from

Farmington curbsides every year have a considerable mineral content. I also generate about 100 gallons (378.5 l) of wood ash each year (from trees that grow on nearby land, if not my own), most of which is used to "lime" the hay fields, but ultimately ends up on the gardens via the compost heap.

Even allowing for all these the amount of minerals I am importing onto my cropland each year is minuscule compared with what most market gardeners add to keep their soils productive, yet a recent lab test shows that my soil is very rich. The humus level is off the chart, which I believe helps account for everything else being so good. I was a bit surprised to see that the pH level was a bit low, though well within acceptable range, and that the potassium was slightly low, probably due in part to the low pH.

How can this be? When I bought the place in 1970, a test showed soil moderately deficient in nearly everything, despite eight years of lying fallow. For a century and a half before that, grain, apples, cream, syrup, livestock, and cannery crops had been exported from the land. Obviously it had been depleted by more than it was replenished. In the course of all these years I have been growing crops on this land while importing so very little, how is it that the minerals have not been (even more) depleted but, on the contrary, greatly improved?

Organicizing Minerals

The key to my mineral management system is not in the minerals I bring from elsewhere, but rather in my not sending them away. Obviously, the lime and borax I mentioned are not all the minerals I *add*; rather, they are the only minerals I *buy*. The rest comes in the form of those things—hay, leaves, chips, et cetera—that are products of the nearby landscape.

In a closed system such as I am trying to create, the minerals that *are* imported (whether from California or the adjacent woodland) are allowed to accumulate. I don't test my soil regularly—the recent test was the first one in decades, mainly for the purpose of this book; the original 1970 test was merely to confirm whether I should buy the place. I stay well on the safe side with minerals by being vigilant about

what departs the place. No, we don't require dinner guests to stay until they've used the privy, and yes, I do plan to be cremated and spread on the hay field (forget about my mind, "a corpse is a terrible thing to waste").

The most effective way I know of preventing loss of minerals through erosion is by organicizing them: tying them up in living or stable decaying matter, be it crop plants, centipede bodies, or actinomycetes. The most vulnerable moment for any nutrient loss is when it is mineralized, or reduced to a small simple molecule. Free ions of potassium or nitrate are easily lost until they are taken up and incorporated into their next life-form. Big complex organic compounds tend to get tangled up in one another and cling to adjacent soil particles, never travelling very far, certainly never leaving your land.

This is particularly true of trace elements, none so much so as iodine. Iodine has the double jeopardy of being both soluble and volatile; it can not only leach away like potassium, but also evaporate like nothing else I know, except maybe its close cousin chlorine. That is probably why so much of the world's iodine and chlorine (as well as bromine) are found in the oceans and are harvestable via seaweed. As with boron very little iodine is needed to keep soils and bodies healthy for a long time, its main role in human nutrition being to supply the thyroid, which regulates the metabolism of other nutrients. A bushel of kelp meal or a gallon of seaweed extract can supply enough for a lot of acreage—*for as long as you can hold on to it*. Avoiding bare soil is a no-brainer, since volatilization is a constant threat.

Even more than the other trace elements the best way to keep iodine with us is to maintain very high levels of soil humus, much higher than is feasible with animal manure. (How to keep humus high? Review chapters 1 and 2.) The nitrogen content of the humus is much less important than the volume of it. The key is cellulose, which in itself contains no minerals, just lots of carbon. Humus needs a modest amount of nitrogen to nourish the universe of microbes that digest that cellulose, but as decay continues new nitrogen appears in the system, presumably from the atmosphere. More to the present point all of this rich and diverse soil biology incorporates unto itself the wealth of hitherto soluble minerals (including iodine and lots of others) where it is stable until those life-forms die and release it yet again. Many of those life-forms are very short-lived (some bacteria have a life span of less than one hour), so the nutrients are returned to solution often. There need to be myriad other organisms waiting to snarf them up before they are somehow lost. The nonliving complement of the humus is no less essential for averting the loss of the soluble nutrients. Its role is to physically attach to the minerals until they are incorporated into the next organism. Humus, humus, and more humus.

One trace mineral that sometimes is just not there, perhaps not even in the parent rocks, is boron. The amount needed is truly minuscule—indeed you must be careful not to overdo it, as it can just as easily become toxic. A deficiency is characterized by brown heart rot in beets and turnips, and apples with translucent interiors (water core), blasted foliage, and witch's broom growths (these symptoms may have other causes, so a soil test may be essential to avoid excessive application).

However much or little boron is in the soil, its availability can depend on soil conditions. For example, in alkaline soils it may simply be locked up in precipitate form, whereas in sandy acidic soils it is prone to leaching downward, especially where rainfall is abundant. As with iodine abundant humus is key to stabilizing available boron. But in cases where boron is just not there at all, bring it in, at whatever expense, from wherever it is found. It is not the importation that is problematic; I mean, a few boxes of 20 Mule Team Borax (sodium tetraborate, *not* the detergent) per acre will resolve any boron deficiency for years to come, indeed indefinitely *if* the escape routes are kept blocked. Keeping it there is the problematic part; boron is most vulnerable to erosion via the marketplace, especially in every bushel of apples or root crops that leaves the farm, never to return. I know of no remedy for this dilemma; it is inherent to the marketplace. Fortunately a very little boron goes a very long way for a long time, but ultimately the only way

I know to hold it on the land is to close the circle by avoiding the marketplace; the only other solution to any mineral deficiency is an external support system.

I do two other things to avoid the loss of minerals (and other nutrients) from the soil: terracing and mulching. Even moderate slopes can lose lots of goodies to runoff, both in solution and in the physical carrying away of particles. Terracing largely averts this by giving water no place to go but down into the soil. (Refer to chapter 13 for a description of how I make terraces.) If rain is so heavy that even high-humus soil cannot absorb it all and it overflows the terrace, the sod-covered margins should trap it there.

Plant "Miners"

Fortunately very few soils are completely devoid of most minerals; rather they may be inaccessible to our plants, for a number of reasons. They may have originally been present in the topsoil, but have been leached downward, or they may be so dispersed in the soil that our crop plants, especially those Band-Aid species, cannot reach or accumulate enough of them. Then again they may actually be abundant, but locked up in insoluble compounds from which crop plants cannot extract them, in which case they might as well not be there.

The downward leaching of nutrients into the subsoil is less of a concern to me. In the course of rotating crops around, some deep-rooted crops will bring stuff back up to the surface. It would be a greater worry if my system were to allow an excess of soluble nutrients to be available at one time, especially in the early spring; again, that is the advantage of ample active humus.

I'd like to tell you more about those deep-rooted crops and other plants that can reach down farther than their neighbours and bring up those minerals to the ultimate benefit of everyone. The "miners" of the plant kingdom use taproots instead of picks and shovels, and their ore cars may come up with treasure from several feet deep. Alfalfa and sweet clover are very effective, but so are many non-legume taproot plants like yellow dock and burdock. The ultimate miners are the trees, especially old-growth hardwoods whose fine root systems reach very deep, probing the boundaries of the biosphere. Ultimately they are the difference between dirt and soil.

Certain other plants have a knack for unlocking rock particles and converting the dissolved minerals into organic forms (for example, chelates), which their less resourceful neighbours can then appropriate. For example, soils that test low in phosphorus may contain significant amounts of it in the insoluble form of apatite, or calcium phosphate, useless to most plants but not to buckwheat, which can "eat" the particles and render the phosphorus useful to the rest of the biosphere. A small number of plants can do a similar trick with silica. Now, silica is the most abundant element in the earth's crust, and many species use it to form a sort of armour against attacks from pests and diseases. Yet silica is among the least soluble soil minerals, so most plants need it to be pre-digested for them. Lichens can do that—that's how some soils are formed—but to an extent so can ferns and sedges and some of the succulent plants, perhaps all of them to some slight degree.

In my younger days as a prospector we once did a procedure called bioassay (although the word seems to have acquired a very different connotation in more recent times): At different points on a survey grid we collected samples of tree clippings—only the last three years' growth. Back in the lab they were burned and the ash tested for various minerals, especially the ones we were after: copper, lead, and zinc. These were relatively shallow-rooted red spruces, yet even they could extract measurable quantities of those metals (of which the main ore body was hundreds of feet below). If you're searching for treasure, ask a tree for directions.

Of course your garden or cropland is where it is, and chances are there are trees, or forest, not too far away. So even though one ecosystem is supporting the other, you have a pretty sustainable arrangement. The ideal agricultural system I envision is one in which the landscape is a patchwork of different uses, cropland here, pasture there, woodlot over there, all within the basic subsistence unit of a homestead.

The level of overall self-reliance is very high; the carbon footprint is very low. (That's how they do it in Esperia.)

Supplying Minerals Through Wood Ash

You will notice that in this entire discussion of minerals I have so far not said much about amendments, or mineral concentrates from off the farm, beyond a small handful of strategic trace elements. I've said nothing of dolomitic lime and colloidal phosphate and Sul-Po-Mag and azomite. I have little or no experience with those, they don't grow here, and they all have a huge carbon footprint. Therefore I don't much use them and I have little to say about them. That's partly because they are from off the land and involve a commitment to the marketplace and partly because they are mined concentrates, isolated from their whole context.

I generally eschew concentrates: I prefer perfumes from flowers to those from distilleries, I get my vitamins from food rather than pills, and my happiness from within me rather than from a bottle. I favour the crude impure matrix over the refined essence. I actually like the fibre, the pulp; it is more than filler; it is the marrow. I relate to the gritty riff-raff more than the cultured elite, and yes, I like my minerals mixed in with lots of humus.

There is one mineral concentrate that does grow on my land, and that is wood ash. I do not concentrate (that is, burn stuff) for the sake of making a soil amendment; rather I burn wood to heat my home and cook my food, and the ash is an unavoidable residue, to be returned whence it came. I must confess that in my early years on the land, I used fire as a tool for getting rid of waste material like brushwood. Brush piles were an unavoidable product of clearing land and harvesting firewood, and burning seemed a logical way of disencumbering the land while "creating" a fertilizing material.

Early on someone pointed out that I was destroying organic matter while wasting the energy therein

Figure 4.1. Wood ash is the one mineral supplement we don't need to import. Photograph courtesy of Yaicha Cowell-Sarofeen

and releasing sequestered carbon. It did bother me, but what to do about it? The idea of chipping and composting occurred to me, but buying such machinery was out of the question. At some point I was able to purchase a 12 hp Gravely walking tractor with a chipper/shredder attachment, and my relationship to my land changed dramatically for the better. I messed around with schemes for composting and mulching with it. I later came across the ideas of Giles Lemieux and colleagues at Laval University regarding use of chipped ramial wood (brush) as a *central part* of long-term soil building. From then on I saw those brush piles not just as a source of ash, but also as a source of minerals in their organic context. Nevertheless, I still have ash from fuel consumption.

This is all hearsay, but lab analyses of wood ash show potassium to be only 10 percent of the total mass, whereas calcium (carbonate) is 25 to 45 percent. This makes me wonder what the *other* 45 to 65 percent is, since magnesium, silicon, iron, and so on, are all trace elements; phosphorus is a very minor component, since much of the original content is supposedly lost in vapourization, as is sulphur.

These analyses are based on samples from biomass incinerators, which may yield a different ash from our own cookstoves and fireplaces. For one thing, that incineration involves higher temperatures than your stove or fireplace; therefore, ash from them may have less loss of those minerals (phosphorus, sulphur, and so forth). Incinerators produce lots of "fly ash"—fine ash that goes up the flue with the smoke, probably due to the higher heat, stronger draft, and particle size. It would be interesting to know the analysis of that fly ash, compared with the firebox residue.

Regardless of any analysis this is what we know for sure: Wood ash is the solid residue of land plants, as opposed to lime, which is from fossilized marine organisms. Therefore it stands to reason that the ash must contain all the minerals needed by land plants in a commensurate proportion. Granted the wood ash is not representative of the whole plant. We know that leaves and ramial chips contain a much greater percentage of minerals and in a different proportion than the trunk wood. Also wood ash is from trees and not annual crop plants, whose needs are presumably not identical. However, as long as we are recycling *all* that matter in some form, whether as ash or compost, then *all* of the minerals are being held within the system that produced them. The only net loss is the heat that warmed us and our food. Oh yes, there is all that nitrogen, hydrogen, and oxygen that went into the atmosphere, along with carbon dioxide unfortunately, but as long as the removed trees are replaced by new ones and the overall forest health is maintained, all those gases will be sequestered again and again, with no net loss. There's a big difference between carbon that is already in the biosphere and that which is sequestered in layers of anthracite or petroleum.

The pH Factor

If you read some of the literature put out by various universities and government agronomists, you might get the impression that wood ash is a toxic waste rather than a precious resource. That is not completely due to the mentality of left-brained, lab-coated technocrats in the thrall of corporate agribusiness; there is also truth behind it. You see, although the elemental ingredients are nothing but beneficial, wood ash is also very alkaline. That is correctly cited as both an advantage (it can replace mined lime as a balancer of overly acidic soils) and a problem (it can make soils *too* alkaline, locking up phosphorous, iron, and other nutrients). It is a question of balance: how much ash

on how much area? One source suggests a maximum of 10 to 15 pounds per 1,000 square feet (4.5–6.8 kg per 92.9 sq m). In addition to the rules of thumb you can always get a soil test; indeed you can get the stuff to test it yourself. Either way it's pretty cheap, especially if all you need to know is pH.

Many years ago we encountered a problem ourselves due to the alkalinity of wood ash. After years of spreading all our wood ash (including from brush pile burning) on our croplands, we discovered that our soil in some places had a pH of 8.3! That's about right for raising tumbleweed and sagebrush, but much too high for any garden crops. In essence we were taking the alkaline residue generated by several acres' annual tree growth and concentrating it on a fraction of an acre of cropland, and doing that year after year.

Once we realized the problem, the solution was obvious. We have several acres of grassland whose growth we mowed and removed every year for compost and mulch. That grassland badly needed some liming, which had not happened in at least 20 years, though hay crops had been removed many of those years. Well, I could not afford much lime, but what about all the wood ash that I could no longer apply directly to my garden crops? If I used it to "lime" the grassland—ash contains nearly half as much calcium as agricultural lime, but many other minerals as well—then those minerals might still eventually end up on our cropland. The difference is that those minerals would arrive there in a vehicle of organic matter (composted hay) that would break down slowly, releasing those minerals in a more sustainable manner. And of course the application of ash to the grassland would increase the overall grass yield, giving us more biomass, more humus to assimilate those minerals. Win–win. (There's more about this in the next chapter.)

I should point out that when I speak of wood ash, I'm assuming clean wood ash, free of wood with lead paint, preservatives, or nails (the iron content in the ash isn't a problem, but it's rough on tires and bare feet). I'm also assuming complete aerobic combustion. Ash from airtight stoves is likely to contain creosote and lots of bad stuff that soil life may not appreciate. By the way, hardwoods produce three times as much ash as conifers and are less likely to generate those toxic products of improper combustion (such as formaldehyde and alcohol). I look for light grey fluffy ash with a minimum of charcoal in it. The charcoal itself is actually a plus, but its incredible absorbency makes it a carrier of many other things, including toxic by-products of anaerobic combustion. Plant matter itself is the result of numerous reduction reactions, such as photosynthesis, that lock up energy as food; burning plant material is largely an oxidation reaction, which releases energy.

Proper use of wood ash is a matter of balance, but how do you determine that balance? There are rules of thumb that can be quite inadequate. Someone suggests 10 to 15 pounds per 1,000 square feet (4.5–6.8 kg per 92.9 sq m); however, is that dry or damp? There is a *lot* of difference. Dry ash may be simpler to spread, assuming a windless day, as it drifts around before settling, hopefully where you want it to land. Dampened ash is easier to control and safer, as you can better avoid breathing it. I have spread dry ash with my bare hands, but I don't recommend it. It combines with sweat to make an alkaline solution that is pretty harsh on the skin (especially any that gets under your fingernails). I usually spread ash with a small scoop that puts it where I direct it, hopefully not on me.

Some sources portray wood ash as having marginal fertility value—I suppose compared with lime or commercial fertilizer—but no one disparages its potency as a pH balancer. The latter is partly because the calcium in lime is tied up with carbonate to render it less soluble, whereas the various mineral elements in wood ash are mainly in the form of oxides and hydroxides, not only very alkaline but quite soluble.

If the calcium and potassium, the two main ingredients, are only 35 to 55 percent of the total content, then what is the rest? The unburned charcoal particles may account for a tiny bit, but not much. I have made lye before by leaching ashes and boiling down the leachate for soapmaking. There is a lot of *something* left, referred to as spent ash, and I find that it is still very valuable as fertilizer. I can only assume that that residue of a residue is largely more of the same minerals, still locked up in a precipitate form and available to plants over a longer time, probably less potent as a pH changer but precious as a mineral source nonetheless. I also know that wood ash contains most of the needed trace minerals, things like iron, copper, manganese, and zinc, in roughly the same proportions that other plants need them, whereas lime has none. I have to wonder if the various lab analyses take the total content into account or merely measure the more soluble part. I appreciate the laboratory as a source of answers, but at best they can only answer the questions we ask them. Therefore we must be very careful to ask the right questions.

CHAPTER FIVE

Grassland Improvement and Management

In addition to using leaves and ramial as mulch and in compost, I employ a yet more indirect way of parlaying forest stuff into human food: by way of the pasture. Despite my enthusiasm for the forest connection, I still intend to have grassland. Given that I have orchard I will automatically have grassland, since that's how most fruit trees naturally grow, not in deep forest.

If you research the topic of grassland, you immediately reach the conclusion that grassland exists for one purpose only: to feed livestock. That might lead you to conclude that if I am not raising livestock, then grassland is of no interest to me. In fact nothing could be farther from the truth. Remember, grassland is second only to forest as an eco-efficient ecosystem. Grass is great for composting and mulching, and it breaks down more quickly than wood while containing more nitrogen. It is an excellent adjunct to the forest stuff, giving us both longer- and shorter-term humus building. And anyway, I don't want to live in the deep dark forest; I aesthetically prefer a mixed habitat with some open land around me.

My grasslands are far more eco-efficient than my cultivated land, yet much less so than my forest. However, while forest materials (tree leaves and ramial wood chips) may be bigger producers of soil-building biomass, my hay field has an extremely important role to play in supporting the ecological diversity that makes my gardens sustainable. Quite aside from the succulent biomass that supplies my compost piles and yields stuff for mulch, the community of pest-eating birds and insects is a vital stabilizing complement to my system that would not exist in forest alone. It has a much shorter decay cycle than woody materials, breaking down and releasing nutrients more readily. Since the "prairie" ecosystem sequesters much less carbon than trees, the proportion of readily available nitrogen is more appropriate for cultivated crops, most of which evolved from grassland species, after all.

In at least one respect my grassland needs are much like any dairy farmer's: I want all I can get. Yield per acre and quality and low inputs, that's what it's all about. By "low input" I emphasize low *off-farm* input; if this part of my food system is to be as eco-efficient as the others, then I must avoid covering any deficits with imports, which would be a support system as opposed to a production system. And regarding quality my criteria are very different from the herdsman, who aspires to a pure stand of grass or grass/legume mix. Plants that are useless for graziers are for me a vital part of my "grassland."

The problem is: How do I maintain the long-term productivity of the grasslands? Because of the great

Figure 5.1. Cutting and hauling hay at the peak of maturity.

eco-efficiency of grass, I do not need to replace all I remove (were that so, life on land would have been impossible). Although I can remove crop after crop without putting anything back, sooner or later I do have to give it something, but what? Couldn't I take the grass from a portion of the field, compost it, and return it to the whole piece as compost? Yeah, but that's a lot of work, and I am both lazy and greedy. I want a greater return for less investment—I want to take *all* that grass and return something of less value in terms of earth-bounty.

The Cover-the-Earth Strategy

You may remember the "cover-the-earth effect" I described in chapter 3. (If not, I suggest you review it; it is one of the most important ideas in this book.) To recap: If you cover the ground closely with *any* material for an extended period (weeks or months),

that ground will become enriched all out of proportion to the inherent fertility value of the cover material itself.

Here's something I did once: A section of hay field was badly run out from repeated mowings with nothing returned to it. Even worse the neighbouring hedgerow was sending its tree roots out there and presumably outcompeting the sod for what nutrients were there. First I tried spreading a thin mulch of ramial chips over part of it, and left it for a year. Not so good—it was enough to nearly smother what little grass was there, and it was questionable how much I really gained. Next (a year or two later) I applied ramial chips to the adjoining area, but *not* evenly spread. I deliberately plunked down a forkful here, another there, leaving the piece pocked with chip-pies every couple of feet. Now where the dense chip-pies lay there was no improvement at all—as before the dense mulch annihilated the sod, which immediately began to decay. On the perimeters of these chip-pies rings of much lusher grass appeared. That grass was benefiting from the moist, rich remnants of its dead neighbours as much as from the woody mulch. As the season passed and another followed, those rings kept expanding outward until they coalesced with one another, leaving the ground either noticeably improved (between the chip-pies) or noticeably dead (under the chip-pies). However, the chip-pie areas were not idle all this time. As the chips themselves began to break down, the mulched bits were re-invaded by the surrounding grass, especially the quack grass; quack grass may be a plague in the garden, but there are few comparable allies in the veganic hay field (more about this later in the *Pasture Components* section of this chapter). By the end of the second year the whole piece was noticeably improved.

Notice I am *not* saying "dramatically improved," as it might've been had I applied a few barrowfuls of chicken poo. In my garden-without-borders I factor in many things: what it cost the land to grow the brush, what it cost me to chip it, how much brush I had, and how badly it was needed elsewhere. The more I reckon up all the externalities on both sides, the better the scorecard looks for ramial. I have parlayed a few barrowfuls of "worthless" woody rubbish into a trailer load of succulent high-nitrogen grasses and pasture weeds, which will soon be further converted by soil alchemy into onions and peppers and beans (oh my!). Just some more gold spun from straw.

There are other things I do to improve grassland with or without ramial chips. The first thing I do is examine the grassland carefully and note which areas are better or worse and why they're like that. For example, I had one area that was low and flat and rather wet; it hadn't been mown for decades but nevertheless was run out. The sod was thatch-bound, so choked by its own buildup of previous growth that only thin spears of grass poked through the thatch. That thatch should have decayed as fast as the new growth formed it, but the area was too wet and matted, almost anaerobic. I scrubbed over the area repeatedly with the hay rake and even with a steel garden rake to pull out that excess thatch and take

it off the piece. Now the wetness of the area became an asset, and it began pumping out grass, which I removed so that no thatch could build up.

Where did I move that grass to? Perhaps 100 feet (30.5 m) away was another area that was steep and dry—*too* well drained. In fact the minerals were leaching downhill, and far from having a thatch buildup the sod was so thin that there were bald spots threatening to grow worse. Any growth there was mineralized before it could form a sustainable humus. So I moved the thatch from the first area over to this one and spread it thickly. There was very little actual grass growing there—rather a sparse stand of goatsbeard and black-eyed Susan—so I broadcast a mixture of grass and legume seeds to see what combination would take best. I would like to have irrigated the piece but had to settle for the moisture-retaining mulch. The grass mix came in thinly at first, but improved as the thatch began to decay. The following season I mowed what grass there was and raked it into windrows, but left it lying there. That had an effect similar to the chip-pies: The narrow windrows smothered the grass under them, but the wide sod strips between them were notably invigourated, and soon the sod areas rebounded with vigour and recolonized the mulched strips.

More Strategies

When I first began to investigate grassland improvement, I tried to build up one weak area by "liming" it with wood ashes alone. Afterward I noticed the yield of grass did not improve, but the colour was a much darker green. That was due to the potassium in the ash. So I could see that applying wood ash to grassland is a great way to create a mineral-rich material in an organicized and stable form that I could then apply to my gardens. But clearly the wood ash was not what the grass needed for *growth*.

It occurred to me then that the grass wanted nitrate, which I didn't have to give it. Clover of course could "make" nitrate where there was none, provided the clover got what *it* needed to function: lots of potassium and other minerals, which were found in the wood ash. So a pound of red clover seed had a profound effect. In fact, for a year or two after I sowed the clover seed in my experimental grassy area, it appeared to have become nothing but clover, save a few spears of grass here and there. Then the clover exhibited something that has been called self-toxicity: It nearly died out, replaced with a luxuriance of herd's-grass (timothy), fescue, and redtop. I am very sure there was nothing self-toxic about the clover; by fixing all that nitrate, it merely gave its competitor—the grass—a huge advantage over itself.

The clover made a slow comeback and henceforth continued as a working component of that very prosperous grass community. Several times since I have reseeded that upper field with red clover, but I now realize it was pointless. There are always plenty of dwarfish clover plants lurking in the grass. All I need to encourage them is some combination of water, ash, or spot mulching, and they will proliferate to the betterment of the entire pasture community, including me.

Let me say this very clearly: There are other, more conventional ways to improve the production of grassland. Someone will point out: But if I put some cow manure on that area and perhaps a spreaderful of lime, the grass will grow lush without messing around with wood chips and ashes and stuff. That certainly is true, but it begs the question of just how much earth-bounty it took to make that manure. If you're figuring that manure to be just someone else's waste, then you're missing the whole point. It is not waste to the earth, any more than meat or milk are; it is a product that cost dearly to produce, and that cost must be factored into the value of the crops we produce from it. To ignore that is to confine your view to the apparent borders of your garden. It is an overly narrow concept of what is organic, of what is sustainable.

Maintaining the Balance

Although the main component of grassland is grass, its other constituents—broadleaf plants—are no less important in maintaining the whole. The same triumvirate of plant groups—grasses, legumes, and non-leguminous broadleaf "weeds"—that rules over a sustainable crop system is also found on our dinner plates as grains, pulses, and veggies. For a sustainable hay field I must maximize all those components. The system largely does this on its own, but there is much that I can do to enhance it. For example, mammals, birds, insects, and the wind will ensure that a highly varied mix of seeds is present, yet I can still add to that.

Due to the eco-efficiency of grassland, adding relatively small amounts of fertility-enhancing material such as ramial chips can generate a disproportionate return in soil-building biomass. However, pasture plants have a self-asserting power far beyond the export of hay, which is, after all, only the top growth of the ecosystem. Grass and all of its cohorts (legumes and other broadleafs) are all the while expanding their root systems, while older root growth is senescing, dying, and decaying. Therein lies much of the sustaining power of the system: It is creating its own fertility from within. However often we may "steal" the top growth by mowing, the underground part continues merrily along, converting, storing, and releasing the energy for its own continuity. *This* is why grazing land continues to improve out of proportion to what is added, despite what is removed. It is not so much what the animal *adds* in the form of manure—in the net account it is adding nothing at all—it is rather what the animal is *doing*: keeping the grass "mown" short, which promotes grass over less eco-efficient crops. We can do that mowing ourselves and convert the growth into human food more directly and with greater eco-efficiency.

Because of this self-sustaining power of grassland, it is easier to maintain than to create from scratch. Whether "scratch" means converting cultivated cropland or newly cleared forest, it is obviously necessary to introduce seed of all the species we wish to have there. Birds and wind will take too long to do the job, and what is already in the soil seed bank is not the kind of species we ultimately need (although those Band-Aid species may contribute to an early beginning). From that initial seeding the grassland will largely determine its own composition depending on many factors, including your management practices.

An elderly friend of mine, Bing Etzell, once explained his philosophy of creating and maintaining grassland in a single word: mow. He showed me a part of his large lawn, which he had wrested from the adjoining old-growth forest many years before. He told me he had cut stumps flush to the ground, but never removed them, just let them rot. He contented himself with constantly cutting any stump suckers that sprouted, as well as mowing any of the forest plants (ferns and the like) that continued to come up. I assume he sowed some grass seed, instead of relying on natural seeding (by birds, wind, et cetera), but his key strategy was to continually repress anything that didn't act like grass—in other words plants that constantly initiated a new growth from the spreen rather than the top. This is mainly a characteristic of monocots, though some dicots like clover will be effectively the same; annual flowering weeds will be choked out.

My friend's lawn didn't become lawn overnight. During several years of nothing but mowing with nothing removed, the erstwhile forest evolved into grassland—a residential lawn—with no fertilization beyond the residual forest humus, which obviously was enormous. My friend was under no delusions about the inferior eco-efficiency of this (he was after all trained as a horticulturalist)—he wanted lawn—but he pointed out emphatically that were he to cease maintaining that artificial ecosystem by frequent mowing, the land would soon follow its own path, which would lead right back to the old-growth forest.

In olden days farmers in some parts of England used a practice called "flowing the meadows" to maintain and enrich meadow pasture. These were lowland fields that were naturally quite level, and a ridge could easily be added around them to allow water from nearby streams to flood the fields, stand on the piece for a period of time, and then be drained off. On drier upland I assume they could have allowed the water to simply drain downward into the soil. It is suggested that the effectiveness of this method lay in the minerals that were borne onto the piece by the floodwater, sort of like Egyptian fields being fertilized by the Nile's overflow. That may be true, though I find it hard to imagine rivers flowing through English

countryside having quite the burden of silt and soluble minerals that we see in the Nile. I rather think the effect was due mainly to the water itself. Grasses in general, and meadow grasses in particular, benefit more from regular moisture than many of the more drought-tolerant dicots. Typically hay fields make their first flush of growth in early spring, when the soil is moist and cool. Warmer, drier weather tends to favour flowering and seed maturing. In hot dry summers grass production shuts down significantly, regardless of the fertility of the soil. Without enough moisture that fertility is inaccessible to grasses, and so dicots see their chance: Clover and daisies and asters and mustard and their ilk proliferate and set seed if allowed to.

I have dabbled with irrigating small areas of my regular hay field in the late season when natural rainfall isn't having enough effect. Where I have done it—moving multiple hoses around from the irrigation tank—the result has been fairly dramatic: Those areas keep pumping out lush greenery when other areas are more or less dormant, growing as best they can with what they have. My attempts at this have been very haphazard; to keep moving the hoses or sprinklers around requires a modicum of timing and attention, which I can ill spare from my cultivated crops. It's not usually a shortage of available water; my system can pump 700 gallons (2,649.8 l) a day when the sun shines (when it doesn't there is less need to irrigate), and my two tanks have 2,500 gallons (9,463.5 l) capacity, ample for the gardens and some to spare. In fact, I am always so pressed for time that my garden crops often sit there languishing while both tanks are full. Although I'm expanding my pumping and storage capacity, it will avail little without my time to manage it all; it is a classic example of the Chinese proverb: "The best fertilizer is the footsteps of the farmer."

Pasture Components

The preferred pasture grasses in Maine, for purposes of feeding cattle, are Kentucky bluegrass, timothy (or herd's-grass), smooth brome, fescue, perennial rye, and redtop. There are some other grasses that are also common, but not so popular. Quack grass may be a major nuisance in the garden, but I know of no other plant quite so useful in the hay field. Its aggressive network of high-carbohydrate rhizomes create a thick sod that does a lot to retain nutrients and moisture. It makes several cuttings per season of high-quality herbage. I occasionally let it get by me through neglect, so there is plenty of ripe quack grass seed in my compost and/or mulch. But I find that quack grass is relatively slow to infest cultivated land *by seed*. Its intrusiveness seems to arise more from the rhizomes in the garden edges and lynchets, which are ever ready to invade the cultivated soil.

I suspect that quack grass has perfected its asexual strategy at the expense of sexual reproduction. I read once that agronomists were experimenting with quack grass as a source of industrial starch (from the rhizomes), but that they had difficulty establishing pure stands! While this news might be greeted with knowing guffaws by those who have contended with this aggressive weed, I am not wholly surprised at the apparent irony, assuming they started with seed, as opposed to using rhizome fragments. Like other species with persistent and high-nutrient root systems (think milkweed and comfrey and terrasols), a single cultivation will only spread the plant still farther, whereas repeated and well-timed cultivation will repress it if not annihilate it. Therefore such management practices can be used to promote a certain species as well as restrict it.

Let's say that a field is infested (depending on your perspective) with both quack grass and foxtail millet. Annual grasses such as foxtail millet are the opposite of perennials like quackgrass; the former make lots of highly viable seed that appeals to birds (especially mourning doves). Annual grasses get around. In this field of quack grass and foxtail millet, a single harrowing will favour the quack grass and repress the millet, whereas repeated tilling will reverse the mix, as the rhizomes of the quack grass are subdued but the dormant seed of the millet is poised to resurge. (Although harrowing, and especially discing, is a very effective strategy for managing grassland, I prefer procedures that don't require a big heavy gas-burning machine.)

The most sustainable and eco-efficient system is the most diversified. A dairy farmer has a different perspective from mine; a pure grass-legume mix is just perfect for domestic cattle. Daisies, buttercups, mustard, and milkweed are seen more as nuisances that don't convert to milk as well as timothy or clover. My own livestock—earthworms, centipedes, bacteria, et cetera—are much less picky. They aren't much concerned about *what* I mow as long as there is *lots* of it. Many pasture "weeds" like goldenrod produce more biomass with less input, but they are not as milk-yielding. One reason why ill-managed pastures, groaning under compaction and close-cropping, come into goldenrod and thistle and Queen Anne's lace is that those species rebuild the overgrazed soil while encouraging the livestock to leave it alone for a while. Although I would be disappointed to see those species dominate in my grassland, I welcome them as useful complements.

Grasses

I confess to being not very up on my grasses. There are several species in my fields that I cannot identify for certain. Although I recognize all of the dicots there, I do not know the nuances of the monocots' behaviour and preferences as well as I ought to. What meagre acquaintance I have with them I will now attempt to share.

Timothy (*Phleum pratense*). A wonderful grass by any standard, timothy is not a New World native, but was brought from England in the 1700s by a Maryland farmer named Timothy Hanson. I first knew it by its local name, herd's-grass, and one version of its story is that it was previously described by John Hurd of New Hampshire, who called it Hurd's grass, hence herd's-grass. From the name you might infer that it is popular with all classes of herded stock (don't try herding cats), but it is also an excellent addition to compost piles. I've plucked the seed heads individually by hand and rubbed and winnowed them by hand. I've never bothered to harvest more than a cup or two—it's one of the cheaper grass seeds—but it would not be difficult to ramp that up given the need. I'm probably unlikely to have the need, however, as timothy is very self-sustaining and reseeds itself easily. My paltry production would be ample to reseed the occasional bare spot or two-year green-manure rotations.

Smooth brome grass (*Bromus inermis*). My elderly neighbour Lucian told me once that brome grass was more challenging than some grasses to get established, but that once established it would stand for several decades. That matches my own observation; down on the intervale (floodplain field), there used to be a lot of brome, which Lucian said had been seeded down over 30 years previous (now mostly grown to alders). Moreover he said its feed value was high due to a large amount of leaf compared with stem. Again, that subtlety may be lost on the red wrigglers in my compost heaps, but even they appreciate its massive production of easily decomposed organic matter.

Orchard grass (*Dactylis glomerata*). One corner of my hay field used to have a lot of orchard grass. It was there when I first arrived and seemed to thrive even more after I set out young apple trees there. Orchard grass seemed like an appropriate name, but why? As the trees have grown bigger the orchard grass has thrived, making me suspect that the grass tolerates or even prefers a light shade, at least in comparison with other grasses. In another part of the field an invading growth of black locust was cleared and for several years thereafter produced a flush of young locust suckers that I mowed (with the rotary mower, which chopped them up) and composted, as the thorns on the new growth were soft and rubbery. Anyway orchard grass had been present there before, and when the locust was cleared the orchard grass responded with the greatest vigour and persists there to this day, although the locust has largely died out. It seems to thrive adjacent to brush piles that are awaiting chipping; that might have implications for its management.

Be all that as it may, one day when haying for my elderly friend Orlando Small, I pointed out how luxuriantly the orchard grass grew in a particular section of his big hay field—in this case far removed from any trees. I supposed that he would exult in its

Figure 5.2. These piles of newly cut hay are waiting to be transported to compost bins.

early lushness, but he instead complained of it. Its earliness only meant that it went by all the sooner, not remaining long in "chewing quality" before running to seed. Even worse all that rank growth was mainly stem, not leaf, giving it less nutritional value for milking animals. That's just fine with me: lots of early biomass to get the compost piles off to a roaring start. One of Orlando's complaints that I can relate to is its habit of forming tussocks—raised bunches of tough sod that catch the scythe point and make it hard to mow short. The rotary mower, on the other hand, tends to scalp those humps and probably harms the grass somewhat. However, even that has some advantages: The tussock-habit tends to leave some open spaces around itself where other species—broadleafs—can thrive as useful companions.

Redtop (*Agrostis gigantea*). Redtop is a very fine-textured species, which Lucian extolled for that very fact: Livestock could chew it easily, and it cured to hay quickly. These assets count little for me, and I do not feel a need to promote redtop, as it seems only moderately productive, though I am content to have it find its own place in the mix.

Various other grasses are all valuable components of my grassland. I don't know them well enough to appreciate their peculiar assets and liabilities, though I'm sure they have both and I should learn them.

Sweeping Up Seed

On a few occasions I have cleaned out the bottoms of the hay barns of farmers I've helped with haying. Perhaps more valuable than the few pickup loads of broken bales for compost are the sweepings from the floor underneath, a thick layer of chaff and dormant seed. This I would scatter in my own hay field, particularly in thin spots, but anywhere I wished to ensure the diversity of the population. If I overseeded, the winners would crowd out the losers and whoever was best adapted for that spot would thrive best. There were probably no species that were not already endemic to my farm, but the more ubiquitous each was, the more opportunity for niche-adaptation.

Broadleaf Pasture Plants

I have often deliberately allowed goldenrod and wild asters to ripen seed (well, not always deliberately, but cheerfully), feeling sure that as long as I mow regularly (not always a valid assumption) they would contribute to the yield of biomass without getting out of hand. And exactly what is "getting out of hand"? There is an area downhill from my root cellar that was neglected for several years and got out of hand: It is almost solid goldenrod mixed with a few ferns. The mowing there is always exceptionally heavy, indeed difficult to get through, and the aftermath (second cutting), if and when I get to it, is even better than any grass. Again that resinous stalky stuff would be unsuitable for any system involving milk animals (they would eat it but it would taint the milk), but it is of great value to me.

One year some salsify went feral from one of my seed isolation plots and promptly infested the adjacent hay field. The *fest* in *infested* means "pest," and the word would apply only if I chose to see it that way. Instead I see the salsify as a strong ally in the maintenance of the grassland soil. You see, it makes a large amount of fleshy taproots that ultimately decay, leaving the soil both enriched and aerated by all the remaining channels.

Other taproot plants can be equally useful in creating underground biomass while loosening the compacted soil and pumping minerals up to the shallower root zones of other plants. Yellow dock is a master at this: While it does not produce a great deal of aboveground growth of itself, it promotes the growth of neighbouring plants that have more biomass. Whenever I spot a dock plant ripening its seed, I am careful to spare it until it is ready to shatter. If it reseeds the area around it, so much to the good, but even better if I can collect it and sow it in a new part of the field.

Comfrey is a formidable producer of valuable biomass, both above- and belowground. Among other nutrients it is an accumulator of potassium, so it is one effective way of cycling wood ash into the garden via the hay field in an organicized form. In one part of the orchard I once planted a lot of comfrey by simply slitting the sod with a spade and dropping in pieces of root cutting. To my delight most of the plants are still producing plenty of compost material nearly 30 years later. Moreover, the grass and other surrounding plants are more productive because of it.

Although I frequently refer to a plant's value in terms of biomass, I must point out that there is a lot more to it than that. Many plants have a powerful life-promoting quality way out of proportion to their quantity. The biodynamic folks focus on that, much to their credit. Being a not very spiritual person I don't hold much stock in some of their methods, but I do think that in their appreciation for the subtle-but-potent effects of certain "minor" materials—a handful of dandelion leaves, a bit of horsetail "fern"—they are acknowledging the intricacy of the living world around us. The hay field is perhaps a good place to observe and utilize that.

PART II
The Seed

Again and again I discover some exceptional new (usually old) variety of crop plant that is particularly suitable for my garden-without-borders and I recommend it to others, who then ask where they can buy the seeds (or plants). All too often it is not available in the general marketplace—either it is the sort of thing that gardeners don't usually grow (naked barley, chickpeas, sugar beets), or it has simply fallen off the shelf due to low commercial demand. It may be that my own source is one not everyone can access for risk of overwhelming the supplier—the National Germplasm System, for instance, or members of the Seed Savers Exchange, which is not set up to function as a seed company. In some cases I may be the *only* source, but while it may seem self-serving to have every reader of this book as a captive customer, that doesn't work. I don't want you all as customers, especially not repeatedly, because I am not in the seed business. So it is with other niche suppliers. Whenever I (or NGS or SSE) offer to share special seeds or planting stock with the general public, it is assumed that if you wish to continue growing them (or expand your planting), you will propagate your own seed stock. I don't wish to make you independent of the marketplace only to be dependent on me.

Fair enough, but what if you don't know how? What if you've never saved salsify seed or hand-pollinated pumpkins or grafted medlars (or don't even have rootstocks)? This section of the book is intended to help you with all that so that henceforth, when you find some great new plant you want to add to your diet (or garden), you can buy or borrow it just *once* and thereafter be self-reliant in supplying your own needs for the future.

CHAPTER SIX

Sexual Propagation: Why and How

I think it is highly significant that every seed is the densest part of its own plant, regardless of size. Even parsnip and dandelion: Remove their unaerodynamic accoutrements and they are leaden. And why wouldn't they be? The rest of the plant tissues are either energy-producing leaf surface or cellulosic support structure; the seed is where it's really at. It is the plant's long-term memory, its operating manual, its family Bible containing all the stories from its own remotest past. Every single event in its ancestral lore, every adventure, every struggle, every triumph, is encrypted in that indelible DNA. When you have to pack all that stuff into a tiny package, it must be dense indeed.

Why Bother?

For many years my Scatterseed Project has been collecting, maintaining, and distributing thousands of varieties of crop plants, many of them rare or endangered. But when I first got into seed saving (and grafting, and so forth), there wasn't much talk of genetic erosion or heirloom varieties; the Seed Savers Exchange did not exist, and it never occurred to me to maintain many hundreds or thousands of crop varieties. I merely wished to be independent of the seed companies—wasn't that a basic part of self-reliance? I fully appreciated the many varieties offered, but my intent was to buy them only once and take it

from there myself. I figured that I could more easily afford to try new stuff if I weren't always having to repurchase the old stuff. I caught on to saving bean and tomato seeds right away—it's pretty straightforward—but what about my carrots and leeks and sweet corn? And what about potatoes and garlic and onion sets? I don't remember when or where I learned about biennials and the tricks to saving their seeds, but almost from the very beginning I started trying to overwinter celeriac and kohlrabi, learning to manage their whole life cycle even when I had barely learned to eat them.

My first awakening to seed saving came when I worked for Orlando Small, an elderly dairy farmer. Orlando had an extensive garden and took just pride in it. To me the most intriguing aspect of it was the several crops that he had just "picked up" somewhere—never bought off the shelf or ordered from any catalogue. There was an odd-shaped purple potato, Cowhorn, which Orlando and his friends called Cowhawn, but you couldn't buy it anywhere. And something he called a lima bean, though it wasn't. Pole limas don't grow to maturity here in Maine; I later learned that it was a white-seeded runner bean. A neighbour of his had picked it up in the Carolinas while driving through to Florida. That was decades earlier, but Orlando had kept it going, as well as an old horticultural pole bean. After he was dead someone said it was also called the "Farmington" bean and was

Figure 6.1. Here is my great-grandmother with my own family seed heritage, lost forever.

an old Abnaki variety. Perhaps, but I persist in calling it "Orlando's Horticultural," as he was the only one who kept it from extinction. That was a new concept for me: that a variety could go extinct, just because folks failed to keep it going from one generation to the next. In Orlando's case his next generation didn't bother to grow them, but I did, so now "the hippie from away" is the guardian of that bit of local heritage. You see, it doesn't matter so much *who* picks up these neglected heirlooms, as long as *someone* does, and that someone might as well be me—or you.

A few years later I discovered a more personal link to this crop heritage business. While helping my father sort through some old family albums, I came across an old (circa 1910) clipping from the *Waterville (ME) Morning Sentinel*, showing my great-grandmother Emma Taylor by their farm's booth in the Exhibition Hall at the Waterville Fair. On display are bushels of potatoes and squash, garlands of flint corn, samples of all the myriad crops raised on that diversified farm on Priest Hill in Vassalboro, Maine. My great-grandma died long before I first breathed, and few alive today recall that there ever was a Waterville Fair. But what about the crops displayed there? The potato varieties are probably still around, but the squash and flint corn and most of the other stuff is most certainly extinct today, replaced by what may or may not be improvements. How I wish I could

step into that photo for a few moments, greet my grandam and snatch a few kernels of this and that to bring back to my own time! But alas, they're gone forever; once the chain is broken it cannot be mended.

It's all well and good to get all fired up about preserving the world's genetic heritage, but perhaps instead of emulating my quixotic efforts you would do well to save seeds of those crops and varieties that you *like to eat*. If you have to save the seeds of a vegetable that you're accustomed to having on your plate, or else do without it, I expect you'll just do what you have to do. Then maybe you'll try other crops and other varieties and save seeds of those you like; you'll probably share those seeds with others and you'll be off and running. You don't need a mission statement, just a healthy appetite! Oh . . . and a few new skills, including how to harvest, clean, dry, and store seeds. Plus, you also need some knowledge about plant reproduction and genetics, so that you'll know how to produce a crop of seed that will be viable and give you the results you want.

There are two basic ways that living things, including crop plants, increase themselves or are increased by human intervention. One involves sex and the other does not. Most folks seem to assume that increasing numbers is the whole reason for sex, but that is just not so. The other way—asexual or vegetative propagation—is so much more efficient for making more plants that it makes sex seem ridiculous. I mean, plants invest *so* much energy in sex, so much "thought," creating all those structures: flowers and seeds, all the elaborate chicanery and negotiation with insects and other species to enable sexual reproduction to happen. And even then the odds are bad: Of thousands of pollen grains blowing in the wind or riding on bee-shins, how many are going to reach their intended goal, a receptive anther of the same species? And of those, how many will mature seed and be borne to a favourable location for survival? Most do not; the few that do must make it all worthwhile, but why? Asexual propagation is so much more sensible. Send out runners or something, make tubers with lots of shoots sprouting from them—so much more reliable, so much less expenditure of energy and matériel! Why mess with sex? Nature is usually so conservative of energy and stuff, it surely would not bother with such foolish extravagance if there were not some compelling advantage. That all-important advantage of sexual reproduction is *diversity*, without which no species can adapt to the world around it. Even if there were such a thing as perfect adaptation, it's a moving target. Circumstances change constantly, and all things alive adjust to those changes or else die.

Plants and other organisms must also find some advantage to *not* changing, to remaining stable and predictable, else asexual propagation would not continue to exist. That's the topic of chapter 7, but for now we're talking about seeds, and therefore we're dealing with sex.

Strategies for Sexual Reproduction

Plants have several strategies for sexual reproduction, each presumably with its advantages and drawbacks. Obviously, plants must have male and female organs in order to reproduce sexually. Rather than use arcane terms like *anthers* and *stamens* and *pistils* and *styles* to refer to those male and female organs, I'll just use *boy parts* and *girl parts*. Flowers that contain both girl parts and boy parts are called perfect, which doesn't imply that they look lovely in bouquets (indeed many are quite inconspicuous); it just means they have all the equipment necessary to procreate. Some types of plants that bear perfect flowers can "do it" to themselves and make viable offspring. (It sounds kinda kinky, but think what you could save just on wedding expenses. . . .) Such plants are called self-pollinators.

Pea is a good example of a self-pollinating plant. Each pea blossom has a keel-petal that folds over the sex parts and keeps foreign pollen from reaching the ovary. The *male* sex parts are also under that canopy, and so the only way that flower is going to get pollinated is by itself. In fact the only way any *other* pollen has a chance of reaching that ovary is if a person were to remove that keel-petal *very* soon after the flower forms and tear off the boy parts, later inserting pollen from another flower. If that other flower came from another plant of another variety, then you have made a cross, which is how plant breeders create new varieties. That's an important distinction to keep in mind: Are you trying to preserve and increase an existing variety or to create a new one? Either may be appropriate, but it helps if you're clear on your objectives.

It doesn't necessarily follow that all perfect flowers can pollinate themselves, though. Many perfect flowers are *self-sterile*, which means that even if both sex parts are there, they are incompatible. It may be a chemical rejection or some other sort of built-in anti-incest mechanism, which guarantees that the plant, as well as the individual flower, will receive its genetic fulfilment (or DNA complement, if that sounds too romantic) from a separate plant. Thus the plant population is forced to share its collective wisdom, rather than each lineage hoarding its own narrow vision of its history and destiny.

Let there be no mistake about this: Every seed is a story, or rather a book containing a story. That story is written in four letters: A, C, G, and T. They stand for adenine, cytosine, guanine, and thymine, the amino acids that form the segment components of DNA. Using this very ancient alphabet every organism records its own personal saga, going back to the very dawn of life. A lot of that story is shared, of course, but the more recent pages diverge in their myriad versions. That's where we distinguish an elephant from a palm tree, but also a Berner Rose tomato from a Mortgage Lifter tomato.

Plants that can reproduce only by cross-pollination (self-sterile plants) are called obligate outcrossers. These include carrots, cabbages, onions, and more. This reproduction strategy can tell us a lot about how to save seeds from those crops.

Let's digress for a minute here to talk about the value of genetic purity in food crops. A *variety* of a crop plant is a group of individual plants that have consistent appearance and characteristics. In crop plants a variety may be a clone (all individuals have identical genotype, or genetic configuration) or as fuzzy as a landrace, which may include dozens of genotypes grouped into one phenotype (similar appearance), perhaps with a single name. In between there are pure-line selections, mass selections, market samples, et cetera.

So why do we care about purity in a variety? I assure you the plants themselves do not care, as is shown by hybrid vigour. They embrace the broadest spectrum of their variability; it's only we human domesticators who are hung up on this purity thing, yet for some very compelling reasons. In the cultivation of crop plants we need two things: uniformity and predictability. Without them it all falls apart. An extreme case: If your sweet corn and popcorn become interbred, the sweet corn will become tough and unpalatable, while your popcorn will not pop. Grown in isolation each is a wonderful food for a different purpose; grown together they are both useless. It is possible to carry this purity concern too far in either direction, but overall, as gardeners it serves our purpose to keep different varieties distinct, with a minimum of deviation from some recognizable norm.

From a plant's point of view, considering the importance of variation for adaptability, the question arises: Why would nature produce a species such as pea that *prefers* incest (self-pollination)? Isn't that counter to evolutionary progress? It is in a way, and yet it, too, has some survival advantage. But before I explain how, I first need to walk you through a little review of some basics of how genes and chromosomes operate. I'll try to keep it unmystical.

A Little Bit of Plant Genetics

If you're planning on saving your own seeds, it's really worthwhile to understand what's happening at the genetic level during pollination of food crops. And in order to explain it I first need to take you through a little review of some basics of how genes and chromosomes operate.

You may recall from biology class that cells contain chromosomes, and that sex cells—pollen grains and egg cells—contain only half the normal number of chromosomes. When a pollen grain and egg cell combine, the fertilized egg has the full complement of genetic material. Genes, those bits of DNA that govern heredity, have certain ways of lining up when the chromosomes in which they are embedded pair off during sexual reproduction. Some, though not all, genes have a straightforward way of manifesting themselves: They either dominate or are dominated, meaning the dominant trait will prevail over the recessive. That recessive trait continues to be present in the chromosome, but it doesn't manifest, or express itself.

Let me give you an analogy. Picture this married couple: The guy is domineering and always talks about world affairs; the wife cares little about that, but listens politely. (They get along fine.) Now there's a second couple, too, but in this pair it's the wife who insists on talking about foreign affairs, while hubby nods meekly. The two couples get together some evening and chat. When the two guys are talking, it's a one-sided conversation about world affairs; ditto with the gals. But if you put the male poli-sci wonk and the female wonk together, they'll hit it off debating the latest UN resolution. The quiet guy and gal will sit placidly in the corner and discover they each have a passion for collecting Jell-O recipes. So there are two very different conversations going on in the room, one of which would never take place if the couples were in their usual guy–gal configurations.

This analogy is stretched somewhat by the fact that neither Jell-O-passion nor policy-wonkishness is genetically determined, at least to my knowledge, but it serves my purpose, so please work with me here.

Maybe there's a good reason for the political types to be the dominant spouses; maybe world affairs are more important than Jell-O. Or sometimes maybe not. Maybe sometime you'll need a good dessert fast, and you're all out of tapioca, so you really need someone who knows what to do with Jell-O. Then you're grateful for those two quiet (recessive) types who hung around waiting for their time to shine. Don't be judgmental about the UN or Jell-O; only time and circumstances will reveal what is worth what.

Back to genetics and two more terms: homozygous and heterozygous. *Homozygous* is the genetic condition of a dominant paired with another dominant (the two folks second-guessing the General Assembly), *or* a recessive paired with another recessive (two folks talking about dessert ingredients). The first pair will never even mention fruit cocktail, and the second pair don't know Ban Ki-moon from Madonna. But when both couples go back home they become *heterozygous*: Each pair has one dominant and one recessive. And all the talk reverts to politics, and you'd never know that the quiet spouse in each couple had other interests and ideas. Think about it: Heterozygous presents the same surface appearance as the homozygous dominant, but underneath it's not so simple.

Many crop plant genes are like that. In peas, yellow seed coat is recessive. Whenever you see a green seed it could be either homozygous for green colour (let's use the symbols GG) or heterozygous (Gg). You can't tell which genotype a green pea has: GG or Gg. But if the pea seed is yellow (or white), you know it is homozygous recessive (gg), because the recessive gene gets expressed only when paired with another recessive.

This is useful knowledge, but it's not the whole story. Nature is often less simple. Some traits are products of two or more genes acting with variable dominance. One example of this is plant height. Pea plants are not simply short or tall. They can exhibit a whole spectrum of heights—depending on how their genes are interacting.

Let's return to the question of why self-pollination might also offer a crop an advantage, even though sex confers on a population the all-important advantage of diversity. The thing is, diversity sometimes comes at the expense of stability, and stability can be pretty important, too. When some rare-but-desirable gene (or gene combination) pops up in a population, you'd like to keep it and increase it, yet especially if it's a recessive trait, it might vanish back into the gene pool and rarely if ever be recoverable. Self-pollination captures that gain in a new combination that is more stable over time, so the plant can benefit from the survival advantages conferred by the gene.

I would point out that when it comes to propagation, you must never say never. Life has such a lust for itself that it will contrive extreme schemes to

Table 6.1. Self-Pollinators and Cross-Pollinators

Selfers	Crossers
Beans, common (*Phaseolus vulgarus*)	Beans, fava (variable)
Beans, lima (*Phaseolus lunatus*)	Beans, runner
Chickpeas	Corn (all types)
Lentils	Rye
Peas	Beets/chard (both are *Beta vulgaris*)
Soybeans	Cabbage family, including broccoli, brussels sprouts, cabbage, collards, cauliflower, kale, and kohlrabi (all are *Brassica oleracea*)
Barley	Carrots
Oats	Celery/celeriac (*Apium graveolens*)
Rice	Chicory (all types)
Wheat	Chinese cabbage, pak choi, mustards (often considered to cross-pollinate)
Eggplant	Cucumber
Endive/escarole (*Cichorium endivia*)	Leeks
Ground cherries	Muskmelons
Lettuce	Onions (all types of *Allium cepa*)*
Parsnips (according to one source, dubious)	Parsley
Peppers (but may cross-pollinate in some climates)	Parsnips (assumed)
Tomatoes (except potato-leafed type)	Radish (including daikons)
Flax	Rutabaga (*Brassica napobrassica*)
	Squash/pumpkins/gourds**
	Spinach
	Turnips (*Brassica rapa*, not including rutabaga)
	Watermelon
	Poppies
	Sunflower

* Cross-fertile within the species; the category "onion," however, comprises at least two species: *Allium cepa* and *A. fistulosum*.

** Cross-fertile within species; however this crop group comprises at least four species: *Cucurbita maxima*, *C. moschata*, *C. mixta*, and *C. pepo*.

surmount any obstacle to proliferation. Thus I cannot insist that peas will *never* cross-pollinate or that carrots will *never* self-pollinate. However, especially with cultivated plants, those respective habits are well enough established that we can certainly work with them, expecting occasionally to be thwarted.

With that in mind I've included a table of selfers and outcrossers for your reference. Some qualification is required. Environmental factors can skew behaviour to a considerable degree. For example, hot temperatures can allow much more outcrossing of beans than I am likely to see in my Maine garden. Although tomatoes are quite thoroughly selfing here in northern New England, one type is a major exception: Those varieties whose leaf margins are more nearly entire (often called potato-leafed) also tend to have blossoms whose girl parts extend beyond the protective cone of the boy parts. This renders them receptive to pollen carried by bumblebees, so that they may be crossed by other varieties, although other non-potato-leafed varieties cannot be crossed by them. The potato-leafed trait is comparatively rare among commercial varieties, but much more common among the heirloom varieties like Brandywine.

The Importance of Selection

Plants in a population (read "variety" if you like) are not genetically identical (except for clones, which we're ignoring for now). They may be similar, but not identical. This means that the fine points of appearance and behaviour of future generations of a variety that you grow in your garden will depend on the genetics of the particular plants you select and save seed from. You might at first decide to save *all* the seed from *all* the plants you grow, but at some point that will become untenable and pointless and you will need to make choices—that's what I mean by "selection."

Mind you, selection will happen whether you give it any thought or not. The fact that you're growing plants in a particular set of climate and soil conditions will certainly put pressure on the population and influence what conditions future generations of plants grown from your saved seed will tolerate. Everything about where and how you garden will inform the evolutionary course of that population. It is part of their ongoing story and they are taking careful notes, all in those "letters" I told you about earlier: A, C, G, T. And since *you* are a part of that story, why not be an active player, making conscious decisions about where the story is going?

Selecting intelligently requires a knowledge of your plants: Just what is it that defines them, makes them what they are? It also requires a knowledge of yourself: Why are you doing this? What are you looking for in a plant population? That may be as straightforward or as nuanced as you make it.

Some examples: There is an extremely early-maturing field corn called Gaspe Yellow Flint, not a big yielder but a very quick cropper. Now, in corn those traits—earliness and yield—are pretty much inverse: You increase one at the expense of the other. If you were to select Gaspe for yield by saving seeds from the larger, more productive plants, it would cooperate nicely. Over a few generations of selection you might double your yield, but the crop might take a few weeks longer to ripen. Is that the result you want? If not, then perhaps you should have selected for the combination of yield and maturity that works for your particular situation.

It is also important to consider the characteristics of the whole plant when you're deciding what to select for. If one tomato plant in your garden ripens its first fruit long before the other plants of the same variety, but the rest of that plant's fruit ripens way behind the others, then that first plant is the worst one to select for seed saving if your goal is earliness. You see, even though that one fruit turned red early by some fluke, all of the seeds in all of the fruits on that plant contain genes for later ripening, including that earliest fruit. There could be a connection between flavour and ripening time, too. Maybe that tomato variety is noted for its exquisite flavour, which takes longer to develop. Do you want to lose that by obsessing over earliness?

This particular example reminds me of a trick I often need to use to ripen radish seed where I live.

The more bolt-resistant (and thus better) varieties may require an excessively long season to mature a seed crop even after they've bolted. Therefore I often start them extra early in 4-inch (10.2 cm) peat pots, thinned to three or four seedlings per pot, so that when I set them out the radishes are already fattening. This strategy is especially helpful with daikons, which I have also sometimes treated as a biennial.

Population Size Matters, Too

Whatever criteria you use as your basis for selecting seed plants, here's another consideration you must add to the others: How *many* plants to save seed from? It's not a simple matter of how *much* seed you need, else you might choose three or four nice plants and call it good. It's also a question of how many plants are required to encompass the genetic base of that variety. All the plants will contain some range of variability, however slight, and by choosing too few individuals you risk chopping off an important part of the spectrum. Just how much risk depends on the plant species, as well as how much variation you had to begin with. For example, many pea varieties began as a single plant, and because pea is a highly self-pollinating species, all the progeny make up what is known as a pure-line selection—not as perfectly uniform as a clone, but highly consistent nonetheless. In contrast, corn crosses freely and there is typically a considerable range of variation, however subtle, even within the kernels of a single ear. To save seed from a single pea plant would preserve a far greater amount of the original gene pool than would saving seed from a single corn plant. Saving seed from only one or two corn plants may give you enough seed for your planting needs, but the population may become weak and inbred, the opposite of hybrid vigour. Repeat that in a second generation, and you will have done great harm to the health and vigour of your corn variety. With corn you need to save some seed from dozens of ears, better hundreds, to avoid the phenomenon called inbreeding depression, commonly referred to as "running out."

Saving seeds from hundreds of ears is easy to do with field corn, where *all* of the seed is going to ripen anyway, so whatever you don't use for seed is food. What about sweet corn, where the ears you save for seed are useless for food? Keep in mind that if you need to save 500 seeds, it's better to have one seed each from 500 ears than 500 seeds from two or three ears. But that could mean having to allow *all* of your crop go by for seed.

One way to approach this is to plant a lot of extra sweet corn one year. You'll end up with far too much seed to plant the following year. But if you store the seed properly it should last for the next several years. So once you've got your stockpile you can reduce your future plantings accordingly, saving few if any plants for seed.

Also keep in mind that if you plant a patch of, say, 500 sweet corn plants and you save seed from only 50 ears, eating the others when they are ripe, you will still get most of the diversity benefit from all 500 plants, since the ears on the 50 plants surely received pollen from all the surrounding plants. (The hitch would be if any sex-linked genes were involved, but I won't bother you with that now.)

Let me note here that in the vegetable garden, corn may be the most extreme example of the need to save seed from a large population. For many crops saving seeds from just a few plants should give good results for home use. But for any crop that benefits from saving seeds from a large population, another good strategy is what I call the community strategy. It can be done with any kind of seed, but my neighbour Lucian explained it to me in the context of flint corn. There used to be an old variety of flint corn (never named) that everyone in the neighbourhood grew—for chickens, porridge, and corn bread mostly. Everyone kept a small patch, though not always sufficient to guarantee an adequate gene pool over time. According to Lucian it was the custom on New Year's for each farm family to visit their neighbours, taking along a scoop of corn seed to dump in the neighbour's seed bin, taking another scoop out for the next farm, and so on. In the course of the day (and over years), the effect was to give every farmer the genetic diversity of hundreds of acres, even if he only grew ¼ acre (0.1 ha) himself. It worked partly because every farmer was growing the same type of corn, though each farm might develop its own special strain.

This community strategy can also be used to get around another potential obstacle that seed savers face, which is the need to isolate different varieties of cross-pollinating crops from one another to avoid unwanted pollination. In this variation on the strategy, two or more people who wish to grow two or more varieties of a cross-pollinating species each agree to grow enough seed of one variety for both of them, ideally enough for more than one year. This allows the gardeners to avoid worrying about whether they can provide an adequate isolation distance to maintain purity. Let's talk more now about what this concept of isolation means for seed savers.

Isolation

When saving seed it's helpful to know what species of plants we are dealing with. I mean, it's always nice to know, but with outcrossers it's critical. Moreover it's not always obvious. For example, if we know that cukes are *Cucumis sativus* and that cantaloupe are *C. melo*, then we know that they are not capable of crossing, because although they are closely related, they are not the same species. (Not that interspecies crosses are impossible, but they are rare.) A different example: Beets are *Beta vulgaris* and chards are *Beta vulgaris*, too, so we know they can cross. We can further classify beets and chard as var. *rapa* and var. *cicla*, but those are arbitrary racial criteria perceived by humans; the plants have no idea what we're talking about. Never mind that we see them as two distinct vegetables; they see each other as bed partners. If we wish to keep these two crops uniform and predictable, we must make sure they do not flower in the same garden at the same time: We have to isolate them from each other.

Likewise the species *Brassica oleracea*: The vegetables kale, cabbage, collards, brussels sprouts, kohlrabi, broccoli, and cauliflower are merely races of the same species, and they will interbreed readily if you let them.

On a different note the crop we generally call squash or pumpkins in fact includes several species. In North America the species involved are *Cucurbita pepo*, *C. maxima*, and *C. moschata*. Now, *C. pepo* includes the summer squashes (zucchini, yellow crookneck, and scallop types) and all of the "true" pumpkins. (Australians use the word for some maximas, such as Queensland Blue Pumpkin—how like them.) *C. pepo* also includes the Acorn types, Delicatas, and Spaghetti Squash. These will all cross with one another, but not with *C. maxima*, which includes most of the big warty, winter squashes with corky stems (Hubbard, Buttercup, Kuri, et al.). The moschatas are basically what we refer to as Butternut squash, with many varieties all having thick fleshy "necks." If you grow two different species of squash, there's no need to worry about isolating those plants from each other. You can grow and save seed from a zucchini and a Buttercup and a Butternut all in your little garden with no concern about crossing or purity. But if you plant a zucchini and a pumpkin together, it's quite another matter. Seeds saved from those plants may not turn out true to type, because they are both the same species and can cross-pollinate.

What complicates this need for isolation of outcrossers is that very few cross-pollinating species rely on random movement of pollen or wind pollination. The exceptions are beets, chard, spinach, and amaranth (all members of the Chenopod family), which produce copious amounts of very fine pollen that wafts well from one plant to another. Most species rely on a less random strategy: hitching a ride on some insect (or bat perhaps) whose destination is another plant of the same kind. Understandably all of those species produce lots of sugary nectar: If you tip the deliveryman well he's more likely to remember the address. So to isolate outcrossing plants it's not always simply a matter of preventing windblown pollen from moving between two different varieties; you may also have to outwit insects that are quite determined to go about their business of carrying pollen from one plant to another to another.

Let's think about an example of two populations (read "varieties") of carrots, each with enough plants

Figure 6.2. Stems provide a clue to the species identity of squash and pumpkins, which is important information for seed savers. This rough-ridged stem is characteristic of *Cucurbita pepo*.

Figure 6.3. A typical *Cucurbita maxima* stem is fat and smoother and spongy and tapers at both ends.

to represent an adequate gene pool. If we wish to keep each of those two populations pure (uniform and predictable), we must prevent the flow of genetic information between them: We must keep the wind, bees, flies, et cetera from carrying pollen from one variety or population to the other. The simplest way of doing this is the way nature does it: by separating the two populations far enough apart that pollen transfer is unlikely. This begets the obvious question: How far is far enough? That answer depends on other questions; for example, if it is a wind-pollinated species which direction is the prevailing wind? Of course that only affects the probability, not the possibility, of crossing, since there's no way to know for sure with so many variables. An hour of crosswind during a key period of pollen shedding could throw out all such calculations.

In the case of insect-pollinated species (which is most of them), the isolation distance required depends on what type of insects are the "designated carriers" for that crop. Fava beans are largely pollinated by bumblebees (at least where I live) and those fat fellows don't travel as far as honeybees, which are the major pollinator of turnips and cabbages. There are other lesser pollinators (flies, syrphid wasps, moths), however, which can further complicate the decision of how far apart is far enough. It is said that hedgerows and tall buildings tend to deflect the flight paths of insects, which may be true, but it doesn't take a very determined bee to find a shortcut through a hedgerow.

What does make a very big difference, I believe, is what other flowers are growing in that distance between two isolated crop plant populations (or "isolations"). If the intervening space is full of wildflowers or a patch of buckwheat, that's a huge help: Every time the bees collect a load of nectar (and pollen), they must return to the hive to empty out before continuing. Because they tend to work rather systematically, by the time the bees find their way to the second isolated plot of the crop, they will have loaded and emptied wildflower or buckwheat pollen many times, and the chance that they are still toting some of the pollen from the first crop patch becomes negligible. By

Spinach Pollination

Spinach is a rather exceptional case in terms of pollination, because spinach plants are monoecious: There are boy spinach plants and girl spinach plants. I believe there are also some hermaphroditic plants. Just looking at spinach plants, though, it's hard to visually distinguish boy and girl plants until they become sexually active. Fortunately we don't really need to know as long as they do; at some point you will see a lot of scraggly-looking seed-loaded vines. What you *do* need to know is that you have adequate numbers of both sexes, preferably heavy on females, but some guys need to be in on this, too. If you have a sizable portion of plants—100, say—you should have plenty of both genders. Some varieties tend to predominate with males (the better for leaf production), so if you saved only half a dozen seed plants, you could have a lonely garden. By the way, kiwi, holly, and yew operate on the same system.

comparison, if the two isolations were separated by a desert or lawn or parking lot, the pollinators would quickly finish collecting pollen from one isolation and turn their attention directly to the other, without wiping their feet in between. That would result in lots of cross-pollination potential between the two isolations.

The other wind-pollinated crop species (beet/chard, amaranth, quinoa, spinach) pose somewhat different problems. Their very light pollen may travel farther than an insect is likely to fly, so what is a safe isolation? For beets/chard I've had consistent luck with an isolation distance of 650 feet (198.1 m), which is much less than most authorities recommend, and I cannot account for my success.

For each crop you want to save seed from, you need to weigh all those factors and decide what distance will give you adequate assurance of purity. I cannot

give you some fixed number and say that a shorter distance is unacceptable and anything above the minimum is safe. The concept of isolation distance is rather a spectrum, and you need to decide where you are on that spectrum. I can only say that, for myself, I maintain a minimum of 650 feet (198.1 m) between all cross-pollinating species (but even greater distance for beets and spinach, as mentioned above). I see little or no evidence of contamination. To assure that I *would* see it if it were there, I usually plant crop types that would clearly show it; for example, a popcorn in this isolation and a sweet corn in that one, or a salad radish here and a daikon there. If any genetic mixing had taken place it would be obvious in the appearance of the next generation of the crops grown from the saved seed. When I don't see any it boosts my confidence. You will find recommended isolation distances in various seed-saving guides, and while I would not naysay any of them, it is important to consider those variables just mentioned. My greatest concern is lest the more extreme distances cause folks to be daunted from saving their own seeds.

Another factor is the degree of purity required. I need more guarantee with the varieties for which I'm saving seed as part of my Scatterseed Project than you may require for seed that's for your use only. If you require absolute certainty of 0 percent crossing, I'd recommend planting one variety in the corner of your garden and the other on Venus. However if you are growing two excellent varieties of a crop—let's say Red-Cored Chantenay and Scarlet Nantes—then a cross between the two would probably be a pretty fine carrot, too. If you isolate them by only 100 feet (30.5 m), much of the seed will be pure, but you will likely also have quite a lot of crossing. Does that really matter, especially if you can always go back to the marketplace and buy some really pure seed if you wish? Only you can decide that. I will point out, though, that it would not be ethical to re-offer that carrot seed, which may contain some impure seed, to others and call it Red-Cored Chantenay or Scarlet Nantes. That impure seed will actually be a bunch of F1 hybrids, which are something else altogether, as I explain in the *What About Hybrids?* section of this chapter.

This discussion is all predicated on the assumption of a need to save two or more varieties of the same species—corn, or carrot, or leek, or what have you. Quite possibly you need to seed-save only one variety of a species, so it's all a moot point, lucky you. But maybe not, because there may be wild (or feral) populations of your crop species growing near your garden, and they could cross-pollinate your crop. The best example is Queen Anne's lace, aka wild carrot (although more correctly termed feral carrot, as it has gone wild from the domestic carrot). These have thoroughly reverted to their wild ancestry and have objectionable traits that will ruin anything they cross with. It is highly unlikely that you can control these by mowing, so if you have them around, you'll need a different strategy, because you probably can't isolate your flowering carrot crop plants from them (more on this later). Some other feral species that you might have to contend with are wild parsnip, wild radish, and pigweed. These are the wild contaminants I know about or have heard of; there may be others in regions beyond the Northeast.

It should go without saying that you must also allow for what crops your near neighbours are growing and allowing to flower. Bees might be deterred by distance, but they don't understand the concept of property lines. If your neighbour grows radishes but pulls them to eat before the plants flower, then those plants pose no danger to your radish seed-saving efforts. (You don't get pregnant just by sitting next to someone on the bus.) However, growing a vegetative food crop such as potatoes is quite different from seed-producing crops, and it does not pose the same problem of contamination.

In addition to isolating varieties by distance we can also isolate them in time. This relates back to growing corn seed by planting enough plants one year to produce seed enough for several years. That same strategy can enable you to maintain several varieties of corn or other cross-pollinating crops such as carrots: You can grow two or three varieties every year for eating, but each year you allow only one variety to produce flowers and seeds for saving. The next year you'll save seeds only from one other

variety and so on. This assumes that you can store viable seed for several years. You can, and I tell you how in the *Collecting and Storing Seeds* section of this chapter.

Isolation Cages

Let's assume you can't save carrot seed *any* year because your neighbourhood is infested with Queen Anne's lace. You're not licked, but you've got to jump through a couple more hoops if you want to save seeds. You can create an artificial or micro-isolation by caging your plants. I've made semi-permanent cages out of pine strapping and nylon or aluminum window screening. The cages were 4 × 4 × 12 feet (1.2 × 1.2 × 3.7 m). (Next time I'll make them 5 × 5 feet, or 1.5 × 1.5 m, so that blossoms won't end up touching the screen.) I staple the screen to the strapping and added a lath tacked on the bottom edge. The screen was a few inches overlong, so I could hill dirt over it for security. I would lift up a corner of this to insert pollinating insects at the proper time.

The cover is a separate piece, also with screening attached, and it is 3 inches (7.6 cm) longer than the cage to ensure good coverage. I also tack the corners and the centres of the cover's long rails in place using scaffold (double-head) nails. It's important to keep the caged area well cultivated up until flowering, after which you cannot open the cages, or insects bearing unwanted pollen might get in. Check to be sure the frames don't warp over time, which can lead to gaps where insects may enter (or leave). If necessary, place a cement block on each corner to anchor the whole cage against high winds.

The cages can be reused for several years if you can protect them from heavy snow in winter; I've yet to devise a collapsible design, which would be a vast improvement. A bit of shingle or tarpaper under each leg of the cage delays rotting from soil contact.

I've also used 1½-inch (3.8 cm) flexible black plastic pipe, covered with polyester row cover (ticky-tacky, too) or better yet some cotton gauze product like Elmer-plantex row covers. The cage should hold at least a couple dozen plants. It is important that the cage be tall enough that blossoms

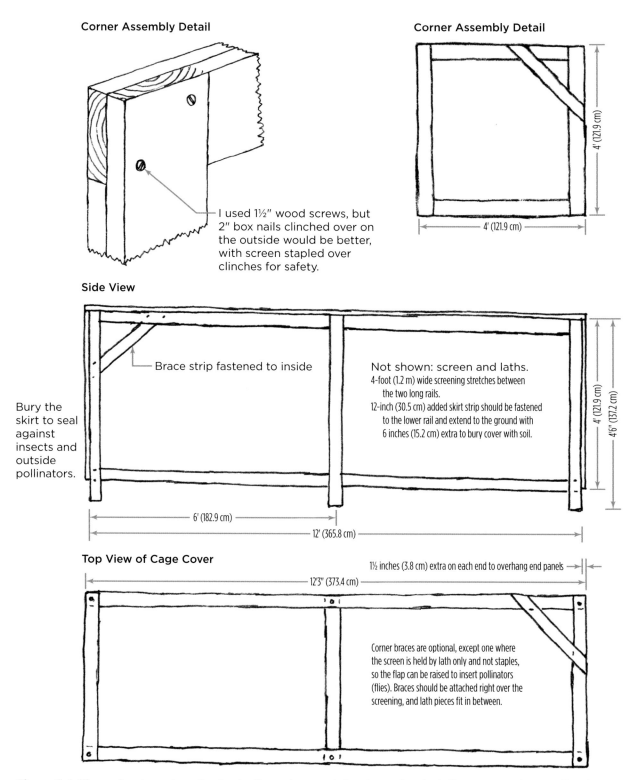

Corner Assembly Detail

I used 1½″ wood screws, but 2″ box nails clinched over on the outside would be better, with screen stapled over clinches for safety.

Corner Assembly Detail

4′ (121.9 cm)

4′ (121.9 cm)

Side View

Brace strip fastened to inside

Bury the skirt to seal against insects and outside pollinators.

Not shown: screen and laths.
4-foot (1.2 m) wide screening stretches between the two long rails.
12-inch (30.5 cm) added skirt strip should be fastened to the lower rail and extend to the ground with 6 inches (15.2 cm) extra to bury cover with soil.

4′ (121.9 cm)

4′6″ (137.2 cm)

6′ (182.9 cm)

12′ (365.8 cm)

Top View of Cage Cover

1½ inches (3.8 cm) extra on each end to overhang end panels

12′3″ (373.4 cm)

Corner braces are optional, except one where the screen is held by lath only and not staples, so the flap can be raised to insert pollinators (flies). Braces should be attached right over the screening, and lath pieces fit in between.

Figure 6.4. These drawings show the basic dimensions and structure of an isolation cage and cover. Staple screening around the entire frame; excess hanging down at the bottom can be buried in the soil.

can't push against the screen, because they could then be pollinated by outside bugs landing on the screen. Of course you can use more than one cage if you don't want to make one big enough to cover a whole planting. It is ideal to connect cages that cover a single variety so that pollinators can move freely between them, optimizing the degree of genetic mixing. I do this by placing the ends snugly together and sealing them with duct tape.

Now the plants are isolated from the larger environment and any stray insects carrying foreign pollen—but that is also a problem, since crops such as carrots are self-sterile and wholly dependent on insects as pollen vectors. If there are no insects inside the cage, there will be no pollination and thus no seed at all! You need to get some bugs in there, only they must have clean tootsies. You could just buy the preferred pollinator insect for your crop species (yes, buy bugs; hold on, it gets worse), assuming you're doing only one.

Since I save seeds from a couple dozen species that I keep in isolation cages, I buy the one insect species that will pollinate all of them, the common housefly (say *Musca domestica*, it feels much better). *Musca* is not the best pollinator for any of the crops, but it's simpler than buying many different insects. Now I'm not suggesting that any of you have houseflies around your homes, but even if you did they would be useless for pollinator purposes. They've been walking around who knows where and are tracking all kinds of stray pollen (and that's not the worst of it). No, we need virginal flies, never stepped on a flower—and those come from maggots just ready to pupate. A very helpful woman at the National Germplasm System explained to me how to grow maggots. It involves rotting meat and soda straws and . . . no, wait, there is another way. Yes, there are insectaries that will sell you fly pupae. They're incredibly cheap, maybe $25 for 1,500 ready-to-hatch, second-day-shipped, never-been-used pupae. You put a dozen or so of the little fellers in each cage (assuming you have several), plus a Dixie cup with a spoonful of molasses for a coming-out party, and they'll take it from there. Really, it's that simple!

Hand-Pollination

Hand-pollination is a method that works well for keeping squash family varieties pure without isolating them. It can also work for other crops, but squash plants are the easiest because of their large flower parts. Squash, pumpkins, melons, cukes, gourds, luffas, et cetera, all have "imperfect" flowers (although they're a lot showier than the perfect ones)—which is to say, the flowers contain only boy parts or only girl parts. So we say that the plants bear male flowers and female flowers, and each plant produces both kinds. The two kinds of flowers look similar, but also different. In general male flowers, which have long and succulent stems, start forming earlier in the season. The females usually have shorter stems and form in the axils of the vines. But the dead giveaway is the ovary, a swollen lump at the top of the stem but at the base of the female blossom. The ovary is the virgin fruit-to-be and also looks much like a miniature version of what you can expect that harvest.

The strategy behind hand-pollination is to exclude bees et al. from the ovaries, so that you can do the matchmaking yourself. The trick is finding a female flower and several male flowers that show signs of being ready to open on the morrow. To do this you need to learn to distinguish male flowers from female flowers well before they blossom; but don't worry, it's not rocket science, because of the presence of that swollen ovary on the female flowers. It's also important to know that squash and other cucurbit flowers are ephemeral: They blossom for one day and are done. If they don't get pollinated on the day they first open, they wither and die. Watching the flower buds from a very early stage, you can monitor their development over several days from small and green to large and yellow orange.

On the day before a flower opens the tip of the bud just begins to spread open a little. That is when you need to "cloister" them; once they open it's too late. I usually do this with little paper bags and paper clips, although some people use tape or twist-ties. With bags I fold in the top of the bag snugly against the stem and secure it with a paper clip. Whether you bag them or tape them shut, it is ideal to have several males for each female. It is also ideal if the male and female are from separate plants (of the same variety, unless you're trying to breed a new variety); however, cucurbits are not self-sterile, so you *can* use flowers from the same plant.

If you were to open one of those male buds the day before the wedding (don't, but if you were to), you would see a very shiny bald greenish phallic structure. If instead you wait until the next morning to remove the bag or tape, you will observe a startling transformation: That male organ is now coated with a fluffy luxuriance of bright-yellow pollen, all formed overnight. Thanks to your good work bagging or taping, *you* are the first critter to have access to those blossoms, even before the bees. For reasons I hope I don't need to explain, the next step is to pick the male blossom and take it to the female—*not* the other way around. You carefully tear away the limp petals of each blossom so you can gently daub the pollen all over the three-lobed ovary. You must do this quite thoroughly, even if you feel the need to recruit more males, and here's why. *You* may think the purpose of these plants is to make food for you, but their take on it is that they want to reproduce themselves, and the fruits are merely food for the developing seed. Therefore, if only a few seeds are pollinated, the plant will probably "decide" that it's not worth wasting all that effort, and it will abort and focus its energies elsewhere. Moreover, if the ovary is not completely "pregnant" (there are many seeds for one ovary), a bee may come along after you leave and drop some stray pollen grains onto those spots you missed. Then your fruit will produce a mixture of pure and impure seeds, and how will you know? An added precaution is to re-bag the pollinated ovary for the rest of the day, after which there will be no further risk of accidental unwanted pollination.

When you've completed this procedure it's important to label that nascent fruit (it will look more and more like a fruit as days pass) so you can distinguish it from the others at harvest. One way to label the fruit is by scratching a shallow X with a

Figure 6.5. Here, apprentice Arica is hand-pollinating a pumpkin blossom.

Figure 6.6. The withered blossom is a sign that this flower has already been pollinated—too late to try hand-pollinating it.

safety pin into the skin of the pollinated ovary; as the fruit grows and matures it will form a corky scar. Or you could tie coloured tape or ribbon securely around the stem. You don't need to hand-pollinate all of the female flowers in your squash patch, only enough for your seed needs.

By the way this all works very well for all squash species; everything I've said also applies to cucumbers and melons, except it's much more difficult due to the smaller blossoms. I rely on isolating those instead.

Another crop species that bears separate male and female flowers on the same plant is corn. The tassel is the male part and the silk is the female part. Unlike squash, corn flowers are not ephemeral; the tassel just keeps cranking out pollen over a period of time, while the silk just keeps growing longer until it gets pollinated. You can use a similar bagging strategy as with squash. I've only had a little experience with this myself and so will not presume to instruct you further. Rather I would refer you to some other source (see appendix B). Since I maintain fewer than 20 corn varieties, I rely on a combination of isolation plots and alternate-year planting.

Saving Seeds from Biennials

Some crop species—the biennials—require that we jump through two hoops in order to grow their seed. These crops are also self-sterile cross-pollinators, so we must maintain purity by isolation or one of the other strategies discussed previously. But in addition to that we must also spend two years growing the crop of seeds, due to their peculiar life cycle.

I tend to think of a biennial as an annual in denial: It thinks it's going to live forever and will insist on behaving like a perennial until some reality check convinces it of its mortality. An annual such as a bean has no such delusions: It knows from the moment it breaks ground that it has only a matter of months in which to grow, flower, get pregnant, and ripen its seed before it all ends in the holocaust of winter. It knows its clock is ticking; it doesn't waste time building surplus structures or saving up for a better time; this *is* its time. In contrast the biennial flatters itself that it can survive winter by storing up lots of food energy in a vegetative underground mass (typically a fleshy taproot) full of antifreeze in the form of sugars, cell salts, and other chemicals. Indeed it

Figure 6.7. Notice how small the flowers on these cucumber vines are (the tomatoes are there for colour contrast). Compared with big squash blossoms cucumber blossoms (male at left, female at right) are much harder to hand-pollinate.

can survive a single winter with that strategy, but no more. The biennial comes out of winter a lot wiser; it now knows it must budget its energy reserves for the desperate dash for posterity. This reality check, technically referred to as "vernalization," is the trigger to flowering; without that stress the plant will just continue to make foliage. This is why if you live in a place where winters are too mild to make the proper impression of urgency (Florida, for example, hey . . . talk about being in denial), you'll never get seed, period, unless you can create an artificial winter. For example, if you trim off the foliage (but *not* the crown) of carrots, beets, and turnips, and store them in a refrigerator (*not* freezer) for several weeks before replanting, they should get the idea and decide to cooperate. Note that vernalization is not always a matter of chilling; in some tropical species flowering and seed production are induced by drought.

It's critical to understand the importance of a biennial plant's crown. Technically called a "spreen," the crown is quite simply where the earth meets the sky, the point whence new growth originates. If the spreen gets damaged or decays it's all over—the plant will die. On the other hand other parts of the plant can sustain considerable damage and keep soldiering on. For example, the foliage can be stripped back repeatedly and more will arise—think Swiss chard. In many cases some or most of the taproot of a biennial can be cut away and the top still survives. The fleshy remnant taproot will typically rot but often not before a new set of shallow feeder roots has been generated to sustain the plant. You see, biennials do not take this mortality thing lying down. I've heard that this procedure—growing a new plant from a mere spreen—is called "snibbing," which sounds like a perfctly logical namc for that sort of thing. Theoretically you could go on the rest of your life eating the same crop of leek plants.

There is a possible use for this snibbing business: Let's say you're a radish breeder who wants to select for nice flavour (really, it could happen). The obvious problem is how to sample the taste of the root but still be able to collect seed from those plants that pass muster (as opposed to passing the mustard,

which is hardly recommended with radish). You see where I'm going with this? Right: You pull the young radishes (they're called "stecklings,"—don't you just love all this new vocabulary?), chop off the bottom halves, or perhaps the sides, of the roots and nibble them, then replant the tops (including spreens) of all the tasty ones, feeding the others to someone you wish to annoy. With care and a bit of luck some or most of those maimed plants will reestablish and go on to make a seed crop—not a great one, understand, but enough to provide foundation stock for continued selection.

Overwintering Strategies

Getting back to true biennials (which the radish is not), the main challenge is shepherding plants through that vernalization experience to the beginning of the second, sexual, season. In areas where winter is chilly but survivable—say Northern California or Portlandia (in fact much of the lower 48)—plants can just be left to overwinter in the garden or field. But here in my winter world of −15°F (−26.1°C), most of those plants would just freeze to mush. A few exceptions—parsnips, salsify, and witloof chicory—can reliably overwinter in the ground here in the Western Maine Mountains, and parsley, leeks, and kale have often come through fine, celery and chard on an occasional fluke. Carrots can manage winter fine under heavy mulch like old hay bales, but rodents will usually wipe them out.

To reliably overwinter most biennials I need to move them into the cellar. Fortunately for me the cellar under my house works admirably, but only because it is rather funky: dirt floor and laid-stone walls, with mortar sealing mainly the area under the sills to avert penetration of outside air. If I had a poured concrete basement with an oil furnace down there, it would be too dry and too warm. I want it to be close to freezing but not quite, and almost dripping damp, but not quite. I know of people who have successfully adapted a part of their basement by partitioning off a room with insulation and covering the cement floor with a layer of wet sawdust. You can also build a separate root cellar outdoors, especially if you have no cellar or no suitable one, but I prefer the convenience of having mine right under my feet.

People often tell me that they store their food carrots in buckets of sand and ask me whether the same method works for seed carrots. My answer is that it is *not* the best way to store *food* carrots but probably *is* an ideal way to store them for *seed*. When root crops are in close contact with sand, sawdust, or another medium, it encourages them to form root hairs, which lowers their eating quality but gives them a head start on replanting. For eating I prefer to store mine in barrels with cushioning layers of maple leaves. The leaf layers hold in the moisture while surrounding much of the carrot surface with air. In fact I also store my *seed* carrots that way, but I can easily believe that sand storage might be an improvement.

I've also been asked whether an unheated greenhouse can be used to overwinter. Yes, it can work very well. Fleshy root crops, such as carrots, beets, and turnips, can be kept in barrels or buckets out of the light, which would break dormancy. As for those biennials that don't make a nice storable taproot—celery, leeks, chard, and the cabbage family—I replant them in soil in buckets and put them in my cellar. I believe these species would be especially suitable for greenhouse storage, as the light would not harm them, provided the temperature stays above, but near, freezing. As damp as my cellar is I often need to add a little water to the soil in mid- or late winter; I try not to wet the plants themselves.

Before bringing bucketed plants into the cellar, I usually remove most of the foliage. With the leeks I snip off anything floppy; with chard, kale, and kohlrabi I remove all the larger leaves, leaving ½ inch (1.3 cm) of stem (don't damage the crown!).

Brussels sprouts are a happy example of having your cake and eating it, too: The "sprouts" will all wither or rot by spring anyway, so you may as well harvest them for food—it is the heavy fleshy stalk where most of the plant's energy is stored. The top tuft of leaves, the tiny fluffy little cabbage, should be left intact, but any remaining side leaves should be clipped off. Special care should be taken in removing the sprouts; next year's flower stalks will arise

from buds under the sprout, so you mustn't tear off the sprouts so as to damage those. If instead you *cut* off the sprouts with a sharp knife, leaving the base attached, that base will just wither off when it's ready and avoid any damage to the bud scar.

As for cabbage you can simply store the whole plant and let the head rot off, but is such a waste necessary? Here's a trick I learned from a Swedish friend, which works reasonably well: Hold the head sideways over a cutting board, taking great care to hold it steady, as the dirt-covered root-ball is still attached. Cautiously slab off one side of the head, then turn it 90 degrees and slab again until most of the "slaw" is removed. I emphasize caution, because you must avoid chopping into the conical tapered woody core,

Figure 6.8. These beheaded cabbage stumps were stored overwinter and are now sprouting flower stalks. After the stalks are replanted in the garden, they will shoot up 2 to 5 feet (0.6–1.5 m) tall and produce flowers and seeds.

which is after all the extension of the stalk, and as with the brussels sprouts each of those triangular bud scars is the source of a new flower stalk. Of course a downside of this method is that you have a lot of chopped cabbage unsuitable for storing. If you can time it with sauerkraut making or further chop it up for the freezer, you will have salvaged much of the food value (except for the hearts, which make lovely pickles, but that's another matter). Here's another reason why that method is handy: We need to save seed from a large number of plants in order to maintain the gene pool of that variety. Thus if you only grow 40 or 50 heads, this method allows you to save seed from all and still enjoy a good harvest.

Yet another method is to simply cut off the heads and overwinter the stumps in buckets. I've had altogether mixed results with this, mostly poor, because the stumps are likely to wither or rot from the cut end. However, some have succeeded, and if the stumps are otherwise useless then what have you lost by trying?

Sometimes when the cellar begins to warm up in spring, stored crowns and plants may break dormancy before it is time to plant. Sprouting too early may cause them to rot or, in the case of rutabagas and turnips, to form long blanched flower stalks that grow into a tangle (they don't know which way is up) and may break off when you take them out of storage. I deal with this by bringing the stored plant materials out several weeks before planting time, when the snow is barely gone. Because I have several varieties of each species, I keep them in woven-fiber feed bags, several to a barrel, each with its marker stake enclosed. I place them (in the bags) on the ground against a sunny corner of the house, where they can get fresh air and a little sunlight. I don't worry that there are still plenty of frosty nights—after all, I almost could've left them in the ground all winter—I want them to begin "hardening up." In particular I want them to develop a little fibre and some chlorophyll, so they won't freak out when they get put back in the real world. Some years I have removed them from the bags and replanted them in buckets (with drainage), so that what does sprout will at least head

upward. By the way those roots could have been left planted in buckets all winter like the cabbages, leeks, chard, and so on, but it would take a prohibitive amount of space.

As soon as the soil can be worked, it's time to get those second-year plants into the ground; indeed it needn't be all that workable, since you just have to dig up the space where each individual plant is going to go. No need for a smooth friable seedbed, as with sowing seeds.

Even in cases where the plants were overwintered in the ground, such as chicory or parsnips, I still dig up the roots and transplant them, for several reasons. Since I usually grow more than one variety of a species, I need to move them away from each other for pollination control. Even if I only have one variety, and it is going to remain in that place, I must still adjust the spacing: A rutabaga plant that required a square foot (0.09 sq m) to grow in its first season now demands at least 3 feet (0.28 sq m) in every direction—some commercial growers allow 5 square feet (0.46 sq m). I'm serious; hitherto you have only related to these plants as adolescents, harvesting and eating them long before they are really mature. Now as adults you will find them much more interesting and much more substantial: waist-high parsnips, beets looking more like tumbleweeds; you need to transplant them with that in mind. In most cases they will want support—yes, you will need to stake up your kohlrabis.

Another reason to dig and move your plants, especially those whose taproot is the crop, is so you can examine the roots. Assuming you have grown and overwintered more plants than you need for seed, now is the time to select which ones are best for seed stock. A forked chicory may not be forked because of its genes, but why take a chance? An exceptionally large fat carrot among a crop of uniform forcing types may conflict with the goal of earliness. Are you trying to breed something new or maintain what you have? Either way you must look at the whole plant, including the root, to make that decision.

Seed saving, especially of biennials, is a classic example of what I mean by gardens-without-borders.

Figure 6.9. A flower gallery for seed savers, *clockwise from the top left*: salsify, onion, broccoli, carrot, beet, and parsley.

So many people tell me, wide-eyed, that they have gardened all their lives and never seen a carrot seed or turnip seed that didn't come out of a packet. Because they've been cake-mix gardening, importing manure, lime, and seeds, it never occurred to them how these things came about and what went into them. When you begin saving your own seeds, you embark on a whole new adventure, one that awakens you, like nothing else, to what's happening in the world beyond that illusory garden border. It's impossible to tell where that adventure may take you.

First-year biennials are like our teenage children, poised to realize their potential as adult parents. How intimately can we know our food crops when they only originate in a packet from somewhere else? How much more fulfilling for us and for them when we get to see some of them in their full life cycle.

Pest Problems

I must point out that there are some problems associated with seed saving. Most are easy to remedy. I can speak only to those that I have dealt with.

Pea maggots. When you grow table peas and eat them in the shellout stage, you may never be aware of pea maggots; they may be present in your garden but unnoticed, because you pick and eat the green peas before the pupae can overwinter and hatch. But when

you save pea *seed* there could well be eggs in the ripening pea seeds, from which the maggots hatch out and continue their cycle. By the way there's a very nice control for this: freezing. Not in your freezer, silly, that merely amuses them. No, if you're blessed to live in a place like Maine, where –15°F (–26.1°C) is not remarkable, you can leave the seeds outdoors in winter. I stash them in heavy steel ammo chests with gaskets (to keep moisture out) in my unheated woodshed; that pretty much annihilates them, at least the bean beetles (the seeds themselves are unscathed by low temperatures; in fact they store much better).

Parsnip webworm. First-year parsnips are largely pest-free, but in their second year they are often attacked by the parsnip webworm, which makes a little cocoon in the umbels. These are easily controlled by just a modicum of diligence: I check the umbels every day or two during the early seed-setting phase and look for any "webbies." The worm is very quick to wriggle out and drop to the ground or lower foliage where it's really hard to find, so I am even quicker to dispatch him (or her) between thumb and forefinger, very effective. I never find more than one per umbel and a couple of quick patrols will usually control them for the season. They seem to be exclusively on parsnips.

Plant diseases. Organically grown carrot crops always have a little alternaria blight in them, but in second-year plants it can be much more severe. Again, organic methods (including enhanced biodiversity) will partially remedy this, but are no panacea. The same applies to many other diseases, including aster yellows and sclerotinia. For example, sclerotinia spore bodies are said to persist in the soil for 10 to 20 years, but that is in "chemical" soils that are biologically dead, whereas in a humus-rich, vibrant soil community, such pathogens are soon whipped into line and rarely cause much mischief.

This reminds me of an analogy: In a discussion of political systems with my friend Denis Culley, he made this observation. "People sometimes refer to the 'problems of democracy' with the implication that it's a good thing if you don't carry it too far, but in fact the solution to the so-called problems

Figure 6.10. Parsnip webworms feed in the flower umbels. It's easy to control them by pinching them between your thumb and forefinger.

of democracy is more democracy." I would likewise suggest that when we encounter problems in an organic system, we should not back off; rather we should seek to be still more organic, and that when diversity presents problems, we should respond by creating yet more diversity.

Collecting and Storing Seeds

The most frequently asked seed-saving question is: How can I tell when seed is mature enough? Very important, since immature seeds will not store well, nor will they germinate. We must remember that "maturity" for our purposes means that the seed *germ* has developed fully; the seed may appear full-sized but still be succulent and underdeveloped (leek seed is an example).

Frequently in seed saving you cannot have your cake and eat it, too—you get seed or food, not both. However there are several exceptions. For example, cukes and zucchinis are eaten in the immature stage, and those you wish to save for seed must be left to ripen way beyond usefulness as food. On the other hand winter squash and pumpkins are eaten at full maturity anyway, so your seed squash are also food squash and vice versa (provided you've kept them pure). In fact the quality of squash seed improves with storage in the fruit—after-ripening, it's called—so it's best if you don't process them all at harvest, but scoop out and save the seed of each fruit as you use it. At some point however, depending on variety, seeds may start to sprout inside the intact fruit.

Working with Dry Seeds

Seeds are sometimes classified according to their processing method: wet or dry. Most are dry and are processed by a combination of methods called threshing and winnowing. For these seeds their degree of dryness is a measure of their maturity. You can purchase devices for measuring the seeds' moisture content, but for our purposes it's easy to determine well enough by following this basic rule of thumb: Are the seed and its surroundings (pod, bract, et cetera) brittle and crumbly? For example, do the pods crack rather than bend like leather? Does all the trash crumble into dust and stuff that can be blown away? If not, leave the plants until they are fully mature. If rain wets them let them dry out again. If frost threatens and you're not certain whether the bean seeds are ready (unlike pea seeds, damp beans can be ruined by heavy frost), then you must make a decision: Can you cover the plants to protect them through that frost or should you harvest them, even though the pods are a bit leathery? If you opt for the latter, it's better to pull the entire plants, roots and all, rather than pick pods. The seeds *may* be able to draw a little more life force from the parent plant, just enough to finish their ripening. To use the earlier analogy, their story in A, C, G, T is already written, but the ink is not yet dry. Next time plant earlier or use some season-expanding strategy, so you will not have this concern.

What about overmature? In general *overmature* is a meaningless term, like *jumbo shrimp* or *modern art*. If seed is mature it is just mature, that's all. However there are concerns lest it should shatter onto the ground, a real issue with brassicas and parsnip, or be eaten by birds, particularly composites like salsify and sunflower. And if seed stalks sprawl over so that the seeds or pods are on the wet ground, they may either rot or sprout. But the seed itself cannot get "too ripe."

I've developed several strategies for avoiding the above-mentioned problems. When I notice brassica seed stalks that have reached the shatter-prone stage, I leave them alone until the next morning. The seed is no less mature then, but the pods will have taken up a slight dampness during the night hours and can be handled with little or no loss. I collect them by stripping the pods from the stalks into a large paper bag or even just clipping off the whole stalks or plants if they're not too badly tangled. I leave this bag opened to the air until the pods are as dry as they were the previous afternoon and then proceed to process them.

When salsify seed is ripening I cover individual heads with scraps of gauze, but only after pollination is long over. In extreme cases of persistent wet weather I have covered plants with some sort of canopy, but that's very rare; what gets wet will dry out again.

Working with Wet Seed

The "wet" seed crops, those that hold the seed within a succulent fruit, like tomatoes, peppers, squash, and cukes, require a somewhat different approach. For each of these fruits there are slightly different clues to determine when seeds have reached maturity.

A tomato should be ripe enough for eating or, better yet, overripe to the point where you might hurl it at an inept performer or political opponent. Very rarely will a fruit ripen so much that the seed sprouts within it. Despite what I said earlier about underripe seed, tomatoes and peppers are somewhat more forgiving. I have regularly harvested green tomatoes, full-sized and shiny but green nonetheless, and ripened them in a sunny window, at which point the germ was fully developed. Likewise peppers: Ideally peppers should have turned their mature colour (usually red, but sometimes yellow or chocolate brown) on the plant, but again, I have been forced by early frost to pick them green or half green and after-ripen indoors, quite successfully.

As for cucurbits obviously a cucumber or summer squash at the eating stage is useless for seed. The cuke must be left to become a deep yellow-orange blimp or better yet brown and netted like a melon. As for zucchini you know those giant inedible prizewinners at the fair? You get the picture.

With cukes and tomatoes you have a clingy goo that consists of amniotic sacs, much like frogs' eggs in function and appearance. You could simply spread the whole business out on a rack or paper towel and it will eventually dry, but that would be a huge mistake. You see, that gelid stuff is designed to retain moisture, which is very helpful for seeds that are expected to germinate in the short term but not for seeds that may need to remain dormant in the packet for several years. It also provides an ideal medium for the growth of various disease spores, such as the fungus that causes damping off in seedlings. Hmm! Might just be a good idea to get rid of that stuff, but how? It turns

Figure 6.11. These cukes are too gone by for eating, but perfect for seed.

Figure 6.12. If they're brought indoors before hard freezes, cucumbers can continue to after-ripen for a few weeks before processing to improve quality.

out that goo is easily broken down by a brief fermentation, which not only doesn't harm the seed but actually leaves the seed coat impregnated with lots of antibiotic and probiotic stuff that actually protects the seed from those bad cooties. How simpatico! Marketplace seed is usually de-gelled by exposing the seed to a solution of hydrochloric acid. That sterilizes the seed coats but destroys all that probiotic benefit. The marketplace doesn't have time to let nature take its course, but perhaps you do. How much time does it take to ferment seeds to the right point? A few days; some say until traces of mould dot the surface, others say when fruit flies start to take note; I say when your significant other yells at you to "Get that stuff out of here!"; that's a pretty reliable indicator.

When your mixture reaches that critical point take the bowl or cup of ugly glop and put it in a glass jar along with several times that volume of water. (I occasionally grow cuke seed commercially, in which case I use 5-gallon plastic, or 18.9 l, pails.) Stir the mix well, breaking up lumps with your pinkies—yes, your hands will smell gross, but it goes away in a few weeks. If you prefer, use a long-handled spoon. As it swirls around in the jar you'll notice some stuff swirling around to the bottom and other stuff swirling around to the top. The bottom stuff is mainly plump seed with some of the denser pulp following it, especially bits of pulp attached to seed. The top scum is mostly gas-filled pulpy fibre. If there are seeds mixed with the floating crud, either they are attached to some lightweight pulp or the seeds themselves are no good, filled with buoyant gas; no loss there. Especially with cucumbers I always have lots of floating seed; it is alarming to see such nice plump-looking

Figure 6.13. Scrape the innards of cucumbers into a bucket or other container in which you can ferment the seeds. Photograph courtesy of Joe Hodgkins

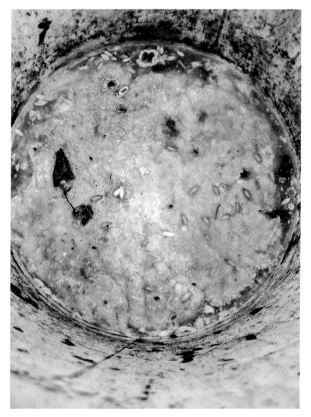

Figure 6.14. This bucketful of fermented cucumber seed is just gross enough to be processed.

seed getting poured off with the scum. If you need reassurance save some out and dry it separately: See how it shrinks to hollow hulls that blow away at a puff. The plumpness, like a baby's smile, was the result of gas.

Don't be fanatic with this first decantation; be satisfied with pouring off half of the cruddy slurry. Then repeat the process again and again; each time the cruddy liquid will become thinner and clearer and the good seed will sink faster (it was slightly buoyed up before by the denser dirty water). When the water is clean enough to suit you, put the seed in a strainer and give it a last cold rinse before spreading it out on a tray or screen to dry. Allow several days to be on the safe side. As with any seeds too long is better than not long enough, provided they're safe from mice. The fuzzy tomato seeds will form clumps that are easily "singulated" by rubbing between your palms. Cucumber seeds will retain a dry residual film that is easily sloughed off and blown away, but beware: The seed ends can jab your palms nastily if you rub them hard. Same with melons, although they do not need to be fermented, nor do dry-processed squash or peppers.

As with tomatoes and peppers the ripe seeds of cucurbits continue to draw some life force from the placenta. Even though the squash fruits may be fully mature at harvest, the seed is further improved if you leave it in the fruit in storage for several weeks before scooping out the seeds.

Viability and Dormancy

Well, there we have it: a lovely harvest of plump cleaned seeds ready for planting next spring. But how good are they? Will they germinate as well as those from the catalogue? Very likely they'll be fully as good, if not better, but if you are uncertain of any point in your process, or just want to feel reassured, why not take a test drive, aka a germination check? The proper way to do home germ testing is to place a given number of seeds, say 20, on a paper towel, which you then roll up, moisten with water, and put in a warm place for a number of days. An ideal place is a box with a lightbulb, much like a brooder

for baby chicks. It's important to keep the toweling continually moist. My problem is that I'm altogether too negligent to monitor their moisture level, and periodic absentminded dry-outs are *not* the way to determine viability. Instead I usually plant some seeds in a tray of potting mix as if I were getting a jump on the season. For some reason I'm better at keeping that watered, plus I'm more confident that the potting soil test gives a more reliable indicator of how the seed will fare in the real world. It's not what I recommend, it's just what I do, and it works well for me. Figuring out percentage germination is pretty easy when you start with 20 seeds. All you have to do is multiply the number of seeds that actually germinate by 5, and you'll have your percentage. Let's say you sow 20 cucumber seeds and 16 germinate: $16 \times 5 = 80$, so that represents an 80 percent germination rate, which is a fair result. (If your starting sample size is only 10 seeds, then of course you have to multiply by 10 rather than 5 to figure out percentage.) Commercial seed is required by law to be above 55 to 80 percent (the specific requirement depends on the species). For your own use you can set your own standards. However, I would point out that low germination also means that those that *do* sprout are probably weaker than fresher seeds and may get off to a slower start.

Seeds that have been overdried (which is unlikely with home methods) may go into a deeper dormancy and may require a day or two of pre-soaking before planting, to wake them up. Large seeds that are old may resist absorbing water (ever cook 10-year-old beans?), but scarifying the seed coat with a file or sandpaper will hurry the process. Some people pre-soak old seeds in flat beer to wake them up (yeah, I know, it never affected me that way, either). I suppose the idea is that an extract of sprouted barley will put them in a germinating frame of mind—power of suggestion, maybe. I used to keep a bottle around for trapping slugs, but felt guilty for corrupting their morals.

Viability of any seed depends on how it is stored. Drier is better than damper, and colder is better than warmer. Seed is alive and metabolizing, albeit at an extremely low rate—dormancy is not the same as death—and the lower the temperature, the slower the metabolism. As far as is known there's no such thing as too cold, provided the seed is adequately dried. If there is too much residual moisture in the seed, the internal water may freeze and expand, which could cause the walls of cells within the seed to burst. The seeds in the National Seed Storage Laboratory in Fort Collins, Colorado, are kept over liquid nitrogen at −321°F (−196°C) and some of them have been there for several decades with no loss of viability. For home seed-saving purposes you could put seeds in tiny ziplock plastic packets and double-seal them with duct tape. Store those seeds in your deep freeze, and you'll probably have to replace the freezer before the seeds. Their viability is at least thrice that of ambient storage, but there is one hitch. Let's say that next spring you want to use some, but not all, of your frozen seed. So you remove the packet from the freezer, open the packet, remove some seed, and then reseal the packet so you can return it to cold storage. Disaster! Inevitably, some moisture from the ambient air will immediately condense on the cold seed in the packet during the brief moment the packet is open. Henceforth, that seed will not store as well. The remedy is simple: Before breaking the seal let the packet sit at room temperature for a few minutes to allow temperature to equalize. Then you can reseal with no problem.

The learning curve for seed saving can be as long or as steep as you care to make it, but the important thing is to make a beginning. Afraid of screwing up? Good, then do it sooner than later, so you'll know how to correct course and get it right. Worried about losing valuable varieties? Then buy some generic pinto beans from the store and use them as your test crop for seed saving. Ready to move on to biennials, but haven't any first-year plants? Just buy some carrots or beets from the produce department and give them a whirl—who cares what kind they are? You may plant them too deeply or too shallowly, you may discover that woodchucks love to eat those roots, but you'll know what's wrong and can seek remedies before next time. You cannot steer a ship unless it is moving, and you cannot run while sitting on your hands.

Seed Storage Expectations

How long will seeds remain viable outside of a freezer? It depends on the species: but here's a rough guide, based on my own experience (predicated on storage conditions of very low humidity and room temperature or cooler):

- Parsnips: 3 years.
- Umbellifera (other than parsnips), lettuce, brassicas, and alliums: 4 years.
- Peas, beans, and muskmelons: 6–8 years.
- Tomatoes, beets, radish, endive, and chicory: 8–10 years.

The benefits of seed saving go far beyond the seed itself. All those flowering plants, especially the biennials, are also great nectar producers, and as such they attract a much broader assortment of nectar feeders than would otherwise be there. Many of these, such as syrphid wasps (*love* parsnip flowers), are at other times voracious consumers of aphids and other plant pests. Also, second-year biennials produce far more biomass than their adolescent forms and thus contribute more to soil fertility.

What About Hybrids?

Gardeners often ask me, "What about hybrids? I've heard you can't save seeds from them." Well, of course you *can* save seeds from hybrids, the question is, how useful will those seeds be? It may clarify the issue if I explain what hybrids are and how they're made.

Most new varieties came into being as a result of a cross of two different parent varieties. The process takes more than one step, but after the first cross the rest of the process is a matter of refinement by further selection. Technically all those crosses are hybrids to begin with, but when we speak of F1 hybrids we are referring to something a bit more specific. If we take two separate parent lines and inbreed them separately for several generations, each line will become highly uniform—the downside of incest here becomes useful. I've written elsewhere of the need for balance between purity and diversity, but in this case "moderation is a good thing if you don't carry it too far." The goal here is refinement, even at the risk of inbreeding depression. That risk is only temporary, because we then cross those lines to create an *F1*, or "first filial," meaning that is it the first generation of that cross. That undoes the harmful effects of incest, while retaining the uniform consistency of the parent lines.

For the commercial grower this breeding process seems like a total advantage, because it results in "hybrid vigour"—a little-understood synergistic gain over either parent line—and the results are consistent throughout the crop; there are no off-types. That benefit is short-lived, however; in fact it is good only for the F1 generation. If we gardeners save seeds produced by the F1 generation and plant them in our gardens, the seeds will certainly sprout and grow, but those next-generation offspring (called the F2 generation) will not show any of the uniformity of the F1 generation. Some plants will retain most of the desired traits of their F1 parents, but mostly it will be a crapshoot: You will have a mishmash of genes from the maternal grandparents, paternal cousins-in-law, old Uncle Harry, and whatnot; generally quite useless.

Well, perhaps not completely: If you find an F1 hybrid that particularly suits your fancy, you could stabilize it, thus creating a new variety of your own. That, after all, is what a plant breeder does. You see, creating an F1 hybrid is in fact only the first step in a breeding project. The breeder has made a promising beginning; now she needs to finish that work by continuing to select for desirable traits in the progeny. This takes several generations of continued roguing of off-types. By the third or fourth generation you will see more uniformity among the offspring, but

not until about the seventh generation will you see enough stability to be able to give the variety a name, and thenceforth consider it as an open-pollinated variety. You should *not* continue using the original F1 hybrid name, as that will only cause confusion. A good example is the onion variety Clear Dawn, which is a stabilized version of the F1 hybrid Copra. Clear Dawn is not as perfectly uniform as Copra, but it seems to me to store somewhat better.

What's the point of F1 hybrids anyway? Granted they have exceptional vigour, but most of that can be gained by the stabilization process just described, so why F1s? After all, hybridization mainly confers vigour, not resilience. In fact the main justification for F1 hybrids is commercial control. If you buy stabilized seeds there's nothing to prevent you from propagating your own henceforth. Now, there probably are not enough of us contemplating doing that to cause the seed companies much loss of sleep, but there's also nothing to prevent *another* seed company from taking that variety and offering it as their own—for example Laxton's Progress #9 pea becomes Morse's Progress #9, although the new company may have done nothing to change it. Indeed most companies just keep the same name if they legally can, so they can capitalize on the variety's reputation. A seed company has two ways to protect their breeding investment: They can patent the variety, which is expensive and often hard to enforce, especially if the thief has changed the name; otherwise they can just offer an F1, automatically giving them exclusive control, as only they have the original components for creating the F1. Other advantages: They save all the great expense of further stabilizing the variety—in essence, they don't have to finish their work—and they get to capitalize on the consumer appeal of the F1 (until rather recently, most buyers have assumed that hybrid meant superior).

What about that last assumption: Are F1 hybrids categorically better than standard varieties, often called OPs or open-pollinated (although that name is technically incorrect)? It is certainly correct that F1s are more uniform, but they are not consistently better in other respects such as vigour, flavour, or resilience, especially if those are not the traits that have been bred for. Many stable tomatoes varieties, including heirlooms, are far superior to any F1 hybrids, especially in those traits for which they were selected, such as taste and resilience. In fact hybrids are not always even unstable, a detail no one wants you to know. Many tomato hybrids in particular will come remarkably true to type, since the parent lines were not very divergent to begin with; for example, Sungold, a very popular yellow cherry tomato, has often been seed-saved by folks who didn't realize it was "impossible" to do so.

That is perhaps the paramount reason for F1 hybrids: to daunt you from saving your own, to keep you dependent on the marketplace. And that is the reason, above all others, why you should grow and save your own seeds.

With a growing number of crop species, especially sweet corn, cukes, and brassicas, hybrids so dominate the marketplace that there is little choice, especially when the few remaining open-pollinated varieties have seen so little improvement in decades, while new hybrids are constantly being developed. If they wanted or needed to seed companies could either look around for superior non-hybrids to offer, or they could stabilize some of these F1s by selecting them to F7 and beyond. But why should they *want* to, if their only payback will be having some customers save their own seed? And why should they *need* to, as long as you the customer are willing to accept what they offer? And what if you *don't*? What if you reject

their F1 offerings, regardless of whether you intend to seed-save or not, and insist on purchasing only stable varieties? Of course it's very important that you *notify* your seed company of what you are doing and why, so that if they notice any shift in demand away from the hybrids toward the OPs, they will understand what's happening and hopefully respond. If they do not (which is unlikely, unless a boycott were to really get some momentum), then perhaps you should look around for companies that *do* offer more non-hybrid choice and reward them with your business, again letting them know why you are favouring them. As long as you keep buying hybrid seed, breeders will have no incentive to complete their work, and you will remain a slave to the marketplace.

Here is a great example of the marketplace failing to meet the needs of a democratic society while it could be fulfilling them. Whose fault is this, if not ours? In our delusions of democracy we somehow expect the political world (government) to regulate the economic world (marketplace); how naive, when it is nearly always the other way around. It may be frustrating to try changing that dynamic, but it's important to remember that, given its vast power, the marketplace itself can be a vehicle for change through your purchasing decisions. You only get to vote at the polls once a year, and no one seems to be paying attention anyway. However, you get to vote at the checkout counter *every* day, and believe me someone *is* paying attention there. Don't throw away your vote.

CHAPTER SEVEN

Asexual Propagation

As I pointed out earlier sex confers on a population the all-important advantage of diversity. However, diversity sometimes comes at the expense of stability, and stability can be pretty important, too. Since I've already explained homo- and heterozygosity, I'm hoping you can see why a population or even an individual whose genes are largely heterozygous might take an incredibly long time to stabilize. When a plant has lots of heterozygous traits, its genes are constantly and unpredictably realigning themselves with every new generation. It's a mess; you never know what you're going to get. That's great for creating diversity, but what if one of those genotypes suits your purpose beautifully and you'd like to grow a whole lot more of the same, *just* the same?

A plant breeder might take that unstable population and stabilize it over several plant generations until most of the genes are homozygous (whether dominant or recessive), so that their offspring will tend to come true to type over time. (This process is described in the *What About Hybrids?* section of chapter 6.) That is what has been done over centuries with most of our annual food crop plants. But not all. Some species, like potatoes, just don't lend themselves to this stabilization, perhaps because they're so plumb full of heterozygosity that getting them to settle down is hardly worth all the trouble. Especially when there is a much simpler way: asexual propagation (aka vegetative propagation, or cloning). If you cut up a potato tuber into pieces, each one having at least one "eye" or bud, you can grow genetically identical copies of the same plant, regardless of how hopelessly tangled its genes may be. Many clonal crops are very hardy and survive well in situ, year after year. Terrasols (aka sunchokes), groundnuts (*Apios americana*), mint horseradish, comfrey, and rhubarb are relatively safe and secure compared with seed crops. And although tree fruits do produce fruits and seeds, many of them are commonly propagated asexually by grafting.

However, there are at least two particular problems with asexual propagation. Because most types of plant material used for asexual propagation are not dormant in quite the sense that seeds are, you cannot simply put them in a packet and store them for several years. For example, I maintain a collection of 1,100 pea varieties, but I do not have to grow seed of every variety afresh every year; I grow a couple hundred varieties each year in rotation, so that every five to six years the entire collection is renewed. On the other hand I also maintain a collection of several hundred potato clones, each of which must be grown *every single year*; the year I fail to harvest at least one tuber of any variety is the year I lose that variety for good.

Another problem with some clones is the accretion of disease; again, potato is a good example. The potato is prone to several systemic viruses, some of which it has acquired from related species (like tobacco) that have crossed its path. "Systemic" means that the virus infects every part of the plant's system: leaves, stems, roots. They are extremely difficult to eradicate. The plant might endure any one of them and produce a fair crop, but the viruses are cumulative. Over time a population of potatoes picks up first one virus, then another, until eventually those potato plants have trouble getting out of bed in the morning, let alone making a paying crop. Fortunately these diseases are not soilborne (like nematodes) or airborne (like blight), but they are transmitted via the mouthparts of sucking aphids. If you control aphids effectively you can prevent or at least retard the spread of viral diseases. If you have infected seed and can replace it with clean seed, then you are all set, even planting in the same soil.

One way to get rid of a systemic virus is to grow out true or sexual seed from that clone, since those viruses cannot cross the placental wall into the embryo. Many potato varieties do not produce seed reliably, but when they do you save the little green fruits (they look somewhat like tiny green tomatoes and are highly toxic when eaten), extract the seed, and plant it in seedling trays, like tomatoes. The problem is that you will not get what you started with (remember heterozygosity?) any more than you or your siblings look like either of your parents. You can grow hundreds of these seedlings and look for some that contain all the traits you liked in the original clone. If you find something great give it a name of its own, because it is a new variety, for better or worse. You are now a potato *breeder*, and with luck you may develop superior varieties, although mostly the opposite. It requires much patience, and you'll soon appreciate why potatoes are usually propagated asexually.

It is much easier to *keep* potato stock clean than to clean it up once infected. If you bring in new seed stock from away and are uncertain of its cleanness, you can quarantine it—grow it in an isolated patch far enough from your known clean plants that aphids are unlikely to move back and forth between them. Once you're confident that stock is unaffected, you can grow it alongside other stock in your garden with no fear of contagion. Although there may be some possibility of contamination from related wild species, like deadly nightshade, by far the greatest risk is from importing diseased potatoes. If you use tobacco products or allow anyone who does near your garden, you will soon spread tobacco mosaic virus (TMV) to your plants. It is nearly ubiquitous with smokers, even clinging to pant cuffs and fingernails, just one more reason to quit.

It is easy enough to propagate herbaceous clones by chopping up tubers, snipping runners, dividing bulbs, splitting crowns, and so on, but what about woody shrubs and vines?

Suckers, Layers, and Cuttings

The cane fruits—raspberries, blackberries, et cetera—are generally cooperative. As they get well established they send out roots that form suckers, usually far enough from the source plant that you can dig them up and cut the connector roots without harming the parent. This also works well with American hazelnuts. The suckers may not be well rooted, since they have been relying for much of their nourishment on their host plant. Even though I try to dig them in early spring while dormant, as they bud and leaf out, there will be too much foliage for the skimpy roots to support, and so they will immediately wilt unless I beat them to it. If there are several stems or shoots growing together, I clip away all but one and cut that one back by half its length before sticking the roots into a bucket of water. I can hold them that way for many days, but I prefer to plant them out as soon as possible. Soon after transplanting they will start to grow more of their own roots, but they'll appreciate regular and heavy watering until they can set up their own arrangement.

I have used layering for woody vines, like grape and hardy kiwi, when I only need a few plants. In spring or early summer I just bend a low vine along the ground and pin it there with a forked stake. Nicking the stem and covering it with soil or compost, so the cambium is in touch with a rooting medium, will encourage it to form roots. Grapes will root at each of the nodes, as long as those nodes are touching the ground. In a year or two I dig up the layered sections of vine, cut them into as many sections as have roots, and plant them where I want them. I may put them into a nursery bed for a year or two until they're better rooted. A problem with layering is that you must be cautious lest you mow the layers in progress that may be hidden in tall grass. Of course that's less of a risk if you mulch them and remove any tall grass.

When I need a large number of plants (I sell hardy kiwis commercially), I take cuttings to root. I cut long lengths of vine and clip them into pieces about 10 to 12 inches (25.4–30.5 cm) long. I like to do this very early in the season, before the buds have broken. With grapes I make sure there is a swollen node at the base of each (the node looks sort of like an old-fashioned thermometer). The nodes on kiwis are close enough together that I merely snip off wherever I wish. I have a special propagation bed for these; it is in the partial shade of a hedgerow and I have dug it deeply, removing every rock and pebble to a depth of at least 16 inches (40.6 cm). That is a major undertaking in my soil, but I use that bed year after year. I also worked in lots of compost and peat moss and thoroughly mixed the topsoil and subsoil.

I stick in the cuttings at least 6 to 8 inches (15.2–20.3 cm) deep, leaving perhaps one-third above ground. I put them as close as possible, usually 1 to 3 inches (2.5–7.6 cm) apart with 18 inches (45.7 cm) between rows, so I can give them intensive care in a small area. It's essential that they be right-side up, and with kiwis that's not always obvious, because the dormant buds are actually down drooping. To be really sure, I lay all the cuttings neatly together as I collect them, all pointing the same way.

Once the cuttings are planted frequent and heavy watering is needed to guarantee a high percentage of takes. I typically get over 95 percent. In late spring I usually cultivate and mulch with shredded leaves. After the cuttings have spent a year in this intensive bed, I reset the rooted plants in a regular nursery bed 12 inches (30.5 cm) apart in rows 30 inches (76.2 cm)

apart. At the end of the second year they are usually ready to sell; otherwise I keep them yet another season. There are many scant-rooted kiwi plants on the market, and sometimes they barely survive. Top growth is pretty irrelevant; roots are everything.

I have found that many species can be started simply by placing woody cuttings in a juice jar half full of water and leaving them in a sunny window until they form little roots. It may take weeks and even then the few roots are hardly impressive, but they're usually enough so that if I set them out in a nice rich nursery bed and keep them moist, they will become well rooted by the end of the season. I've mainly used this method for currants, gooseberries, elderberries, and grapes.

Using Willow Water

Here's a trick that will greatly increase root formation and help you work with hard-to-root species: Exploit the root-promoting properties of willow. Willow is notorious for its ability to root from a single twig stuck into wet ground. Less well known is its ability to promote root growth in neighbouring plants. A willow growing near a dug well can impart a disagreeable taste to the water. Apparently this is due to exudate from its roots, which acts as a rooting hormone. You can purchase root hormone products for starting cuttings, but you can also make your own. Take cuttings from a willow (any species) and soak them in a jar of water for a few days, as though you were trying to root them. Then add cuttings of whatever plant you wish to propagate. I rubber band them loosely together to keep different types separate. Sometimes I leave the willow in the jar with the cuttings, sometimes I remove it; I'm not sure it makes a difference either way. What I *have* seen for sure is that many species that don't root well in plain water do much better in the willow infusion.

Intrigued by this phenomenon, I also tried using the willow extract as a root dip for annual garden transplants, like tomatoes and cabbage. So far, nothing conclusive, but I mean to continue the experiment, including willow-watering seedlings in the tray before transplanting.

Grafting

With an actual tree, as opposed to vines and shrubs, it's not usually practical to layer a young shoot. And tree cuttings are more difficult to start in willow extract. I suppose this difference is because woody vines have more adventitious or undifferentiated cells in their bark tissues. These are cells that haven't decided what they want to be when they grow up, so they loiter around until they see what's needed, at which time they might suddenly turn into a twig or a leaf or a root cell. Trees also contain these cells, though not as many. I guess most tree cells are more set in their ways and less flexible about midlife career changes. The fact is that most tree cuttings have difficulty holding on to life until they can grow their own roots and be off on their own. One method of helping the cuttings establish themselves has been to "borrow" some already established roots (although the loan is usually permanent). This is grafting, and it's an ancient technique; the Christian apostle Paul of Tarsus gives a rather clear explanation in his "letter to the Hebrews."

To understand how grafting works it's important to understand the stem structure of a woody plant. If you slice a young twig neatly across the grain and examine it closely, you observe a thin light-green ring layer just under the main or outer bark; that's cambium. Cambium is the special type of tissue that can undergo cell division and produce new growth. There are several methods of grafting, but they're all based on the fact that the cambium (inner bark) is the source of all new growth in a tree. Got that? The wood will never produce new growth; nor will the pith or the outer bark (it will crack and spread, filled in with new growth from within).

When we graft two pieces of tree material together, hoping their tissues will unite into one, it is the cambium that will accomplish this feat or it won't happen. As every cambium cell divides and grows it either moves inward to build up the current season's growth of wood, or pushes outward to form bark. Also if the bark is injured the cambium will grow and spread in every direction until it meets itself, eventually healing the wound. That is more or less what happens with a graft, except the meeting cambium tissues are now from two separate organisms: generally compatible, but genetically distinct. Remember this: When you make a graft, whether barks or piths or woody rings align or not means nothing; but if the two cambiums are touching, there is hope.

Cleft Grafting

My preferred method of grafting most species is the simple cleft graft. In this type of graft you will slip a small piece of the variety you wish to propagate into a slit in a prepared rooted base of a different variety.

To make a cleft graft you need a rootstock, the "borrowed" roots onto which you will graft. The other part you need is a twig—called the scion—from the variety you want to propagate. You'll also need some simple tools and some grafting wax.

Procuring rootstocks. The rootstock, though genetically different from the variety we are putting on it, has no real effect on the fruit that the grafted plant will eventually produce. Fruit quality will be determined by the "scion," the piece of twig we graft onto it. What *may* be determined by the rootstock (at least in part) is hardiness and tree habit. For example, some rootstocks are selected because they determine the amount of growth hormone being sent to the top of the plant. They may cause the top to have short internodes (the distance between buds) and thus to be dwarfed. But the fruit will be normal size in shape and taste.

Curious fact: There used to be a variety called Virginia Crab that was popular as a rootstock because it had a significant dwarfing effect on whatever was grafted onto it. However, Virginia Crab was infected with a systemic virus. Eventually a way was found to rid it of the virus. Great, but now the rootstock no longer had the same dwarfing influence as the original and was generally useless for anything.

If you want rootstocks that will confer specific properties, you may have to purchase them. But if you are less choosy you can often grow your own rootstocks. For example, I've planted seeds from my spent cider pomace to grow apple seedlings for

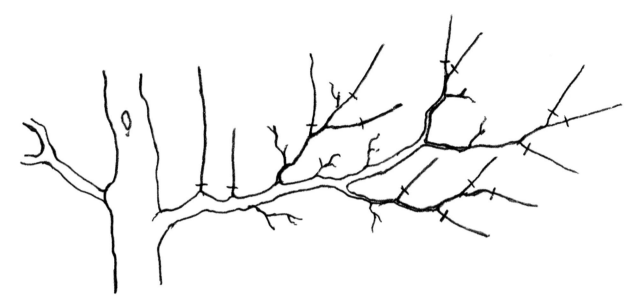

Figure 7.1. Scionwood should all be cut from smooth, straight (new) growth. Suckers are ideal.

use as rootstocks. Apple seeds, like many other tree species, will not sprout unless they've been stratified. That means they've been through some winterlike experience of freezing and thawing, which breaks their dormancy. You can store seeds in your freezer to do that, but I simply spread some pomace (the residue left from pressing cider: skins, cores, seeds) on the ground under a thin layer of soil and a square of ½-inch (1.3 cm) hardware cloth (for contact and to deter the deer) until it germinates. I'm careful to watch and remove the screen as soon in the spring as the seedlings appear; if they leaf out with the screen in place, it's hard to remove it without damaging the seedlings. In a year I can dig out hundreds of apple seedlings to line out in my nursery rows until they reach pencil size in diameter, ideal for cleft grafting.

Procuring scions. Now for the scion, a twig from a tree of the variety you wish to propagate. By far the best scionwood is smooth unbranched wood of last year's growth (as shown in figure 7.1). Although I have successfully used older wood it is not so open-minded. Pencil-diameter twigs are ideal, but I have managed with much smaller and much larger. An advantage of the cleft method is that the size differential between scion and rootstock is not too important.

I get my scions from various sources: fellow seed savers, the National Germplasm System, and the Scion Swap, an annual event held at the Common Ground Fair exhibition hall in Unity, Maine. I also collect scions directly from old trees that I know of in my area. Regardless of whence I get scions I immediately put them in a plastic bag with a damp paper towel, close it with a twist-tie, and keep it in a cool, dark place. My root cellar works fine; the refrigerator would probably be better if there were room. They mustn't dry out.

Preparing to graft. I make cleft grafts in April, when the ground is bare but buds haven't burst yet. Some folks prefer to graft earlier when snow is still on the ground; however I prefer dry knees and warm sun to keep the grafting wax workable. That compound, by the way, is a blend of beeswax, pine tar, and linseed oil. You can make your own (I don't know a recipe), but Trowbridge's makes an excellent mix that costs less than a nickel per tree unless you're profligate with the stuff. When the air is chilly and your body heat is challenged, the stuff is impossible to work, so I choose a sunny day when my spirits are high and the wax is supple. Leaving the open block of wax on the windowsill until I'm ready to go outside and start grafting renders it almost runny.

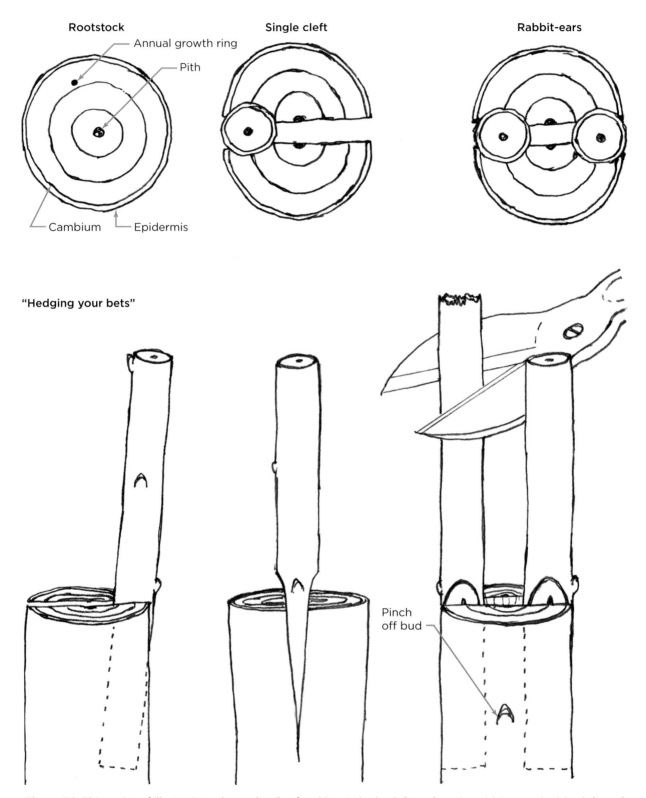

Figure 7.2. This series of illustrations shows details of making a single cleft graft and a rabbit-ears double cleft graft.

The scion and rootstock are not a perfect match (see figure 7.2). Obviously their cambiums will not touch everywhere. Small matter, provided they touch *some*where. Therefore they must not be centred on each other unless they are perfectly matched in size. In most cases they are not, though it is excellent if they are; they will start off with two points of contact instead of one, and very soon they will grow into a continuous cylinder. What matters with the ill-matched pieces is that their perimeters must be tangent at *some* point. Since they grow at slightly different rates, that initial contact *point* will expand to become an entire *ring*, forming a complete cylinder. Over time the graft junction may disappear altogether.

Cutting the rootstock. It is wise to make the split in the hub (or rootstock) first. I always cut each scion for each rootstock as I'm doing it. If I try to do several at a time, the fresh-cut surfaces are apt to dry out before I set them in place. Partly for that reason I greatly prefer to do this with a helper, and so it's not necessary to clean the sticky wax off our fingers for each graft. One does all the cutting and inserting and tagging; the other does all the sealing (and cussing). If I expect that I or anyone will have use for the left-over scion stub for another graft, I hasten to put it back into the plastic bag and in the shade of my body. I always cut the rootstock before the scion, since that part has sap welling up behind it and can better stand to wait for a minute or two until you can complete and seal the graft. That split should be right in the centre of the hub, and you should control your knife hand so the split doesn't go too deep—say, 2 inches (5.1 cm) or so. If the cut is too deep then there may not be enough tension in the rootstock to pinch the scion snugly in place. I have corrected floppy slits by just clipping the hub back slightly, so the slit is a bit shallower and firmer, but often that just mashes the split and makes things worse. In such cases I simply clip off the split portion and start anew, assuming I haven't cut it back too short to begin with.

Cutting the scion. The strongest way to join a scion and rootstock is to shape the scion into a long skinny wedge and insert that wedge into the split in the rootstock. Shaping that wedge is about the only

skill involved in this operation, but it pays to do it well. I use a stiff, straight-bladed grafting knife, but in the past I have muddled along with jackknives and box knives and fared much better than I deserved. Obviously the proper tool is best. Notice (figure 7.2) that the scion cut is not completely tapered; it "shoulders out" a bit near the bottom, since the scion doesn't need to be inserted all the way into the cleft.

It is ideal if you can whittle the scion so that the lowermost bud ends up near one edge of the wedge, where it will be closest to the rootstock cambium. This is to preclude the possibility that the side of the wedge opposite the bud might hold the split open in

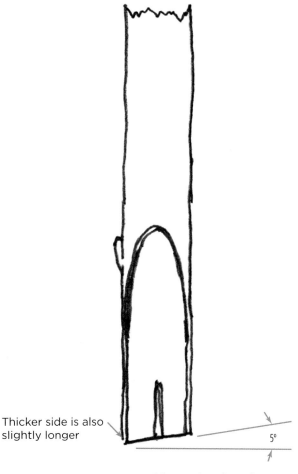

Thicker side is also slightly longer

5°

Figure 7.3. If you must err while cutting the scion, cut the wedge not quite parallel with the first bud on the outer side.

such a way that the bud-side cambium will not come in proper contact with the rootstock cambium. If a rootstock is not too much larger than the scion, then it will spread open evenly, but if the size difference is too great, the scion may twist the split open unevenly so that the outer (away from the pith) part of both cambiums cannot meet. By making the two surfaces of the scion wedge slightly off-parallel (with the bud near the fatter side), this should not be a problem. Remember, if the inner wood of the scion fails to meet that of the rootstock, no harm is done as long as the outer cambiums of both are matched for as much of their length as possible.

First I whittle one side with as few strokes and as smoothly as possible; then I flip it and do the same on the opposite side, so that it comes to a pointed wedge at the end. The trick is to have it end up not like a tongue depressor, but an actual sharp wedge that you can insert into a tight cleft. I do that by giving it enough under support with my left pointer, but hopefully not so much that I scrape myself too hard. Not only is it socially gauche to bleed while you're grafting, but blood is very salty and will ruin the cambium cells. Sweat and saliva aren't too good for them, either, so try to keep a lid on all that if you can; just basic decorum involving body fluids.

Joining the graft. Once you have shaped this scion to your satisfaction, you should insert it into the rootstock pronto, before it gets too dried out. While grafting I try to keep my bag of scions in a box out of direct, drying sunlight. The directions that follow may seem complicated on first reading, but I'd say three minutes per graft is a relaxed pace once you're practised at it.

If your rootstock or hub is stout enough that the sharpened scion is difficult to insert in the cleft, you can hold the cleft open with the knife blade or a narrow wood chisel. I'm usually able to easily and gently slide the wedge-shaped scion itself into the cleft and push it gently in until it holds firmly.

If you manage to insert the scion perfectly into the cleft, then the cambiums will be touching over the whole length of your cut—excellent! But what if not? What if you insert it a tiny bit too far inside or outside,

so that it's not quite touching anywhere? Well, that's not too likely, but it is possible, so some folks hedge their bets by placing the scion as correctly as they can, and then they tip the scion slightly outward (so the bottom twists *in*ward). The idea is that the scion and rootstock cambiums won't have perfect contact everywhere but will certainly be touching somewhere (see figure 7.2). This philosophy reminds me of a very fine watch I once owned; it kept incredibly accurate time, supposedly lost one second every millennium. Still not satisfied with that, I took a ball-peen hammer and whaled on it. Now twice every 24 hours it shows the absolute correct time.

When I insert a scion I withhold it just enough that little parabolic areas of the cut surfaces show above the hub. If I force it in too far the shoulders of the scion would spread the cleft too far open so there would be poor contact.

If the scion is less than half the diameter of the hub, you can double your odds for success if you place a second scion in the same cleft, aligning its cambium on the opposite side, hence the name *rabbit-ears graft* (figure 7.2). If both grafts take, you should clip off the weaker one by the second year, else it will make a weak crotch (figure 7.2).

Sealing the graft. Having done all this it is vital to seal up all those junctions, so the cut surfaces, especially the cambium, do not dry out. That's where the wax comes in. I prefer to have a helper when I'm cleft grafting so I don't have to clean wax off my fingers (a major operation) between every graft. Stretching out the wax like a taffy pull, the helper wraps a band of soft compound around the joint, covering the exposed area from the parabolic cuts all the way down the sides of the rootstock to the bottom of the split. A bit farther, in fact; when rainy weather makes the rootstock swell with sap, the split may try to open a little more, leaving it exposed. I make sure that every cut surface is well covered, leaving no openings for rain or drying air to enter. With the rabbit-ears graft I take special care that wax is gently worked in between them. If the compound is too cold or stiff, there's a risk of tearing off the papery outer bark, or epidermis, as you work. If that happens, exposing the cambium, it's not too serious so long as you replace the epidermis with a "Band-Aid" of soft wax.

With practice you can make a neat waxing that covers everything with a minimum of waste. Gorming it all up with superfluous compound is not such an issue economically—so the tree costs a dime instead of a nickel—but it's good to avoid smothering the buds unnecessarily. They can break through a thin layer of goop, but there's no point in strangling them.

When you're done with waxing clip off the excess scionwood, leaving about three buds. That way if the bottom bud fails—the one in the wedge—you still have two other backups. If they all take, fine, but at some point, maybe next spring, you should clip off all but one strong shoot, so you end up with a tree instead of a bush.

After grafting I often take the cut-off top of the rootstock and stick it in the ground close to the newly grafted plant. All you need at this point is for a big fat robin to come along and decide to perch on your work, possibly knocking it askew. The old top forces him to perch on that instead or go elsewhere.

As the weeks pass the buds will burst and leaf out and you'll suspect you have succeeded. Probably, but not for certain; the scion can survive a little while on its own resources, but hopefully sap will be flowing from the roots before hot, dry weather sets in. Be sure to pinch off any buds that may emerge from *below* the graft (figure 7.4). Those are rootstock and will try to compete with the grafted variety.

Some variations on cleft grafting. In general apple scions can only be grafted onto apple rootstocks, plums onto plums, and so forth. But there are several exceptions. Pears can be grafted onto quince, which makes dwarf pears possible. Peaches and plums can be grafted onto *Prunus besseyi*, or western sand cherry, to create dwarf versions of those fruits, but with poor results.

I graft medlars onto wild hawthorn when I haven't medlar rootstocks. The former may become incompatible over time and reject the other. Therefore I do what I call a nurse graft, as low to the ground as I can get it. If it takes, the following year I transplant the new plant where I want it, but I either set it in a

deeper hole or mound up soil around it, so the graft union is *below* ground. Eventually the medlar top will send out its own roots, so if the foster parent rejects it, by then it will be independent, which we call scion-rooted. That can be done with other species, too—for example, apple on apple—in case you fear compatibility issues; however, if the rootstock was selected for a particular purpose, say dwarfing, you would lose that effect as the rootstock is bypassed.

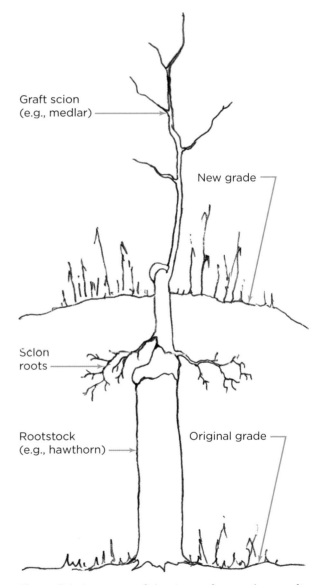

Figure 7.4. A nurse graft is a type of secondary graft to avoid incompatibility of scion and rootstock.

Figure 7.5. There are four apple varieties on this topworked tree.

By the way, some nurseries combine traits by using a rootstock that may be very hardy but non-dwarfing, then grafting on a piece of a dwarfing inter-stem, then in turn grafting on the scion variety. I assume this is done by bud grafting (later on that), but I see no reason why it wouldn't work with cleft grafting.

If you have room for only one or a few trees, and you really want several varieties, you can graft more than one variety onto a tree. It's called topworking; I let the original scion grow several feet tall (wait a few years) and then graft other varieties onto as many as four lower limbs, treating them as separate horizontal rootstocks. Make sure those limbs are situated high enough that you won't regret it someday when they interfere with your mowing. With just one original variety you can always prune out a lower limb without chagrin, but if that limb is a separate graft then you will sacrifice that variety. Of course you can also train a lower limb upward by pruning. One concern with multiple grafts is that if you lose a tree to borers or mouse girdling, you've lost several varieties at once.

More Grafting Projects

How big can a rootstock be? I am often asked this question by someone who has a 4-inch (10.2 cm) diameter apple tree of no value because it never was grafted. Theoretically such a tree can be used as a rootstock; in fact with that huge root system under it, the new scion is likely to make explosive growth, perhaps 4 to 5 feet (1.2–1.5 m) in a single season. An obvious problem is the difficulty in opening a cleft to insert the scion wedge. There's another way around that, which I'll describe anon, but a bigger problem is the great discrepancy in size. Despite the scion's accelerated ("forced") growth, it will be many years, if ever, before it catches up with the rootstock and makes a continuous cylinder of cambium. Worth it? Not to me.

What is more feasible is the topworking method mentioned earlier. You can cut the tree back to a hat rack shape and graft each limb separately, to the same or divers varieties. An advantage of this is that because you have left the basic tree frame intact, it will take much less time to regrow and bear; the individual grafts are nearer the same size, yet it has all that tree mass to support and nourish it. A disadvantage is that any new growth sprouting from the main tree will be useless and competing with the graft. You must watch and prune them out as they appear. In time it will become more obvious and easier to distinguish.

If you insist on grafting onto a stump that's too big for a cross-cleft, you can do something called a bark or veneer graft. As with cleft grafting you cut the tree off several inches aboveground and then make a slit in the bark down one side of the stump. Make the slit about 2 inches (5.1 cm) long. Next cut a scion as for cleft grafting but with one flat side only. Insert the scion into that slit, taking care that the scion's cambium is against the outside of the stump's cambium, but pressing against it. The flaps of the outer bark should more or less hold the scion in place, but you must reinforce it with something; I recommend masking tape, which will degrade by the time it's not needed. Then you must seal the whole operation, just as you would with a cleft graft. However this may require a lot more compound, depending on the size of the stump. There is another product, a black petroleum-based material, that you can paint onto

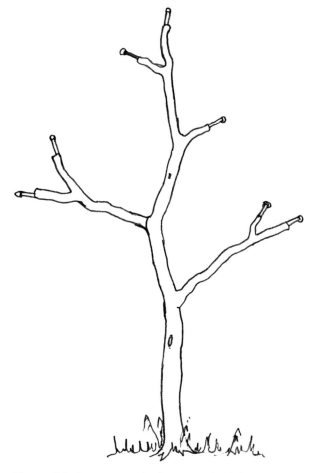

Figure 7.6. On a topworked tree, each of the separate limbs is grafted with a new variety. Any limbs that are not grafted are removed. It's important to wrap each new graft as it's completed.

the surfaces. Because of its fluidity many prefer it for cold-weather grafting. I disprefer its ingredients to the wax-based compound. I've done this kind of graft, but I don't recommend it because the large exposed stump surface, plus the disparity in sizes, assures that it will be many years, if ever, before the graft heals over to make a single strong cylinder. The scion is fairly likely to break off in the meantime. This is mainly useful as a temporary holding place for varieties when you haven't enough young rootstock available.

A popular grafting technique that I have used but little is whip-and-tongue. Its big advantage is that

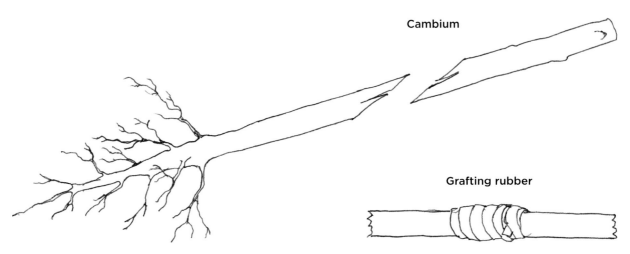

Figure 7.7. When making a whip-and-tongue graft insert the slits so that both cambiums match up all the way around (left). When you're satisfied with the match wrap the graft with a grafting rubber (right).

you can dig up the rootstocks in late autumn (or order them) and graft them indoors in winter. The bare-rooted plants are bundled and stored in a cold, damp place and brought out at your convenience to graft in a comfortable shed or cellar. Because this type of grafting is usually done on a workbench, it's also called bench grafting. After grafting the plants can be rebundled (no wax) and kept until spring.

To make the actual graft you simply cut each scion and rootstock into a long one-sided taper. Of course they need to be held together somehow, so you make a slight slice near the centre of each piece (see figure 7.7). These are inserted into each other and wrapped with a grafting rubber, basically a strip of degradable elastic band (they must be quite well matched). It is the rubber that does most of the holding, the slices serving more to keep it aligned. No wax is needed at any stage. This is much stronger than it looks.

Perhaps the most bizarre grafting method I've encountered is something I found mentioned in *Facts for Farmers* by Solon Robinson, a magnificent two-volume work from the 1860s, no less fresh and inspiring today, if you can find it. Robinson describes a technique used in a region of Tennessee: You cut an apple branch several feet long, roughly an inch (2.5 cm) in diameter, and with a common wood chisel and mallet make a crosswise split clear through the

branch; repeat these splits every foot (30.5 cm), all with about the same alignment. Then you whittle a number of scions and insert one into each split, so that each scion passes right through the branch. Each scion is clipped to several inches long, and the whole limb is buried horizontally in a trench in good soil, in such a way that at least a couple of scion buds on each scion protrude aboveground. No waxing involved; indeed I assume that it's important for the cut surfaces to be in close contact with the soil. I would emphasize that there are *no roots* involved at this point, only the naked limb!

If successful these "grafts" will take and the scions leaf out, drawing their whole life support from the woody mass of the limb. However roots will begin to form at each graft junction, and the following spring you can dig up the limb and saw apart each of the grafted sections, replanting each in its own place to become its own plant. Weird. Well, I've tried it once or twice without notable success—it's hard to find a long straight apple limb, and the splits are pretty ragged—but the idea led me to try something else.

It had occurred to me: What if I tried it with a lateral *root* instead of a limb? I was reluctant to dig up and remove large roots from a perfectly good apple tree, but chance brought another possibility to my attention. Anyone who has ever dug in the vicinity of a mulberry tree knows that they send their bright-yellow roots extremely far in every direction (more like mushrooms perhaps). I have dug up finger-sized roots in a garden over 40 feet (12.2 m) away from an 18-foot (5.5 m) tall tree. So I tried the technique with the mulberry in late spring. Most of the buds were too far opened, but mulberries tend to bud over a considerable time span, so I was able to find some scions that were fairly dormant. With nothing to lose I tried this Tennessee graft on a few of the excavated roots and stuck them in my nursery. The scions were cut from the same tree as the roots, so in effect I was grafting the tree onto itself. Most of them soon croaked; however, a couple of them have most definitely taken. How they will survive remains to be seen, but I am more encouraged by that success than I am discouraged by the other failure, since mulberry rootstocks are hard to come by and seedlings are incredibly variable.

Stooling

If we propagate fruit trees vegetatively in order to get uniform tops, what about the rootstocks? If they also have special attributes, like dwarfing or hardiness, should we want those to be uniform as well? In some

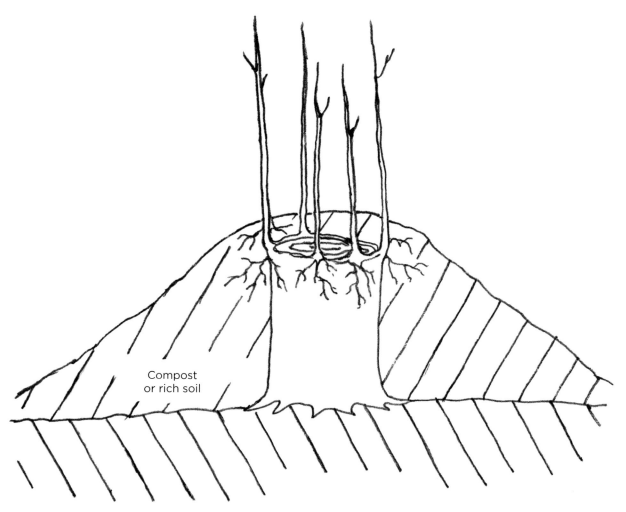

Figure 7.8. Hilling rich compost or soil over a stump can stimulate new shoots to form.

cases seeding rootstocks are significantly uniform, but if not, how can we propagate them vegetatively? One method is called stooling, or stool-bed propagation.

Start with an original rootstock from a nursery supply company, and cut it back to within a few inches of the ground. Or take a grafted tree whose top failed (or make it fail by cutting off the top, if it's important to you) and cut it back to *below* the graft union to be sure you have nothing but rootstock, and within a few inches of the ground. From this "crown" you've created should sprout several shoots, which should grow several inches tall. (I keep saying "should" because I have had variable success with this; mostly it works.) When hilled up with fertile soil each of these shoots should form little roots near its base, just above the crown. You can greatly increase this "should" by first making little nicks in the bark to put some cambium in contact with the surrounding soil.

Next wrap a twist-tie tightly around the point where the shoot arises from the crown; this slowly girdles the swelling shoot, giving it greater incentive to form its own roots above that point. Then apply some commercial rooting hormone (such as Roo-tone) to the lower bark before burying it. I never did that; you can also soak the hilling medium with the willow extract described earlier. I haven't done any of this in years, but as I recall the extract seemed to make a difference. The next spring you unearth the crown and clip off all the shoots that have rooted.

Transplant those to a rich nursery bed so they can have a year to get better rooted before you graft onto them. Some sources say you can continue to cut new rooted suckers every year, which I believe but have never tried. I've only done the stuff experimentally and partly by accident. Personally I'm not so excited about dwarfs, as they tend to be shorter-lived.

Although most of these forms of asexual propagation, including grafting, are really quite simple and should "take" most of the time, there is some degree of knack involved. If at first you have less than 50 percent success, by no means should you feel discouraged. In the beginning I chortled with pride when I got one out of four. Now if I have less than 95 percent success, I consider it cause for some soul-searching. As with everything practice makes perfect, or at least better. When teaching apprentices I always have them rehearse a bit with fresh prunings before I let them work with scarce and valuable scionwood. After all today's youth did not grow up whittling doodads out of box elder. But when you can create your own orchard full of rare fruits for less than a dime per tree, it's well worth a little practice and patience.

PART III
The Crops

The whole point of all this soil building and land shaping and weed and pest control is to produce food crops. After several millennia of the shared human horticultural experience, the diversity of food plants available to modern cultivators is truly impressive, not to mention the many plant species our ancestors ate that have fallen into disuse. How sad, therefore, that most of us are accustomed to eating only a tiny part of that food heritage, millions of acres devoted to a paltry selection of same old crop species. Along the paths that lead *away* from the marketplace, we can find a much more varied and interesting diet, whether new crops or new varieties and uses for existing crops. Let's discuss those now, but warning: The subject of crops is far too broad for me to be comprehensive.

CHAPTER EIGHT

The Veggies

Instructions for growing vegetables abound. I want to share with you my experiences and ideas about growing these crops, especially those that are less commonly grown, but I don't want to fill pages describing common techniques that can be easily found elsewhere. Instead this chapter focuses on those areas where my approach is unique or not widely known.

I should also point out that I've included information on growing vegetables in other parts of this book as well. In particular, because I like to combine different crops together in a bed, the sections on intensive cropping and companion planting in chapter 14 are filled with information on veggies. There's also chapter 15, *Pests and Diseases*, and I discuss many common vegetable pests there.

Therefore in this chapter I've devoted very few words to some vegetables, because what I have to say is not new or because I've written quite a lot about them in the section on plant companions. I've written quite a bit about other crops, those that are little known or underexploited and those on which I have a particular slant I want to share.

As I prepared this chapter I was confronted with the problem of how to organize the discussion: Should I discuss crops according to their botanical groupings or according to utilization or, as in seed catalogues, alphabctically? In the end I found myself arranging the crops into natural groupings with a

minimum of overlap or redundancy. Thus you'll find sections on the major family groups of vegetables and also some sections about individual crops. It may be harder to follow than a nice straight line, but in a garden-without-borders things are rarely if ever arranged quite as neatly as we might prefer.

Veggies are the crop group that gardeners usually think of first, although they are perhaps the least important for fuelling our bodies. They give us relatively little starch, protein, or fat, yet without them we would be as unhealthy as grain-fed cattle. That is to say, we need plenty of bulky, hydrolyzed fibre, which creates spaces where our digestive juices can access the nutrients in the denser foods (like grain) and convert them to forms our bodies can store and use. Your system can actually extract more protein, fat, starch, and minerals from the richer foods if they are dispersed among large amounts of succulent fibre. Also, the bulkier veggies help to assure that the richer components do not spend excessive time loitering in the gut, forming toxic compounds.

Indeed it would be unfair to ignore the amounts of those nutrients in the vegetables themselves. For example, the percent of protein in veggies may be much lower than, say, soybeans or wheat, but if it is more hydrolyzed then it is more easily assimilated. I see no point in eating either to the exclusion of the other.

Although elsewhere I rant on about how most of the vegetable crops are very eco-inefficient

("Band-Aid species"), I do not mean to imply that we should not eat them; indeed their highly digestible, relatively low-fibre succulence makes them well suited for our picky stomachs. I'm merely pointing out that they are not great soil builders; that function needs to come from somewhere else, like grass or trees. I would also point out that Band-Aid species have shallow roots (when compared with trees even parsnips are total wimps) and are thus susceptible to trace mineral deficiencies, absent the generous addition of forest leaves.

In terms of eco-efficiency, it would be a great thing for all of us to eat more directly from the forest, and so we should. But let's face it, I'm going to keep on eating pretty much as I have been, and so are you, I suspect. I think, therefore, that it's worth splashing a little ink over the subject of particular vegetable crops.

Vegetable Legumes

Vegetable legume plants—beans and peas—are distinguished by their tendency to host certain soil bacteria that have the magical ability to convert atmospheric nitrogen (N_2) into nitrate (NO_3). Legume plants can use the nitrate to build proteins and stuff. Those plant proteins (which I tend to think of as dietary nitrogen) in turn feed us; indeed some of their amino acids are not adequately found in grasses or other plant foods. I think it is only too cool that we benefit from this major source of protein from plants that do not even make the protein themselves, but rely in turn on yet another level of life. We may be a predator of the legumes, but their relationship with the bacteria seems somewhat more cooperative. The rhizobacterial (that's techno-speak for "root-cooties") colonies are at first glance parasitic—they feed on surplus sugars in the root sap—but they pay very well for their keep by performing that nitrogen trick. Both parties seem well content with the arrangement, so who are we to naysay?

In order to make this miracle appear out of thin air, the bean, or rather the bacteria, must have access to that air, and since they cannot climb up out of the soil to get it, the air must be brought to them. Therefore abundant organic matter in the soil is critical, not so much because of any particular ingredient in it, but for its role in allowing air to enter the soil and remain. If, however, you reduce the humus level and add lots of nitrogen fertilizer, you will gain nothing and lose a lot. Legumes in general do not appreciate being spoon-fed; they prefer to make their own.

Despite their importance in the nitrogen/nitrate/protein connexion, not all of the vegetable legumes can be properly thought of as protein sources. In the case of green beans and snow peas, the pod is the food part, and that part, I suspect, is not much higher in protein than, say, a pepper. As I understand it, it is in the mature dried seed—the pea or bean or lentil—that the concentrated protein is found, and those I discuss in detail in chapter 10. But we also eat the full-sized but immature seed of many legumes, such as table peas, shelled beans, runner beans, favas, and vegetable soybeans (aka edamame). These legumes are still among the most nutritionally dense vegetables. They may have less protein than their mature forms, but that protein is much more hydrolyzed (chemically bound up with water) and thus much easier to digest and assimilate.

The combination of comparative richness with easy digestibility makes most of the legume veggies uniquely suitable for making dips and spreads resembling guacamole, and I describe how in chapter 17.

The Old World legume crops—peas, chickpeas, and favas—are extremely ancient domesticates and all very cold-hardy. Like all legumes, they enjoy lots of alkaline minerals: potassium, calcium, and magnesium. However, because I plant my vegetable legumes in among other vegetables that I feed with high-mineral compost, the legumes get all the minerals they need.

The cold-hardiness of the Old World legumes makes them particularly suitable for companioning with compact, equally cold-hardy vegetables. By "compact" I mean the largely biennial alliums, brassicas, umbels, and composites (lettuces, chicories, salsify). These all mesh very well with the legumes in terms of fertility and spatial demands. The taller

New World legumes—pole beans and runner beans—are very frost-tender and shade casting, so they work better as intercrops with sunflowers and amaranth.

I use some legumes in two or three forms—as both fresh vegetables and dried pulses—and I risk some redundancy in describing separately how I grow each type, but that being said let's now consider each vegetable legume in its turn.

Peas

Peas are mainly known as a fresh vegetable, one of the foods kids love to hate. Actually even as a kid I had really perverted tastes, because I loved peas, spinach, lima beans, and broccoli whenever they were served; I even liked canned peas. After I went "back-to-the-land" (not that I came from there in the first place), and had no ability to freeze stuff, fresh peas became a seasonal treat that we longed for during the rest of the year, like corn-on-the-cob. We could and did blanch and dehydrate fresh-shelled peas and snow peas. Frozen snap peas just don't cut it, so we still enjoy those in season only.

Our great-grandparents had a similar experience: Before Clarence Birdseye pulled a fish out of Newfoundland waters and discovered flash-freezing, most pea eating was in the form of split pea soup, with the special luxury of fresh shelled peas in July. In fact many of the older shell pea varieties, like Early Alaska, were originally round-seeded field, or soup, peas, which also worked well if picked immature. However, those varieties are quicker to go starchy, so modern tastes have leaned toward the wrinkle-seeded types. The older types have not gone away but have found a niche as canning peas, as they hold up better through processing.

At the opposite extreme are the petit pois or "little peas." Now I learned *petit pois* to refer to any wrinkle-seeded green shell pea, as opposed to the field peas; however it has come to mean a more specific type of tiny tender pea that is used quite immature and whose seed is squat and flattened like a tuna fish can. These are all the rage in the sniffy gourmet set, but I prefer the more robust flavour of regular fresh shell peas like Green Arrow and its many selections.

A word about the "leafless" or tendril-type pea varieties: There are a number of varieties (such as Novella) that have been bred for an exceptionally high number of tangly tendrils. Some folks like to grow these, figuring that the tight-tangled mass of plants will support itself (which it will) and allow them to say goodbye to fussing with hex-wire fence or weaving brush. The problem is that all those tendrils are modified leaves, not true leaves—they grow at the expense of chlorophyll, and that spells reduced sugar. Unless you prefer starchy peas to sweet, "leafless" peas offer no advantages. But if a pea named Bikini turns you on, go right ahead and grow all you want.

Edible-podded peas consist of two types: snow peas, whose immature pods are eaten before the peas begin to fill out, and snap peas, a more recent innovation that can be eaten when the peas are full-sized but the pods are still tender and sweet. All edible-podded peas lack the parchment or papery layer that lines the pods of common peas and makes them inedible.

Snow peas. The snow peas are further divided into the small-podded varieties, and the carouby ("pod" in French; cf. "carob") or cabbage peas, which can grow quite large and still be excellent eating, although they may develop a dorsal string that wants removing; the pod remains quite fibre-free except for that string. Given the fact that cabbage peas can grow much bigger without becoming useless, why would anyone grow the small-podded types? In my experience the latter produce a lot more pods, especially if you keep them picked, plus they can be cooked and eaten whole, whereas the large ones are best cut up like green beans. I don't prefer one over the other; I grow them both and use them in different ways.

I do prefer tall peas over short, for the sake of yield, although shorter varieties are a few days earlier. Since I now have a freezer, yield is more important than season extension. The things that makes peas so cold-hardy—all those sugars and cell salts—also make them freeze superbly. At this point my favourite cabbage pea is Carouby de Maussane and my favourite small-podded variety is China Snow.

Most snow pea varieties, on the other hand, have smooth seeds and can be salvaged for soup peas, though that is not their best use. Many of the snow pea varieties have beautiful purple blossoms and olive-brown seed with purple speckling. These pigments (tannins) impart a strong flavour when cooked as soup peas, which some may find either pleasant or objectionable.

Some friends once boasted that they only grew snow peas because they felt that shell peas are too wasteful. They used snow peas for all three uses: When the pods got too old they shelled out the peas, and if these went by they used them as dry peas. Well, I cannot impugn their taste—I mean, taste is taste, after all—but it seems to me they were settling for second best (or third . . .) in using varieties for their unintended use. One of many reasons I like to avoid the marketplace is so that I can afford to eat the very best, or what I like best, rather than what some system can offer me.

Snap peas. Snap peas seem in some ways superior to snow peas in that you have the filled-out peas, and thus presumably more protein. They are big and dense and seem less wasteful; they are a delectable raw snack that even pea-hating kids love. All good reasons to grow them, but no more than you will use fresh, for they do not freeze or can well; whatever you don't use fresh will go to waste, beyond what you need for seed. Whereas some shell pea varieties (like Alaska) have smooth seeds that go starchy and therefore can double as soup peas, snap peas are wrinkle-seeded and useless for anything else.

Figure 8.1. This trellis of woven alder brush works well for peas, and it can also be used for ornamentals such as sweet peas, morning glories, and scarlet runner beans, where labour efficiency is less important than with food crops.

Growing peas. Studies have demonstrated that peas sown too early will mature early (obviously) but will yield somewhat less for the rest of the season. Peas planted a week later when soil is a bit warmer may crop slightly later, but not necessarily a week later, as the warmer soil enables them to catch up somewhat. But the important difference is that the later peas will yield better overall. At some point the advantage disappears, because late peas suffer from various forms of heat stress. Of course, successive sowings are the best way to extend fresh-pea season, but as the weather turns hotter and drier, blossoms tend to abort and pink root rot may set in. Heavy mulch may help the latter problem, but it doesn't affect air temperature much. I have used heat-tolerant varieties like Wando, or early-planted late-maturing varieties like Tall Telephone, but mostly I just shift over to eating green beans and green runner beans.

As for fall cropping, planting shell peas for a fall crop doesn't work well for me; the timing is wrong and the plants usually suffer from powdery mildew. Snow peas will grow well in fall for me, however, perhaps because I can plant them as late as July 10 and still expect a fine crop, ripening in cooler weather.

For supporting my peas I used to build a trellis of woven alder brush, very attractive, very organic, and very effective at holding vines upright with optimal spreading for maximum production. (You'll find instructions on how to make a woven trellis in chapter 14.) However, since I'm often trying to do too many things and have too little help, I have "relapsed" into using chicken-wire fencing. That works almost as well; it just isn't wicked cunnin' like the woven trellis.

Fava Beans

For much of the world favas are a dry bean, particularly the "minor" race, which are small and off-round, looking much like a peanut. The "majors" are large and flattened, swollen at the hilum (eye) end. Those, too, are often used as dried beans, though in many countries, especially Europe, they are used mainly as a green shell bean, like table peas or green limas.

Favas are extremely cold-hardy; they can be sown at least as early as peas and will usually be ready to eat as early. Or earlier; we rarely have peas on the table before July 4, whereas we have eaten favas in June. Moreover, extremely early sowing doesn't reduce later yields. Regardless of when they are sown, they seem to know when they're good to go.

Although favas have modest nitrogen needs, they respond well to high tilth in the form of abundant humus. Since I companion them with other vegetables (usually early lettuce or multiplier onions), which are always grown with compost, the favas get plenty of humus anyway. I have had fava plants grow to be 5 feet (1.5 m) tall, and so they can work well as the centre row of a bed with shorter veggies on either side. My problem with that is that I consume proportionately more favas than that arrangement allows for; moreover, favas seem to prefer being in blocks, so I usually plant three rows per bed (rows #1, #3, and #5) and plant companion veggie crops in rows #2 and #4. Fava plants cast a dense shade, and most companions would be suppressed; but early lettuce transplants work especially well, as they are harvested before the favas get too tall. Egyptian onions are another crop that meshes admirably with favas. These are the type of bunching onions that propagate from tiny bulbils, which form where flowers should be. I get a head start by sowing the Egyptian onions late the previous fall. They just barely get established before winter, but in very early spring they make rapid growth and are ready to harvest in early June, before the fava plants can dominate them. (For more about using the versatile bunching onions, see page 143.)

Fava bean stalks are relatively weak and succulent for their height, especially the large vegetable types, so I usually set up some sticks with string for support. I insert a stick every 10 feet (3.0 m) and run a single length of binder twine down each side of the row. The string is a slight nuisance and I suspect a short length of branched brush stuck in every 2 or 3 feet (0.6–0.9 m) might work as well or better; I'll have to try that.

The tall plants begin flowering from the bottom and continue to yield successive pickings until they begin to lose vigour. You can delay that senescence by topdressing with compost when the plants are half

Figure 8.2. These favas Kindle is holding are at just the right stage for use as green shelling beans.

Figure 8.3. It doesn't take long to shell out enough favas for a feast.

grown or after the bunching onions have been removed (careful, though; the fava plants break over easily). Much of the benefit is from the cool moisture provided by the mulch, as well as the nutrients released with every rain (or watering, if rain is deficient—steady moisture is essential for optimal and continual bean production). Lacking that, they will tend to discontinue pod formation and concentrate on ripening what they already have, which is fine for dry fava production but not for vegetable favas. Therefore topdressing is a perfect situation to use rough, coarse compost from a batch that wasn't quite ready for spring application. It is biologically stable but still not fully broken down; perfect.

Here is a feature of favas that seems to be little recognized and even less utilized: If you remove the older stalks as soon as they begin to senesce—signs of this are bronzing or spotting of the foliage—a second flush of stalks will sprout up from its base, usually from the crown itself, but often from the lower buds of the smaller stalks. By cutting back to these you can force a small crop of late beans (actually I don't know why we call it forcing, as it's quite in keeping with their own desires). The topdressing you did earlier will greatly encourage this. The second crop will be much smaller than the main crop, but here's the tickler: Late-sown favas *never* do well. Later sowings are likely to fail because they are vulnerable to heat stress, aborted blossoms, aphids, chocolate spot, and more. For some reason, however, the second flush, sprouting from established roots, is much less fazed by the stress factors and will produce a modest crop of fresh green favas in a season when you could otherwise expect none.

I should mention that some folks consider fresh immature pods of favas to be tasty, much like green beans. I've tried them and they certainly are edible, but you may have my share; I'll wait a couple of weeks more for the fat shell beans.

I must also mention that there is a very small percentage of humanity that finds fava beans quite inedible, indeed downright toxic. Those people have an inherited gene for favism, a condition that causes an incremental, cumulative, and irreversible reaction sometimes resulting in paralysis and even death. It's curious that the gene is only found in ethnic groups of the Mediterranean basin, although those groups may be as diverse as Greeks, Jews, and Libyans. If you belong to one of those groups, that doesn't mean that you must be susceptible to favism; the gene is rare even among them. It's just that if you are, say, Italian, and your first taste of favas leaves you feeling sort of tingly and weird, perhaps you should lay off, at least until you consult your doctor and determine that it isn't something else, like maybe falling in love.

Some folks are put off by the rather thick skins of green fava beans, and they insist on *peeling* them. Maybe it's a delicacy thing, maybe they also peel their peas and their grapes, who knows? Shucks, I eat beets and carrots and tomatoes skin and all, love to crunch on grape seeds and such (you can keep your seedless watermelons). I love the light chewiness of fava skins. Also, how do I know that those skins aren't loaded with some little-known trace element (like pandemonium) lacking in other foods? No, you must excuse me if I eat them whole; it's just one of my little vices.

Edamame

Early on I learned that we could grow soybeans in Maine: you know, to make our own soy milk, tempeh, burgers, and so on. It occurred to me that maybe you could eat them at the immature stage. We shelled out a small batch by hand, like table peas—no small chore that—and gracious, weren't they tasty! What a marvellous discovery I had made! Only later did I learn that the Japanese (and other Asians) had been doing that for centuries, serving them boiled in the pod and letting the eater shell them between clenched teeth. Furthermore, we learned that there

Figure 8.4. These green soybeans are all shelled out and ready for the freezer.

are soybean varieties bred specially for that purpose: larger, more tender, and holding longer in the green shell condition. They're called edamame, and any self-respecting natural foods eater knows all about them; even I do now.

I mostly grow Shirofumi, though this year I tried Midori and liked it very well. Edamame are sort of like kisses that way: Some are better than others, but none of them is bad. I never waste precious garden space growing just soybeans, not when I can grow them in my corn with a net gain in the yield. The soy yields somewhat less than if planted by itself, but since the corn is in no way impaired, it is a win–win game. I described this strategy in more detail in chapter 14.

I will just point out here that I grow dry yellow soybeans in my field corn and popcorn, and edamame in my sweet corn. The wherefore of this is that I am in the sweet corn every day or so to pick fresh ears, and likewise the edamame, often at the same time; no point in tramping around in one crop when the other crop would just as soon I stayed out.

The main reason I am often picking them at the same time, aside from their parallel maturity, is that I often eat them together. I shell out the edamame by steaming the whole pods for a few (10 to 20) minutes before cooling them and squeezing the beans out between fingers—it only seems tedious until you have tasted them; thereafter you will do it with the bated breath of anticipation. Meanwhile I steam the corn-on-the-cob, cool it, and chop off the corn. Then I dice up one or two ripe-red bell peppers and cook them all together. We call it Chinese Succotash; after eating this stuff sex will seem boring.

Like peas, shelled-out edamame stores very well in the freezer; also like peas, they can be dehydrated (see chapter 17). Although dried, when rehydrated they will still be an entirely different food from ripe yellow field soybeans.

Runner Beans

I always liked green lima beans as a kid, and I still would if I could grow them, but the season here is just too short and too cool. Fortunately there is a bean that tastes exactly like limas (maybe even better) but matures much earlier and loves cool weather: the runner bean. Most gardeners know about scarlet runner beans, though usually as an ornamental and to attract lots of cool moths and hummingbirds. Some folks, especially the British, like to eat scarlet runner beans as well, either as green beans or as shelled beans. Those are okay, say I; that is, I wouldn't sniff at them if I had nothing better, but you see, I do. There is a white-seeded variation on the purple-seeded scarlet runners, which to me is in every way superior. The purple seed coat pigment that the white runners lack imparts a peculiar flavour I disprefer.

The flowers, and thus the pods, of both types form in clusters on long racemes; even the white flowers are attractive. You see, common beans (*Phaseolus vulgarus*) and true lima beans are largely self-pollinating, and so they feel no urge to stick their sex parts out

where just anyone can see them. Runner beans, on the other hand, are promiscuous crossers; they strut their stuff for any passing luna moth, bumblebee, or rubythroat—it's all the same to them. Not that they can't do it to themselves—that's what sets them apart from the obligate outcrossers like carrots—but they do seem to prefer mixing it up, which also explains the general dominance of the bright-red, attention-getting flowers, an advertisement for pollinators. (For more on this topic refer to *Strategies for Sexual Reproduction* in chapter 6.)

Runner beans of any colour are distinct from other beans in that they come up headfirst, whereas the others stick their neck out first and then pull the head up after. Runner beans are also distinguished by the size of their seeds, as big as any lima but twice as fat. Because of that plumpness they are not very tedious to use; very few minutes are needed to pick and shell out enough beans for a kettleful. They cook up very quickly, too.

Scarlet runners are hard to mistake for anything else, but it's easy to see how some gardeners might take the white runners for limas, even though the runner beans are much fatter. That confusion is reflected in the names of many white-seeded heirlooms, like Small's Carolina Lima, Whatcom Lima, Oregon Lima, Polish Lima, Christmas Lima—all runner beans.

White runner beans are truly the northern gardener's lima bean. That may seem strange considering that runner beans originated in Central America. However, whereas common beans evolved in the Mexican lowlands, runners are from the tropical highlands or cloud forests of Guatemala, whose winter is comparable to our summer. Indeed the main place where scarlet runner beans have caught on outside their tropical homeland is England. There they are appreciated as green beans (as opposed to what we call green beans and they call haricots). In their defence runner bean pods are larger, meatier, and heartier-flavoured than common green beans.

The English are all about pedigree breeding, whether it's people, livestock, flowers, or vegetables, and when you see the runner bean varieties they've produced, it is obvious they have given this species some serious attention. There are rulerlike pods and beans as big as your thumb, bred to perform in a land of peat fires and spirit-quenching fog. A "best of show" runner bean breeder flaunts his blue ribbon like the crown jewels and for better reason.

If I'm growing white runners as a dry bean, I'll often grow them on sunflowers (see chapter 10), but if I'm using them as a vegetable—that is, in the green shell stage—then I want easier access for frequent pickings, and so I usually plant them on poles down the middle of a bed, with shorter crops on either side. Not bulb onions, which mind the shade, nor beets, which don't get along well with runner beans, but I suspect many others would work very well. I currently like early (Jersey Wakefield) cabbage with multiplier onions.

Figure 8.5. The ornamental red-flowering types are the source of the name *scarlet runner bean* as well as the botanical name *Phaseolus coccineus* (*coccineus* is Latin for "scarlet").

Figure 8.6. If white runner beans are the northern gardener's lima bean, then calendula is her saffron.

Although I can ripen runner beans even where I live, I can get an earlier crop of green shell runner beans if I start them in peat pots two or three weeks early and tear out the pot's bottom when I set them out. Of course with a freezer there's less of a rush, since I can still enjoy last year's crop; however freezing hasn't always been an option for me, nor do I assume it always will be.

My Chinese Succotash (see page 124) is none the worse if green shell runner beans are substituted for the soy, in which case onions, celery, and tomatoes are a welcome addition.

Green Beans and Green Shell Beans

Green beans with their barely formed seeds are strictly vegetables, like snow peas. It's nice that their nodule-laden roots produce nitrate for the soil, but unlike dried beans, that doesn't carry over to protein on our plates. So what? We appreciate them for various other things they contribute, more "minor" nutrients perhaps, but important nonetheless.

When I was a kid we always called them string beans or snap beans, and many folks still use those terms, though they no longer apply. Older green bean varieties had a tough string running down the dorsal seam; you removed that by "snapping" off the end and pulling away the "string," thus justifying both names. Today's varieties lack that string (unless they're overgrown and tough) and you can either snap or cut them into shorter pieces for cooking; it's all the same.

My hands-down favourite green bean is Jeminez, a three-purpose pole bean with robust flavour and tender quality well after the pods are overgrown. (See chapter 14 for more about how I grow this bean.) I'm not sure I would even bother to grow another green bean, except that Jeminez is a bit late to come into production; some bush beans are two weeks earlier. To get an earlier crop I always plant a row or two of a bush green bean variety, but once the Jeminez come into bearing, the bush beans are likely to be ignored. Ignored but not wasted, *if* I choose the right bush bean variety. Most green beans make rather poor shelled beans and worse dried beans,

but some varieties are fine for both or all three uses. I avoid varieties with long slender seeds because they are harder to shell, and especially black seeds, which have a coarser flavour. My current favourite is Green Crop, a green bean with plump dull-white seeds, somewhat like a cannellini bean. Then I don't need to worry much if I plant too many green beans, as the gone-bys (dare I say "has-beans?") are always welcome for soup or even baking.

I mulch nearly every crop I grow, but especially green beans. My friend Betty McConnachie has discovered that she can work in her mulched beans at any time without incurring rust, and my own experience seems to confirm that. I generally avoid working around most crops when they're wet to avoid diseases, but green beans need to be picked when they are ready and sometimes the weather doesn't cooperate. Betty's theory is that the rust is in the soil and rain splashes spores onto the plants from the soil, and mulch prevents that.

My late neighbour Lucian and his father used to plant an acre or two of green beans for the cannery, leaving his mother to pick them all. Even though every year they insisted that they would help, they would instead go off to haying and she would be stuck with the work. One year she made a point of going out early in the morning when the plants were all dew-laden—a big no-no—and thrashing about with her skirts as she picked. The whole crop was rusty, and the men decided that if they weren't serious about helping, they'd best quit planting green beans. If Betty's reasoning is correct, the beans would still have been rust-free if only they had been mulched—quite a job for one or two acres.

One type of common bean (*Phaseolus vulgaris*), called horticultural beans, are specifically used for green shell, *horti* referring to the garden as opposed to the field crop or dry beans. The word *cranberry* is also applied to some types of these, particularly in New England, though it only adds to the confusion, as most horticultural beans look nothing like cranberries (the one that does is superb). Typical horticultural types have maroon streaks on a buff ground, and sometimes the streaks dominate. In fact, any bean, including overmature green beans and immature dry beans (like kidneys or pintos), can be used as green shell beans, though they are not all ideal for that purpose. The best horticulture varieties are those with big fat seeds that cook quickly to a creamy texture; they often have a higher proportion of soft starch to protein and thus are less dense. For that reason Tongue of Fire is a marvellous horticultural variety but indifferent for baking.

It doesn't matter that most horticultural beans are frost-susceptible until dry-ripe. I intend to sow all of my beans, especially pole beans, as soon as possible after Memorial Day, to be assured they'll mature in time. However if they are horticultural beans then I can risk much later plantings: mid-June in the case of most pole beans and bush beans as late as the Fourth of July. I can do the same with dry bean varieties that can be picked for use as shell beans if necessary. Occasionally I've pushed planting dates too late in the season and had a crop get caught by an early frost with lots of beans still in the green shell stage. Once heavily frosted those beans will not mature—in fact they will soon begin to turn brown— but if I pick and shell them out immediately, I can package and freeze them (unblanched) for an easy supply of quick-cooking beans.

If I have a surfeit of shell beans and a dearth of freezer space, I often just spread out the shelled beans on screen racks to dry (see chapter 17). Depending on how ripe they are they end up looking much like dry beans, only wrinkled, and they can be stored and cooked up like mature dry beans, only they'll cook up quicker and creamier. This only works if they are mature enough to have full colour; if the beans are still green they may turn brown on the drying racks, though they are still fine for the freezer. Of course none of these is usable for seed for future planting.

Crucifers

The Brassicaceae (formerly called crucifers, a term I persist in using) are such a diverse group, and so digestible and nutritious and flavoursome, that it would be remarkable indeed if a number of them didn't find their way to our tables in some form or other. And how many forms!: the succulent taproots radish, turnip, and rutabaga; the oily seeds mustard and canola; the leafy greens kale, collards, bok choy, and arugula; the inflorescences broccoli, cauliflower, and rapini; and the various morphological variants of leaves and stems including cabbage, brussels sprouts, and kohlrabi. Even condiment specialties— think horseradish, yellow mustard, pod radish, and shepherd's purse.

Members of this group are largely, if not wholly, of Old World origin; they are highly cold-tolerant. Indeed most of them are vulnerable to heat stress and drought. They all share the same basic flower shape: four bilobed petals configured like a Maltese cross, hence the old family name Cruciferae: cross-bearing. They all contain a group of compounds that give them a signature pungency. They all grow happily in my garden, and several of them are among the weeds of the neighbourhood.

The crucifers, more particularly the Brassicas, are quintessential Band-Aid species; they thrive in disturbed soil and are rather quickly marginalized by more eco-efficient species. Nevertheless they are succulent and digestible, and they loom large on the list of things we think of when we hear, "Eat your vegetables."

The genus is taxonomically complex, with some disagreement over just how they should be classified. Part of the problem is their ability to sexually transgress what we see as obvious boundaries. This is typical of a genus that isn't fully speciated—it hasn't yet become reproductively isolated among its members. This plasticity and instability helps account for the huge diversity in plant forms and is useful to plant breeders seeking to develop new crops. At least two species, rutabaga and Ethiopian mustard, are interspecific hybrids "created" by humans.

Brassica oleracea

The biggest single group of cultivated brassicas is the species *Brassica oleracea*, which we think of as several distinct vegetables, although they are technically one. These include kale (var. *acephala*), cabbage (var. *capitata*), broccoli (var. *italica*), cauliflower (var. *botrytis*), brussels sprouts (var. *gemmifera*), kohlrabi (var. *gongylodes*), and Chinese broccoli (var. *albogabra*). I'll discuss each of these crops here, but note that you can also find information about their companion crops in chapter 14 and their pest problems in chapter 15.

The wild form of *Brassica oleracea*, colewort, is found today in Western Europe. Being somewhat salt-tolerant and very calcium loving, it is often found on limestone cliffs on both sides of the English Channel and elsewhere on the European coasts. There is another distantly related species, *Crambe maritime*, or wild sea kale, that is also found on the European coasts and may be used for greens. Its tall stalks may grow to 7 feet (2.1 m) and become so woody that folks in the Channel Islands often use them for walking sticks (I'm trying to picture myself fending off a Jersey bull with a dried kale stem . . .).

I'm telling you all this stuff to give some idea about how to deal with these plants in our gardens. They love ashes and lime, they don't like heavy wet clays, and they're best grown early or late, not so happy in the dead heat of summer. The most heat-loving forms, broccoli and cauliflower, were developed in the Mediterranean where winter is to laugh at. I find cauliflower to be too finicky (it just knows this isn't Italy) so I generally don't grow it, especially when broccoli is so much more nutritious and reliable; it doesn't go all to pieces when the days heat up a bit. Cabbage and its close kin brussels sprouts saw their greatest development in Germany, Denmark, Holland, and Belgium, lands with lots of big cows and, oh yeah, *lots* of manure. Those two crops are very heavy feeders, requiring not only high humus levels but also nitrogen-rich humus; it's just what they have always known. Yet they also crave calcium and phosphorus, which are fairly abundant in maple leaves. As mentioned elsewhere leaves are a decent source of nitrogen *if* you use enough of them, which their great abundance makes possible. *If* you use enough leaves to meet their nitrogen demands, you will also be adding lots of calcium, phosphorus, magnesium, and humus, their other needs. Animal manure (including humanure) doesn't do that quite as well: lots of nitrogen in proportion to organic matter. You may be adding more nitrogen than they require, yet too little of certain other nutrients. This is true of most crops, but perhaps none so much as cabbage.

My friend Eliot Coleman once told me of an experiment where he spread cow manure on part of one plot and tilled in oodles of leaves on the other. The poo-fed cabbage was badly infested with a root disease while the leaf-littered plot yielded a perfectly healthy crop. My own experience confirms that. That being said, I also apply well-matured general compost to my cabbage beds. In conjunction with the leaves it assures me of adequate nitrogen plus everything else without introducing the disease spores so prevalent in manure; in addition, the high humus level avoids leaching and volatilization. Whenever I have any seaweed (which is seldom, as I'm 80 miles, or 128.7 km, inland) I apply it to cabbage as an undermulch. Due to its maritime origin I suspect cabbage minds a dearth of iodine more than most crops. I always overmulch with hay or shredded leaves, as iodine is so very unstable and easily lost to the atmosphere.

That's the trick, you see: If some trace mineral was simply not in the soil, maybe wasn't even in the rocks that made the soil, you've just got to bring it in from somewhere else. No big deal, if it is a small amount, but then you have to *keep* it, hold it in the soil as it moves from one life-form, one chemical form, to another; keep it in the circle, don't let it out. You can send it off to the marketplace, but they won't ever send it back.

Cabbage. I basically grow three types of cabbage: Early Jersey Wakefield for early fresh use, a savoy cabbage variety for fall and early-winter salads and stir-fries, and Danish Ballhead for main crop storage and sauerkraut making. Sometimes I also slip in some red storage cabbage, though I disprefer it. The reds don't perform as well for me, nor do they store; indeed I wouldn't grow them at all except for an interesting discovery. One fall I had very little Danish Ballhead and a fluke surplus of some red variety that Molly insisted upon. She suggested I use it for kraut, which I was loath to do, as I expected the same ugly grey-blue colour one gets from red cabbage that is overcooked or left in an iron pan. I didn't allow for the fact that the red pigment acts like the pigment in litmus paper: It turns blue when alkaline and red when acidic. That batch of sauerkraut, with alternating layers of white and bright red, was the prettiest stuff I ever made.

I grow the Early Jersey Wakefield differently than the other types. It stands much more crowding, at least as much as broccoli. For that reason I sometimes plant it in a five-row bed configuration (see *Crowding* in chapter 14) with Capucijner peas (see chapter 10) down the middle, Wakefield cabbage or broccoli in rows #1 and #5, and multiplier onions in rows #2 and #4 (they come out before shade takes over). I plant the crop at 18 inches (45.7 cm) asunder, which is unnecessarily wide in the beginning, so I put some kohlrabi plants between the broccoli plants. At 9 inches (22.9 cm) between *Brassica oleracea* plants, it looks awfully crowded until I harvest the kohlrabis, which is always in plenty of time for the cabbage or broccoli to take over their space. Sometimes I substitute early lettuce plants for kohlrabi if the latter will give me more than I can use. On the other hand, whereas surplus lettuce is useless, I can always dice up surplus kohlrabi and freeze it mixed with broccoli. More on that later in *Companion Planting* in chapter 14.

I have always used summer cabbage (Wakefield) only as fresh slaw or in stir-fries, so I often had a surplus of that as well, but my friend Elise Glinsky mentioned that she used it for krauting. I expressed skepticism at its suitability, and she assured me that it worked well. I tried it and it did indeed. The only drawback is that sauerkraut made in late summer won't keep well without refrigeration, which is my main reason for making it. I don't can my regular kraut until the cellar starts to warm up in April, by which time much of it has been eaten. However an advantage of Wakefield is that it crops more reliably for me than the late hard types. If I plant some Wakefield in July so that it heads up in October, the cellar will be cooler when I move the finished kraut into it. Plus the later-grown cabbage is sweeter, which makes better-quality kraut. At worst Wakefield cabbage is a reliable backup in case my late main-crop variety disappoints.

Here is an excellent tip for preventing overgrown cabbages from splitting, if I just remember to do it in time. If I grab the growing head and give it a bit of a twist so that the roots are slightly stressed, it will pretty much be stopped in its tracks and wait patiently until I harvest it, which should be as soon as possible.

Brussels sprouts. I don't plant many brussels sprouts, as they seem to demand so much and give so little, yet I do love them, especially in a vegan cream sauce with pepita cheese. I'm the only one in the household who cares much about them, so a dozen plants is ample. Two things are necessary if you even hope for a crop: Prune off the side leaves to let sunlight at the stem (not all at once, but working upward as the plant grows taller); also, at some point in late season, the top tuft of leaves needs to be pinched off, so the plant will decide to put its stored energy into those side sprouts. The spacing needs for brussels sprouts and late-heading cabbage are similar—at least 24 inches (60.9 cm) in each direction—so I usually just put them together, finishing off a cabbage row with brussels sprouts.

Kohlrabi. As for kohlrabi the usual summer types are small and quick to mature, almost like turnips, so I usually just stick plants in between later crops, as I described with broccoli. However there is another type, the winter kohlrabi, that's quite different. Gigant is the variety most commonly found in the United States. These can grow as big as the biggest rutabagas and store nearly as well if packed in cushioning layers of dry maple leaves (other leaves may do just as well, but maple leaves are cool and crisp and fluffy, perfect for this purpose). They are much slower to develop fibrous interiors than the summer types, so it is not as critical when you harvest them, just sometime before they've had too many heavy freezes. They are superb used in any way you would use rutabagas, having a slightly milder flavour. We often eat them raw cut up like carrot sticks and served along with those.

Keeping in mind that kohlrabi is merely the swollen stems of *Brassica oleracea* brings me to a new food from an old plant. When I was in college and trying to squeeze every penny's worth out of store-bought food, it occurred to me that a head of cabbage contained a lot of waste. Once you chopped off all the slaw, there was still that big core, too fibrous to use, or not: When I peeled away that fibrous outer stuff, there was a crisp sweet core that was surprisingly tasty (this was before I'd ever heard of kohlrabi). Now whenever I make sauerkraut I have a substantial amount of those cores, and aside from steaming them whole with carrots, they also make a delicious pickle. Of course, "pickled cabbage cores" sounds too much like a salvage item, but since we have learned to call them "dilled cabbage hearts," they seem far more elegant.

Kale. Kale is quite something else. As you might expect from a plant that is closest to its wild ancestry, it is the least demanding and most nutritious. It will make what it can with what it has. Like all of the brassicas, you must coddle it past the early flea beetles, but henceforth it soldiers boldly on. The very first time I sowed kale, never having tasted it, I didn't get it in until mid-July, and then almost as an afterthought, a belated experiment, probably doomed. I had little time to spare for it, with a baby on the way and a house under construction. We had to leave for

Figure 8.7. These winter kohlrabis in a box in my root cellar will be excellent for fresh eating long after the cabbages are gone.

the hospital just as the first heavy frosts hit, so when I returned several days later with an enlarged family and low expectations, I was shocked to see the kale looking as robust as ever and tastier than before. By a fluke of ignorance I had done everything just right. Since then I prefer to sow in mid- to late June for more and earlier pickings, but the sweetest kale isn't ready until mid-October.

I pointed out to a friend in the seed trade how I was disappointed at the diminishing selection of kale varieties. He protested that there were more than ever: Russian Red, Ragged Jack, Hungry Gap, Lacinato, et cetera. I conceded that those were all very nice, but I meant true kales, *Brassica oleracea* var. *acephela*, whereas those others are technically not kales but mustards. They are beautiful and pungent, unlike the milder crinkly leafed real kales like Konserva or Squire. Seed companies are notoriously careless at making that distinction in their catalogues, which can cause confusion and woe for unsuspecting seed savers who might separate them unnecessarily while allowing the turnip-kales to be contaminated by something else (possibly wild mustard or pak choi).

Chinese broccoli. The least known *B. oleracea*, at least outside Asia, is var. *alboglabra*, also known as Chinese broccoli. Although its botanical name sounds like some magical incantation, it is a very down-to-earth flavour-packed vegetable. I've only grown it once or twice, but its resilience and no-nonsense taste—sort of a combination of broccoli and kale—makes me want to grow it again. The seed I got was from Thailand, but some US seed companies carry it.

Other Brassicas

Outside the species *Brassica oleracea*, there are several other brassicas that are important vegetables—for example, the various mustard greens, Asian cabbages, and others whose taxonomy is more complicated. That doesn't matter much unless you're saving seed and need to avoid crossing, in which case it matters very much. I grow most of these rather haphazardly, knowing that when kale is ready, other brassica greens will probably be neglected. The logical way of

Figure 8.8. I often sow kale seeds in a nursery bed, where they can be growing to transplant size while another crop, in this case bulb onions, is maturing elsewhere in the space that will later be kale's.

growing most of these is as a late summer or fall crop. They are mostly annuals that respond to the waxing summer heat by bolting, whereas the gathering chill of autumn makes them grow tender and mild.

The one brassica I do often grow for early spring greens is something I discovered almost by accident. Many years ago I grew a peck or two of rapeseed, with the intention of pressing our own oil. Well, the oil idea never went anywhere, but while growing the crop we had a ton of thinnings that seemed too valuable to compost. We tried cooking them as greens and were well impressed with their mild turnipy flavour. The next year we sowed them specifically for greens, thinking we'd made a clever discovery. That summer an aunt from Maryland visited, and we showed her around the garden. "Oh, good gracious!" she gushed. "I haven't seen rape greens in the market for years! How I love those!" So much for "new discoveries." The main advantage I see in rape greens over others is their very quick growth. While we never made oil from our rapeseed, we found another fine use for them as sprouting seeds. To learn about that refer to chapter 17.

Turnips. Turnips have a "good-for-you" reputation, but apparently it is the greens that deserve most of the credit. Nevertheless I enjoy a few messes of the early young turnips when they are not too pungent. Of course the sweetest mildest turnips are those sown for a fall crop, but those must compete for our affections with rutabagas, which is a totally unfair match. Only in moments of foolish passion would I opt for the turnips; it's really no contest. Rutabagas are just so warm and welcoming, so productive and enduring, so sensible and . . . well, so Swedish. Yep, that's where they come from, and that's why Britons call them Swede turnips or just Swedes. It's really no insult to either Swedes or swedes, they can be proud of each other. The story of how rutabagas came to be the way they are, indeed how they came to be *at all*, is a fascinating bit of lore, and I'll digress to share it with you.

The true turnip, *Brassica rapa* (that Latin root *rapa*, meaning "turnip," pops up in words like *rape*, *rapini*, and *kohlrabi*) is native to Central and Southwest Asia, where you find annual and biennial, domesticated and wild forms of it. From there it spread east and west; in Northern Europe it gained popularity as a forage crop. In the Baltic region it encountered *Brassica oleracea*, probably the wild colewort that would have grown around some of the turnip fields. Colewort has 18 chromosomes and turnip has 20, so the chances of them crossing seem negligible. Negligible, but not impossible; say, one in a million. However, when tens of thousands of wild weeds are visited by millions of bees carrying billions of pollen grains to thousands of flowering turnip plants, we must realize that "one in a million" is really saying something like "several times every season." At least once (though surely much more often, especially since both species are self-sterile, even among themselves) this improbable-seeming cross became a certainty and one or more plants appeared with 38 chromosomes. I am assuming that this cross was in one direction only: colewort on turnip (or turnip × colewort), since a cross in the reverse direction would be off in the weeds somewhere, unlikely to be noticed by the farmer. But it is hard *not* to notice when some of your turnip plants are several times larger than the others and have the peculiar blue-green foliage of the oleraceas. After propagating and tasting a few the Swedish farmer would surely have realized he had something special.

So if someone says "turnip" when they really mean rutabaga, don't correct them; they're right as far as it goes. But if someone says "rutabaga" for a *Brassica rapa*, do straighten them out, they are just downright wrong. Not that anyone is likely to; in my youth I had countless New England boiled dinners with "turnip," though I'd never tasted a true turnip, nor heard of a rutabaga, until my mid-20s.

Considering the narrow genetic base of this fluke creature, it's not so surprising that rutabagas display much less variability than either of the parent species or of most other domesticated species. I have grown many dozens of rutabaga varieties, and the shape varies only slightly; some have purple shoulders, though some have green; nearly all have yellow flesh, though a few have white, while turnips are generally the other way around. The older, less refined heirloom rutabagas such as Gilfeather and Waldoboro Greenneck are often more irregular-shaped and crack-prone, with coarser flesh. The white-fleshed varieties are more pungent, or turniplike, while the yellow-fleshed ones are milder and sweeter after their wild kale ancestry, but to paraphrase Will Rogers: "I never met a rutabaga I didn't like."

Turnips may be either spring- or fall-grown, but rutabagas are strictly a late crop. I never sow them before mid-June, often early July.

I should mention that in Germany some people make "rubenkraut" from shredded true turnips, somewhat more pungent but otherwise much like sauerkraut. We've made it and liked it, though I would say that comparing it with sauerkraut is analogous to comparing turnips and cabbage. I haven't heard of krauting rutabagas, but since rutabaga keeps so well and tastes so great, I feel little incentive to experiment.

Radishes. What I grow in my not-for-market garden includes everything that will reliably mature here and that I like. The latter criterion is not very exclusive, but it generally precludes radishes of all types. I don't hate them at all, but just can't get interested enough to spare the space, not that they require much. I won't deny they are fascinating to grow; I have cultivated hundreds of varieties from all over the world and the variability is much greater than you'll see in any US garden seed catalogue. They are truly exciting, but not on my plate, please.

For those who may feel differently about them, I'll share what few tips I can. Even more than other crucifers, radishes do not abide hot weather well. They're best planted very early, so that you can harvest some sizable and mild-tasting radishes before they crack and become too pungent and maggoty. It's a race against the summer solstice. An easier way to win that race is to not enter it, but instead sow an autumn crop, say in August, so that as the radishes develop, the days are growing shorter and cooler, ideal for mild radishes. Mulch makes a huge difference with all radishes, keeping the soil cool and moist and deterring root maggots. (A light dusting of wood ash on the surrounding soil will also deter them.)

European-style salad radishes do not store very well, but daikons or other Asian-type winter radishes store almost as well as rutabagas, provided they are not bruised. Unlike rutabagas, they don't take kindly to being stuffed into feed bags and piled into barrels. Rather they prefer to be gently layered with dry leaves in a lidded trash can.

In Central Asia, the radish homeland, there are varieties that grow as big as soccer balls and are used as livestock fodder, grated and used as a relish, and salted and fermented like sauerkraut. In India one type is grown not for its puny roots, but for its long slender seedpods, which are used in stir-fries, like radish-flavoured string beans. In Germany a shorter-podded type (Münchenbier) is pickled and served with beer, like pickled eggs and pickled beets (just in case you don't get pickled enough from the beer).

The Umbel Family

Umbras are shadows, "*umbrellas*" cast a "little shade," *umbellifera* are "umbrella bearers." If you glance at the flower heads of any members of this large and diverse family that includes both vegetables and herbs, you will immediately understand the name, although it seems that most of them look more like upside-down, or blown-out umbrellas. The classic example is carrot, although many folks may be more familiar with the flowers of carrot's wild or feral form: Queen Anne's lace.

All umbel family members are quite cold-hardy, which is important since most of them are biennials. Some that are not biennial—dill, cumin, fennel, and coriander—are hardy annuals, whereas skirret, lovage, angelica, Korean pigplant (wild celery), and sweet cicely are cold-hardy perennials. Since most umbels have very succulent stems and leaves, the sap needs to contain some pretty powerful antifreeze, and it does: various combinations of the cell salts (those alkaline minerals again), sugars (especially in carrots and parsnips, the hardiest of them all), and various terpenes. Those terpenes are the hallmark of the family. In fact, the spice members of the family—anise, fennel, caraway, dill, cumin, coriander, and more—owe their distinctive flavours to the particular terpenes in them. Paradoxically, these terpenes are pungent and resinous (as in turpentine) and some of the family really overdo it: I find lovage about as palatable as pinecones. Yet in the vegetable forms the delicate nuance of terpenes is what makes carrots taste like carrots.

All umbels, except for the perennials, have a pronounced taproot. In the cases of carrots, parsnips, celeriac, and root parsley (which I call parsliac), the swollen taproot is the crop. Perhaps that taproot is a hint that the plant has modest nitrogen needs, which are easily filled by a fine network of shallow feeder roots. The plant's more insatiable need is for the alkaline minerals calcium and potassium, and for those it has learned to delve down deep and fetch its own. They may appreciate some help with this in the form of added lime or ashes, but surplus nitrogen is more of an impediment. Of course if the soil's alkaline minerals have not already been depleted via leaching or the marketplace, they should not need to be added, at least not regularly. The taproot simply pumps up from the lower layers what is already there. And of course loose friable soil with abundant humus makes its digging easier (and ours when we harvest it), while retaining enough moisture to keep those minerals in solution. Obviously those scanty surface rootlets are not extremely efficient at taking up water, so the taproot's ability to suck up water from the depths is especially important. Which helps explain why young umbel plants are particularly vulnerable to drought stress, whereas older plants with established taproots have considerable drought tolerance: The taproot both delves for and stores water.

Another hallmark of umbels is their ferny foliage. Parsnips are somewhat exceptional, but the group's members as a whole are rather frilly, which must convey some advantage. Like all the Band-Aid species the umbels must compete, or at least survive, in a grassland ecosystem, biding their time until some disturbance of the sod by fire, ploughing, or what have you gives them quite literally their moment in the sun. I suspect that light foliage helps them to fit in with the grasses and legumes without undue competition. Likewise the taproots, which reach down below the shallow sod and pump up minerals that eventually become available to their neighbours. Gardener, take note: These things all have implications.

Carrots

From the beginning I have always found it easy to grow a decent crop of carrots. However "decent" isn't what you should be aiming for, but rather something more superlative. For example, if you want mostly large, well-formed carrots, the soil should be deeply dug and free from stones, because stones in the soil cause forked and twisted roots. Seedlings should be thinned to about 2 inches (5.1 cm) apart when less than 3 inches (7.6 cm) tall. Thinning later than that will be less effective and may disturb the remaining plants. I try to keep the newly seeded bed well watered at least until the seed has germinated and usually for a week or two after. With earlier plantings nature usually takes care of that, but I often plant as late as summer solstice, when the soil may be too hot and dry for prompt and consistent germination without supplemental watering. Drought in the early stage can set back carrots badly. High humus levels help assure ample and consistent soil moisture. Other than that the only fertilizer I add is some wood ash (perhaps 2 to 3 gallons, or 7.6–11.4 l, per bed). Much nitrogen should not be needed and in fact may be counterproductive. Nitrogen acts like a growth hormone, and while carrots in very rich soil may grow large quickly, they will often have poorer health, poorer flavour, and poorer storability than those carrots provided with proportionately more of the alkaline minerals (especially potassium). You see, adequate potassium is needed for cell wall turgour; without it tissues become flaccid and weak and prone to invasion by most everything. Alkaline minerals are also linked to the formation and storage of sugars. Therefore carrots with ample minerals are sweeter, crisper, more disease-resistant, and longer keeping. Carrots grown with excess nitrogen tend to be resinous-tasting (due to terpenes) and quick to shrink in storage.

The part of the carrot we eat is not the whole taproot; the little threadlike root that extends beyond the enlarged root is vital. It reaches for water that's way beyond the storage part. This delicate root is easily damaged by burrowing insects or rodents, and when it is broken the plant loses its sense of direction: Without that radical dominance the root grows in every direction, producing the often banal-looking forms you see at fair exhibitions. The plant may produce just as much root, but it's more difficult to wash and use. That is why taproot plants like carrots, parsnips, beets, and turnips are so notoriously difficult to transplant: Even if the seedlings survive, the resulting roots will usually be deformed beyond use.

In recent years there seems to have been an emphasis on breeding early forcing varieties for the fresh market and for novelty varieties of diverse colours. From my limited experience most of those latter are mediocre in taste and very poor for keeping. These are new only to the US market; they have been common in places like Cyprus, Syria, and Pakistan for centuries. The French and Dutch have done a great deal to advance the carrot to its modern splendour; I see no point in moving backward. My own priority is storage quality, and Scarlet Nantes fills that need to my satisfaction.

Figure 8.9. Carrots being trimmed for the root cellar.

Figure 8.10. Packed between layers of dry maple leaves, these carrots will keep in fine shape until spring.
Photograph courtesy of Isaac Maxham

I don't harvest carrots until mid-October, partly because they make some of their finest, sweetest growth in those late weeks, but also partly because I store them, like my beets and rutabagas, in barrels of new-fallen maple leaves. Probably any leaves would suffice, but maples are crisp and clean like crumpled paper and they make a nice cushion without imparting any flavour. Also, by then the root cellar has begun to cool down for better dormancy.

Speaking of carrots it has intrigued me how vulnerable to freezing they are in the garden compared with the gone-wild carrot, Queen Anne's lace, in the neighbouring hay fields. I wonder to what extent the latter is protected by the sod and how much is merely the resilience of wildness. Carrot deteriorates, or reverts, very quickly out of cultivation, which may demonstrate the superficiality of some of our domestication.

Parsnips

Parsnips are a food people seem to either love or hate; I see it as an indication of proper upbringing. In defence of those who aren't enthusiastic they may have not had them properly cooked; parsnips should be steamed in just enough water until the water has nearly boiled dry, leaving a sweet syrup. To say that the flavour resembles a blend of celery and licorice doesn't quite do it, but close. They originated in southeast Europe or the Near East, but today parsnips seem to be most commonly grown and eaten in England and New England, which should help to redeem the culinary reputations of those peoples. There are quite a few varieties of parsnips (I have seed for over 80 varieties), but the range of variation among them is not very obvious. My preference is for parsnips that hold their fatness without tapering too quickly—it makes a huge difference in the yield—and that do not have a very deep or hollow crown. American Hollow Crown, condemned by its very name, is a fine-flavoured variety, but its shape makes it difficult not to waste a lot of the flesh, plus it tapers immediately. I greatly prefer the variety Kingship; Hungarian is also superior.

The trick with parsnips is to sow them early, along with the earliest peas. They are absolutely cold-hardy and will germinate while carrot seed sits there waiting for better times. You mustn't be daunted at the seed packet statement that they require 150-plus days to maturity—that does *not* mean frost-free days. In fact it takes lots of cold to make them tender and sweet; I only dig a few in autumn and wait for early spring to dig most of them. Technically, the seed packet should say "365 days until harvest."

Some people like a parsnip stew with potatoes and onions—I suppose, but for my taste parsnips go best with parsnips. We've also used dried shredded parsnips instead of coconut in a millet cookie recipe; "parsnip macaroons" isn't half as bizarre as it sounds. Some winters I get a craving for parsnips in February, which is not so unrealistic when the ground has thawed under the heavy mulch of snow. The obvious problem is finding them, which is why I usually mark the rows with 4-foot (1.2 m) sticks before winter sets in. Shoveling away a couple feet of snow to reach them gives you some idea how much I value them as a winter treat.

Celery and Celeriac

I've never had much difficulty growing celery, though I've never bothered to blanch it, and I don't even try to grow those huge stalks you get by hyperfertilizing them. Since I'm not aiming for the marketplace, I can just pull off individual stalks as they mature, which seems to produce more than a onetime harvest, and since I can't begin to use it that fast, most gets chopped into the freezer. Before we had that option, we used to dehydrate it; sometimes I still do. Not wishing to waste the green tops, I dehydrate some of those to crumple and sift them into a powder for soup broths and popcorn dressing, which is described in chapter 17. We used to grow a "cutting celery," Zwolsche, which is easier to grow, sort of like parsley, but I really like the larger stalks for stir-fries.

We realized early on that if we wanted to have celery we could store fresh, then we needed celeriac. That gives a more or less compact root-ball that will keep like a beet. I'm still not satisfied with my method; some years I wind up with small knobby things with lots of gnarly side roots and a thick fibrous rind. I had always believed it was because we didn't have a high-nitrogen animal manure to feed it, but sometimes that doesn't make so much difference. I now feel the greater need is for more water, more regularly applied, than I've given it before, especially in midsummer. High humus levels help to stabilize moisture, but those critters just need lots of water and pretty often.

Of course celeriac cannot be used for ants-on-a-log (nor would I eat such kid-fare), but it does lend itself to stir-fries and even more to soups and stews.

Parsley

Some people think of parsley as an herb, but I sometimes use it more like a vegetable, as in Canadian potato soup, where it is used by the cupful. I've also grown Berlin, or root parsley, which I call parsliac, as it is quite analogous to celeriac: You can store it for use in soups and stews. It's a bit bland for me, and I usually don't bother with it.

Like all umbels, especially celery, parsley can be very difficult or slow to germinate, particularly in hot, dry weather. In its Near Eastern homeland, nature has trained parsley to remain dormant until cooler, wetter weather arrives (their winter). By the time hot, dry weather returns it has sunk a deep taproot to seek water and store it. When mulched the shallow soil moisture is more reliable, but since I don't mulch until plants are 1 to 2 inches (2.5–5.1 cm) tall, I sometimes have to provide supplemental water until they are ready to be mulched. I usually thin to about 2 inches apart, but no farther, as a little crowding gives smaller, more tender stems, which I prefer.

There are other food crops among the umbels, which we use as herbs and spices. I talk about those at the very end of this chapter in Condiments and Spices.

Alliums

The alliums are almost unique among the vegetables in being monocots, as opposed to dicots. (The other monocots are sweet corn and asparagus.) These crops keep growing from the spreen, or crown, whereas dicots grow from the most recent growth.

Hey, I don't tell you this stuff just to generate copy; it has implications, some obvious and some that may occur to us as we go along. In this case it means that if you find a (dicot) tomato nipped off by a cutworm, don't waste time waiting for it to come back; it won't. Go ahead and put a new plant in its place (after dealing with the cutworm, of course). If an onion plant gets nipped off, on the other hand, it is a setback, but it is not the end of the onion. There is a new baby leaf curled up inside the crown, just waiting to take its place. In fact it was going to grow anyway and create the next inner ring of the onion bulb and another after that. The cutworm has cost you the outer ring, not the whole onion. Actually, the first few leaves weren't going to form rings anyway; they support the growing bulb and eventually senesce or die back. Think of the outer papery leaves of the bulb and those that don't even get that far. The real damage is in setting back the plant when every day counts in its race toward summer solstice (Banner Day in Esperia). We'll talk about photoperiodism in the Onions section of this chapter.

There is something truly mouthwatering about the alliums. How often have I heard someone say: "Hmmm! Smells like something delicious is cooking!" when the dominant aroma in the air was onions or garlic frying or leeks stewing, not potatoes or squash. What makes the alliums so aromatic is a group of sulphur-based compounds, which, paradoxically, we often dislike in raw food. They may be downright irritating; no one cries when peeling squash or dicing potatoes.

Just how addicted we are to the flavour of alliums is reflected in the story of the Exodus: The people of Israel are trekking around the Sinai, having recently escaped from a land where they endured slavery and genocide. Does anyone say: "Hey, guys, this manna stuff really hits the spot"? No! They keep kvetching about how much they miss the onions and garlic and leeks they had back there! The alliums, ya just gotta luv 'em.

In the Ayurvedic tradition of India garlic and company are viewed as being associated with low spiritual attainments. Boy, they got that one right: I eat tons of alliums and I'm an atheist.

Chives and Garlic Chives

This time, let me begin with the minor species. Chives (*A. schoenoprasum*) are one of the least useful (in my opinion), yet easiest to grow. Because chives are perennial you can start them either by planting seeds or by finding someone who grows chives and dividing a clump. Your fellow gardener should be delighted to share, as chives become self-choked and run out if not occasionally dug up and divided.

Chives are nice finely chopped in salad, especially potato salad, though I prefer scallions. You can chop and dry chives for off-season use, although their flavour is weakened by drying.

Another perennial allium, gow choy (*A. tuberosum*), is also called garlic chives—quite appropriate since it grows in grassy clumps and has a distinctive garlicky flavour. It is flat-leafed whereas chives are round. We used to have a lot of it in the herb garden, as it is so easy to grow. However, like chives, its use is limited since it loses much flavour when dried or cooked.

Chives and gow choy are both considered herbs, and they don't really fit in with annual row crops in the vegetable garden. They'll grow better in a perennial flower garden where their roots are less likely to be disturbed. With their distinctive monocot (grasslike) shape and flowers (white for gow choy, purplish blue for chives), they are more decorative than nutritious.

Leeks

Leeks are probably the most European of all alliums. Onions and garlic were brought in from Central Asia at some ancient time, but leeks started out right there, and the wild descendants continue to dwell there, often in vineyards and olive groves. There it often acts like a perennial, forming little bulbs like a multiplier onion and rarely flowering. Some ethnobotanists claimed that leeks originated in the Mediterranean basin; that is supported by the present distribution and the fact that the ancient Egyptians had them (although biblical and other references could be referring to a particular type called kurrat). Others insist that they originated in Northern Europe and/or the British Isles; even today they are the national emblem of Wales. I see ample reason to assume both stories are correct. You see, there are two types of leeks: the long narrow summer leeks that closely resemble the wild (or feral?) leeks of southern France, and winter leeks, the short fat type popular in Britain and Nordic countries. The latter type are extremely cold-hardy, indeed they do rather poorly in the hotter regions of Southern Europe.

One of the oldest varieties sold in the United States is American Flag. It sounds like a reference to Old Glory, but it was originally called London Flag, and it had no patriotic connotation. Rather *flag* is another word for certain members of the wild iris family, which leeks somewhat resemble.

Leeks require a long season to fatten—not necessarily a frost-free season, however—so I always start them indoors sometime in March. I've discovered that their young roots are much more vigourous than onion plants are, so I start them in 3- to 4-inch (7.6–10.2 cm) deep trays, so they can develop fully. I set them out in early to mid-May, regardless of their size. They are picky about moisture at the outset, but once established they are remarkably resilient. Like most of the European veggies, leeks like ample humus, reasonably rich in nitrogen and abundant moisture if they are to size up well. If they lack that you will not have a crop failure, just somewhat smaller leeks—which, if you are not bound for the marketplace, are quite acceptable. I set my leeks out 3 to 4 inches apart. I have no pest or disease problems of which I am aware. The thrips that sometimes damage onions seem to have little impact on my leeks.

The traditional way of growing leeks is to hill them up with soil to blanch the shaft, but I've never seen any point in that. I have no problem with chlorophyll and anyway the shafts are largely self-blanched by the outer wrapper leaves, which most people trim off and discard. I, too, trim off the floppy green leaves, but I certainly do not discard them. Again, I value the chlorophyll and all the other goodies that are concentrated there and not in the pale shaft. I rather chop and dehydrate them, and then grind them in my Corona hand mill to a fine powder, which I mix with dried powdered celery leaves and make an herby oil infusion especially good on popcorn.

Leeks are very suitable for dehydration or freezing. I used to prefer the winter type because they hold longer in the field, but I've come to the conclusion that summer leeks, despite their name, are quite hardy enough to last until I can get them into storage of some sort. In fact, I often lift them in mid-October or later. By "lift" I mean to gently loosen the soil beneath them with a spading fork, then tug them out by the shafts. They have very strong roots and won't be just yanked out like onions or garlic. Sometimes I store them in a washtub with most of the dirt left on the roots, so I can deal with them at my convenience, maybe weeks later. In fact I have stored them

thus through the winter, but they usually need to be watered in midwinter to prevent withering, especially those I'm saving to grow seed.

The etymological roots of *leek* popup in other alliums and in other languages. For example, chives in German are *schnittlauch*, or "finely divided leeks"; onions are *luki* in Russian; and *garlic* comes from Anglo-Saxon *gar*, a lance or spear, and *laek*, or leek. Speaking of which let's talk about garlic.

Garlic

Garlic's species name *sativum*, which means "cultivated," attests to its central place among the alliums. It has been gathered and eaten for much longer than it has been cultivated. Wild forms still abound in its

Figure 8.11. In the garden or on the table kale and leeks just belong together.

native homeland of Uzbekistan and southwest China, where my friend John Swenson helped to collect dozens of feral and domesticated garlics.

I first grew garlic by purchasing some bulbs in the grocery store and planting them, along with my tomatoes, in early June. What fun; they grew beautifully, and that fall I harvested a bunch of uncloven bulbs, round in shape but not much bigger than what I planted. You see, associating them with Italian and Spanish cuisine, I had assumed that they required a long hot season. Silly me, I soon learned that they are extremely cold-hardy and can be planted as early as the ground thaws, although that may still not yield bulbs as large as if planted the previous autumn. Some sources say several weeks before the first fall frost, but I have planted them so late that I had to bust through the frost-bound crust to reach soft soil, and they still did okay. They like to be planted deep—2 to 4 inches (5.1–10.2 cm)—so they can still break dormancy and begin to root even under the snow. Supposedly it is best if they do not sprout and emerge in autumn, lest they be winter-damaged, but since I always plant mine sometime after the equinox, they never try to do that, but instead focus on underground development, getting poised for an early-spring dash. On the other hand they are ready to harvest in late July, so there is a lot of season left over to grow something else in that space, like kale or lettuce from transplants.

If I have neglected to mulch garlic heavily with shredded leaves in the fall, then I do so in spring, very early for maximum weed repression and before the lettuce (or whatever I intercrop with it) gets big enough to be messed up by it. I give garlic a heavy dressing of urine after planting in late fall and/or early spring. In the latter case I take care to dilute it heavily and apply it on the mulch, which readily absorbs it, eliminating odour.

There are basically two types of garlic: soft-neck and hard-neck. I prefer the latter, as it is hardy and individual cloves are fewer but larger and easier to peel. Hard-necks are reputed to be stronger-flavoured, but I find them milder. I often steam them whole with carrots and other root crops for a subtle

pungency, not at all overpowering. Hard-necks are also reputed to be poor keepers, though I have found the opposite. I'm accustomed to referring to all hard-neck varieties as rocamboles, but I understand that porcelains and purple-skinned varieties are distinct from those. I mainly grow Sarah's (probably synonymous with Russian Red) and Rowe (aka Phillips), a local heirloom originally brought to midcoast Maine by Italian stonecutters. Sarah's is larger but the poorer keeper of the two, Rowe is smaller but holds dormancy very well. When Sarah's begins to show signs of sprouting, if not before, I dehydrate the rest and focus on eating the Rowe. Usually I have a surplus of both, and they both get dried and ground for garlic powder.

Only hard-neck types form a scape—a succulent flower stalk that has turned into a dry roll of papery leaves. But you see, it is a flower stalk without flowers; garlic is so committed to asexual propagation that it has largely forgotten what sex is. In fact some research indicates that all of the hundreds of garlic varieties are really only a small number of genotypes, whose variability is largely the result of superficial mutation. Mutation is what some creatures do instead of sex to create genetic diversity; don't knock it until you've tried it, I guess.

Instead of flowers hard-neck garlics form bulbils: little clusters of baby garlic cloves at the top of a tall snaky stalk (the botanical name for this subspecies, *orphioscorodon*, means "serpent," and hard-necks are often referred to as serpent garlic). This gives them an additional advantage in self-propagation in the wild: Although the basal clones (in the bulb) can divide, they cannot *go* anywhere; the top-set types can and do tip over and replant themselves, travelling a foot or two every time. There is a parallel in the type of onion called Egyptian, or top-set, onions. They are sometimes also called walking onions.

The problem with the garlic bulbils is that they draw energy from the main bulb, which should be our priority, as the bulbils are not especially useful. I break off the scape as young as possible, before bulbils have really formed. Fortunately the scape is not useless. At that stage the snaky top part is still quite tender and mild, an excellent accompaniment to other steamed vegetables at a time when no other fresh garlic is to be had.

As garlic plants start to senesce (foliage turns yellow, then brown), it's a good time to pull the bulbs as soon as possible. Leaving them longer will not ruin them, but the bulbs tend to open and separate, making them somewhat harder to store.

Here is some interesting hearsay: Garlic is supposed to contain lots of selenium—well, as much as anything contains "lots of selenium"—and that is supposed to prevent aging. Of course, cyanide prevents aging, too; eat enough of that and you'll never grow old.

More hearsay: A scientific study found that garlic is reputed to lower blood cholesterol levels, but another study says that's all a lot of bunk. Yet more hearsay—and some scientific truth that is wholly supported by my own personal empirical direct hands-on experience: Garlic does *indeed* repel vampires, just like they say. I eat it every day and have never even been approached by a vampire. Mosquitoes don't seem to mind it, though.

Here's another interesting item that may interest more advanced gardeners, though it is purely hearsay. It's not wholly true that garlic rarely engages in sex. When you look at a cluster of bulbils, especially very early in their formation, you'll often notice a tiny floret or two scattered among them. These are actual sexual flowers, and might ripen and produce viable seed (of a new variety even), except that they are soon overwhelmed and aborted by the developing bulbils around them. *If* you can spot these sexual and asexual organs when they are first developing and gently excise the nascent bulbils (with a tweezer perhaps, or a very sharp carpet knife) before they can compete, there's a chance, albeit slight, that some of the sexual flowers may get pollinated and grow to mature seed. Alliums are generally self-sterile, which is probably why the rate of takes is so discouragingly low. So I assume that if you can introduce pollen from a similarly treated plant of a different variety, then the odds of success would be much greater, *and* the resulting varieties would be much more vigourous

and useful than a mere inbred line. Now understand that I have no time for this sort of nonsense; I'm far too occupied as it is. I only urge *you* to dabble in this, in the hope that you will develop something new to share with me and everyone else.

Onions

Some people tell me they have dismal luck growing onions: They plant nice big sets, put tons of chicken doo-doo on them, water them every day, talk to them, sing to them, and just get little onions. I answer: "Forget all that. *When* do you plant them?" "Why, in late May, when I plant my peppers and tomatoes." That's much too late to plant onions, due to a phenomenon called photoperiodism—a word some people like to say because it's more impressive than *day-length sensitivity*. It's what happens to some creatures who live too close to the equator, where days and nights don't vary all that much year-round. As you head towards the poles winter nights and summer days get a lot longer. This is why onions need to be planted oh-so-early. Until June 21, they think they've got forever, but as soon as the solstice passes and they sense a shortening of daylight (you learn to really notice the light when your life depends on photosynthesis), they get spooked, even though they have months of warm weather ahead. They start to "bulb up," putting energy into something they'll have to show for their summer's work. That's just what we want, except that if they were only planted a few weeks earlier, how much energy can they store? Remember, they need a leaf for every ring, not counting the first few for outer wrappers. Sure, they love extra fertilizer and water and company, but neither they nor you can argue with the calendar. All the alliums have some degree of sensitivity, but none like the common bulb onion.

That is also why onions should get plenty of light, plus the fact that sunlight on their shoulders helps them to fatten up; that's not quite the same as photoperiodism, which tells them *when* to fatten up. It's a reason why bulb onions don't work that well for companion cropping, as I'll explain anon. It is also why you should not plant them deeply.

I say "plant them" instead of "sow them" because direct-seeding almost always fails where I live. I always plant seedlings or sets (more on that, too, later). Sets are certainly easier, as you hardly need to plant them. My late neighbour Edgar Keyes used to make a shallow furrow and walk down the row dribbling sets into it with no care for which way they landed. He covered them little, if at all, knowing that when it rained their roots would find their way down into the soil, and as the bulb expanded, they would automatically right themselves if they lay wrong. I am far more meticulous than Edgar was, but I don't see that I get much better results. Even when using my own plants I try to keep them shallow.

Early on I felt a concern about using onion sets, which came from the store when I was trying to be self-reliant, growing my own seed. Growing onion seed was easy enough, but then I was limited to grow-

Figure 8.12. These late-planted onion seedlings will form sets for next year.]

ing my own plants, which is more time-consuming than using sets. Moreover, my favourite variety, Clear Dawn, was not available as sets, so I've tried a few times to raise my own, and although I've had mixed results it is definitely doable. The main trick, which I'm still trying to zero in on, is timing; again, it's the day-length thing. Clear Dawn seems more day-length neutral than many onion varieties, but it is an issue nonetheless. I sow them late enough so they will not take themselves too seriously and try to make large bulbs, yet early enough so they will make something more than a mere seedling. A mid-July sowing yielded very small sets, usable but pretty tiny. The problem with growing them too large, as you may have noticed with some purchased sets, is that they think they're ready for adulthood and try to skip adolescence altogether, bolting to flower without even making a usable bulb. The best sets are small and uniform. One way to achieve this is by sowing them thickly in very rich soil. I've tried using the short end of a bed (roughly 4½ × 4½ feet, or 1.4 × 1.4 m) for a nursery, including onion sets, but also kale and lettuce for transplanting as space becomes empty in other beds. I don't thin—plants are ½ inch (1.3 cm) apart or closer—and not necessarily in rows but rather in broad bands. This crowding saves much space and restricts size. It works fairly well, but again I need to fine-tune the timing. In addition to uniformly small size, I need them to die back and harden off for winter storage.

I learned a completely different approach to starting onion plants from my friend Amy LeBlanc. She does indeed start her plants from seed sown indoors, but she avoids sowing them closely in trays, which inevitably causes them to be set back by the transplanting. Rather she uses 2-inch (5.1 cm) cell trays (#72 Speedlings with tapered cells for easy removal), sowing or thinning to three or four plants per cell. When she transplants her plugs into the garden, she is in effect planting in hills, like corn or pumpkins, but in this case the "hills" are only 10 inches (25.4 cm) in each direction (if the alternate rows are staggered, this crowding is effectively reduced; I prefer 12 inches, or 30.5 cm). An added incentive for this system is the ease of cultivation: It is much easier to work a hoe in the wider space before mulching. I've tried her method a few times with good results; I should try it more often.

Bulb onions are ready to begin using in late July, long before they're ready to pull and cure for storage. We like to harvest them when the tops still have just a little bit of green succulence in them, so they can be braided. Molly usually does this, spreading them out on a chicken-wire rack so that any clinging dirt can fall through and be gone. A length of sisal binder twine is integrated into the three-ply braid as new tops are worked in, often to 3 to 4 feet (0.9–1.2 m) long. These onions are fresh-pulled, not cured at all, so we tie a loop in the string and hang them from nails in the shed rafters for several weeks until they are well cured and the cellar has cooled down. If the onions are pulled a bit late and the tops have grown dry and brittle, they can be sprinkled briefly to restore pliancy. When properly cured I carry all the braids down cellar and rehang them from nails in the overhead joists. At 20 to 30 pounds (9.1–13.6 kg) each and reaching halfway to the dirt floor, it's no easy chore, but it is a fine way to maximize use of space while keeping every onion ventilated and accessible. If any should rot or sprout, I can spot them quickly and snip them before they can work mischief on the others. The ventilation is even better than in net bags.

I like Clear Dawn because of its globular shape, which wastes less than flatter types like Stuttgarter, plus I find that it keeps better. Even then, as March draws to its close and their clocks start going off, they commence to soften and sprout. By this point I always start chopping them to dehydrate or freeze, reserving no more for fresh use than I'm prepared to risk losing. By mid-April they are decidedly poor in quality, yet the new crop is months away. Then, more than ever, I turn to green onions.

Multiplier onions. What I like most about green onions, or scallions, is that they fit: They fit into spaces and time slots where bulb onions wouldn't make it, either because of the photoperiodism thing or because a neighbouring crop casts too much shade.

By "green onions" I mean any onion that is used as a scallion. There are lots of words for green onions, some of which are overlapping or mutually exclusive, so I'll try to sort it out. A scallion is any onion that's pulled and eaten in the green stage. Some people even plant their bulb onions extra close, then thin them later for "scallions." Usually *scallions* is used to mean "multiplier onions"—any onion that divides to form a clump (yes, yes, I know: How can we call them "multipliers" if they propagate only by dividing? What next, square roots?), or they may flower in their second year and produce true seed, which can be planted.

A further subtype is onions that instead of a flower form a cluster of tiny bulbs, or bulbils, on top of a "flower" stalk. Sometimes one of those larger bulbils sends up a stalk of its own, with a second generation of bulbils on it. These are called Egyptian onions, though I don't believe they have a proven connection to the Nile region. They are also called nest onions, top-set onions, tree onions, and walking onions, after

Figure 8.13. These late multiplier onions were planted directly from seed after early peas were removed.

Figure 8.14. Egyptian onions grow their own sets.

their various habits, that last because they lay over and self-plant the bulbils, each time "walking" 1 to 2 feet (0.3–0.6 m). All three types of plants only form green onions; they never make a bulb of any consequence. A yet further subtype of multiplier onion is the potato onion, which likewise divides at the bulb, as well as forming top sets. However, unlike the others, it forms a bulb of some consequence, not as large as regular onions but sometimes up to 2½ inches (6.4 cm), and good keeping and eating.

There is considerable variation in the size and vigour of these top-setting varieties. How is that possible if they are clones? It seems these vegetative onions, much like garlics, have not completely given up on sex. In among the top-set bulbils you often see

florets, which typically atrophy because the bulbils hog all the energy. As I described with garlic it is possible, though difficult, to carefully excise all the bulbils as they first form, taking care to leave any sexual flowers that are among them. This is a great way to create new top-setting varieties—for example, pollinating them with a red onion, or with a regular bulb onion to further increase bulb size. Moreover, clonal onions sometimes become run out by accumulating systemic viruses or other diseases. Sexual seed bypasses the disease, and may confer renewed vigour just by cleaning up the variety. These are all *Allium cepa*, so they should cross readily.

There is another species of multiplier onion, *A. fistulosum*. They are often called Welsh onions, which is a misnomer, or Japanese onions, which is right on target—they are about as Welsh as umeboshi plums. These also make excellent scallions, whether grown by divisions or seed. It has been conclusively shown that the aggregatum group all originated from a long-ago cross of *A. cepa* with *A. fistulosum*.

Shallots are a distinct species (*A. ascalonium*), though the word is sometimes used imprecisely for scallion.

The reason I am slopping so much ink on these green onions is that they are so easy to grow; they are more reliable than the regular bulb onions and they fit into the garden in so many ways, depending on the type. For example, seed-grown multipliers can be planted in early spring with another companion crop, whereas bulb onions would do poorly there. They are ready to harvest in early fall, often while the bulb onions are curing in the braid.

I'm always looking for another way to integrate scallions into my garden's calendar, and one of my favourites so far involves Egyptian onions. I don't have to mess with delicate seedlings, I just collect the clusters of bulbils and crumble them apart if they are dormant; if they are themselves sprouted, I carefully tease them apart or more often plant a whole cluster—half a dozen scallions in a bunch is no problem. I keep a separate nursery bed in the herb garden to supply me with bulbils. Those "mother plants" are 10 to 12 inches (25.4–30.5 cm) apart, but in the garden I sow them thickly, as close as an inch (2.5 cm) or so for individual bulbils, since they will be harvested as soon as the tops size up, long before they try to divide. If they were left to divide, they would also try to bolt to seed, becoming tough and inedible. Since I am planting those for eating, not for propagating, I merely drop them into shallow furrows from a standing position—let them land where they may. Then I cover them haphazardly; I prefer to sow Egyptians in late summer or early fall following some earlier crop, so they can just get established. As soon as the snow melts, they'll be off and running, and I'll have a crop of fresh green onions by early May, just when the bulb onions in the cellar are soft and useless.

Composites

A sunflower is an iconic flower that's also an ironic flower because it is really not a single flower at all. Rather, a sunflower is a classic example of a composite flower, composed of tiny florets arranged on a bract or disc. Indeed, the true flowers are not conspicuous at all; what makes sunflowers stand out are the big yellow "petals" that surround the discful of florets. Those false petals, also called ray flowers or ray petals, catch the attention of pollinating insects until they close in and discover the nectar. Although they are largely cross-pollinating, especially those with conspicuous blossoms, they are not self-sterile—they can pollinate themselves with no help from the flying yentas. Even those that do outcross have much to gain by being self-fertile, since their disc configuration guarantees that much of the pollen will land on neighbouring florets on the same plant.

It must be a very effective adaptation, this composite head, because there is an enormous and highly diverse family of plants that bear composite flowers. Thistles, dandelions, goatsbeard, and black-eyed Susans teem in our weedy lots and hay fields, daisies, asters, zinnias, dahlias, and chrysanthemums adorn our flower beds, while lettuce, endive, radicchio, and terrasols (aka sunchokes) fill our plates.

Here I'll focus on the vegetables of the tribe. For more on sunflowers, see chapter 11.

Lettuce

All of the veggie composites that come to mind belong to a subgroup characterized by milky sap (no, milkweed is not one of them). That's also where lettuce gets its name: *laitue* in French (*lait* is the French word for "milk"). Lettuce is one of the few self-pollinators in the family, and not surprisingly it has the least conspicuous flowers, pale-yellow brushes that any bee might easily overlook.

I tend to be dismissive of lettucy things because of their lightweight nutrition, but they do play an important role in our diets: Among all the high-protein, high-oil, high-starch things on our plate, it is important to have things that are "merely succulent"—rich not just in water, but in emollient water, which holds the other denser nutrients in suspension so that our gut juices can better digest and assimilate them. Cucumbers and celery are also great that way, which I guess is why I like all of them in summer potato salads, to lighten the starch and oil.

Recent years have seen a gazillion varieties of lettuce of myriad colours and leaf textures. Most of them, or so it seems, are the work of my friend Frank Morton out in Oregon, who is as much an artist as a breeder. That's really annoying, because when you grow out his stuff, you have to choose between eating it or mounting and framing it. I'm not saying that food shouldn't be beautiful, but sometimes Frank carries it a bit too far. Fortunately the stuff all tastes great, too, so I suppose one must endure it.

Actually my all-around favourite is an old English loose-heading variety, Webb's Wonderful. I've yet to see anything hold up to heat as well as Webb's does (probably well adapted to the deserts of Yorkshire). I've even eaten Webb's Wonderful after it had bolted and found it surprisingly bitter-free.

One thing I really appreciate about lettuce is that it does well stuck in odd spots where another crop just came out or is yet to go in. Because I sow so much lettuce in the open gaps around my gardens, I invariably have too much of it reaching eating stage at the wrong time, but I don't much care; excess lettuce can always go into the compost. There is often time to mature a crop of lettuce transplants late in the season, whereas perhaps a direct-seeding won't make it. For that reason I try to always have a nursery bed for frequent small replantings. It is usually about 4 × 5 or 6 feet (1.2 × 1.5–1.8 m), and lettuce seedlings share the same block with kale seedlings or onion sets. Whatever isn't needed I heartlessly pull out and compost; then I replant the nursery bed with more seed (when you have a good supply of home-saved seed, that's not such an extravagance). There's not much that lettuce doesn't companion well with; it even tolerates some shade.

Endives, Escaroles, and Chicory

The genus *Cichorium* includes endive, escarole, and the chicories (whence the name), which includes radicchios, sugarloaf chicory, chicory root, and witloof chicory (also called Belgian endive). Endive and escarole are both *Cichorium endivia*. The chicories are all *C. intybus*, even though they are quite distinct in appearance and use. For seed-saving purposes it's important to remember that all chicories are biennials and will all cross-pollinate. In fact, *C. intybus* and *C. endivia* (which is a self-pollinating annual) will occasionally cross, though only in one direction: chicory by endive, never vice versa.

Let's talk first about *Cichorium endivia*, endive and escarole. The main difference between them is simply appearance: Endive is frilly while escarole is broadleaf, more like lettuce. Here I'll discuss them as one crop: *Cichorium endivia*. The flavour is like lettuce with a touch of bitterness. The nice thing is that in hot weather, it still has just "a touch of bitterness," whereas heat stress makes most lettuce go ugly-bitter. It is also a bit tougher-textured than lettuce, not being quite so domesticated. I'm not too big on toughness, so I chop it rather fine for salads. And in salads is how most people use it, although Italians (who invented escarole; doesn't it sound it?) use it as cooked greens with pasta and beans. Well, think about it: It's very similar to dandelion, only milder. I've tried it and it's quite all right. Endive/escarole will not transplant well, at least not while remaining palatable. The toughness and bitterness are greatly mitigated by eating them a bit before they are full-grown. My favourite so far is Grower Giant, a big industrial-strength variety bred for the factory farms of Central California; it's big enough that you can afford to pick it young and tender, much more vigourous than the French and Italian varieties.

Among the chicories, perhaps best known in the United States are the radicchios (pronounced *rah-DEE-kee-ohs*, though it sounds ridiculous), which have seen a real surge in popularity, especially in cutting-green mixes. Some types are grown like lettuce; others are cut young and allowed to regrow, the second growth forming a loose red-tinged head. If sown to mature in cool late autumn, radicchios are milder-flavoured. They are undeniably beautiful, but you may have mine. The sugarloaf types look somewhat like the conical Bordeaux escarole or even the Wakefield cabbage. As with some other chicory types, growers sometimes cover the maturing heads for a few days with a bucket to exclude light and reduce bitterness. On the few occasions I have grown it, I sowed it for fall harvest and didn't cover it.

I'm scarcely interested in the Magdeburg types, though you may be. Those are the long fat roots that are roasted and ground to make a trendy-chic coffee additive; whether they are an adulterant or an enhancer depends on your perspective. As long as there is fruit juice, I see no need to make beverages out of scorched beans and roots.

The chicory that interests me most actually derives from that ersatz-coffee type. Only a century ago a Belgian gardener was pulling some roots out of cellar-storage for roasting when he was dismayed to find that they were all sprouted. In the dark they had formed little blanched heads. He tasted them and pronounced them delicious. Well, actually he pronounced them *witloof,* which in the Walloon dialect means "white leaf," and ever since they have become a high-priced gourmet delicacy, sold in upscale restaurants as chicons, cut in half lengthwise and dribbled with vinaigrette or hollandaise.

You can grow chicons yourself in your own garden. I like to companion them with Jersey Wakefield cabbage and with a row of Capucijner peas (see chapter 10) down the centre of the bed (on support). Is it so surprising that these three work so well together, or is it coincidence that they all originated within a few miles of one another? Witloof (or whitloof) is not overly demanding; if the bed is rich enough for the cabbage, it surely is good enough for the chicory.

In the late autumn I dig the roots and store them upright in buckets, with sand packed around them and lids on the buckets. In the cold (mid-30s°F, or about 1–4°C), dark cellar, they stay dormant until I want some salad material. I bring the buckets upstairs and put them in the cabinet under the sink. It is cool there, too, but much warmer than the cellar. It's also dark, so I can remove the lid if needed (they may have started growing already). I've always had

Figure 8.15. These fresh-dug witloof chicory roots are ready to be stored in the cellar for forcing in midwinter.

Figure 8.16. Blanched heads of witloof chicory ready to feature in a winter salad.

100 percent emergence (although you should never count your chicons until they . . . oh, never mind), and within a few weeks, depending partly on temperature, you'll have a whole passel of those highly prized yuppie treats that really hit the spot amid the deep snows of midwinter. I am not a big fan of blanched delicacies, but in this case you *have* to exclude the chlorophyll-activating light, or you will end up with gut-retching bitterness that will make you curse me for telling you about this stuff.

By the way, these things are fussy to grow only if you're aiming for the marketplace, which demands tightly formed heads delivered on a rigid schedule. But for my own not-so-picky use, even those heads that have grown a bit loose and may fall apart are perfectly good. Also, when you grow cichons for market, once you cut the heads off, you must throw the root away. But I don't: Those harvested roots will continue to force a *second* set of sprouts, which are even looser and even less salable, but no less delicious—I love to chop them up in a yellow potato salad; let the wealthy drool, only I can afford this.

Salsify

Before I went to farming on my own, I believe the only place I ever heard of salsify was in a Brer Rabbit story. It was associated with words like *persimmon*, *sassafras*, and *calamus*, all of which belonged to a faraway world of rural southern black people like Uncle Remus. Those crops couldn't make it in my

Figure 8.17. *Scorzonera* roots will store well for winter stews and stir-fries.

Maine world, all except salsify, that is. Salsify is not native to my region, but it has gone wild from my garden and become a complement of my hay field, where it is welcome.

Let me clarify that *salsify* refers to two different species, both in the composite family, and both similar in culture and use. True salsify (*Tragopogon porrifolius*) is a white root with purple flowers (in its second year that is; it is biennial). False or black salsify (*Scorzonera hispanica*) has a black-skinned root and yellow flowers.

Salsify is very cold-hardy and slow growing, so I plant it pretty early to get the largest roots. Even at that, first-year roots are rather small (hold that thought). I have sometimes sown it thinly among my carrots, because it's reputed to deter root maggots. That's mainly hearsay, however, as I don't normally have much trouble with root maggots anyway, so I can't verify its effectiveness. Planting salsify to deter root maggots may be like the wooden grating that I installed by the back door to keep anteaters from getting into the house (I call it my avant-garde aardvark guard). People use the grating to stamp mud off their shoes before entering, but I've noticed quite a lot of dirt gets tracked indoors anyway (mostly by me). However, ever since I put in the grating, not a single aardvark has gotten past the door. So perhaps salsify works just as well as a maggot deterrent.

Though salsify is a dicot, its narrow leaves have a rather grassy look, which explains its German name, *haferwurzel* (root crop that resembles oats). It is sometimes called oyster plant, due to the alleged resemblance to the taste of oysters. Well, I am relieved to report that I see no resemblance whatever. Salsify has a distinct mineral-y vegetable flavour. I like it thinly sliced in stir-fries or cut into short chunks in baked beans or stews. My mom grew some once and batter-fried it, trying to get the oyster effect. Happily she failed; they were very nice. Don't get me wrong, I love seafood; a bit of dulse or nori can really add interest to land-based veggie dishes.

The only problem I have with salsify is that it doesn't always size up well, especially the scorzonera

Globe Artichokes

Globe artichokes and cardoons are not easily grown here in Maine, so I have only a little experience with them—and even that is very negative. One day long ago I was hitchhiking down a country road in Watsonville, California, with vast fields of globe artichokes to either side. An aerial crop duster swooped overhead without making the slightest effort to avoid spraying me, though I was on a public highway. To this day the sight of artichoke hearts reminds me of stinging eyes and airplane exhaust.

type. Also, you really have to scrub it to remove all surface grit. That's fine when they are carrot-sized, but when they're smaller, it hardly seems worth the effort. Then I discovered a fascinating fact about salsify, or at least scorzonera: Although they are supposedly biennial, making seed in their second year, if you leave them in the ground the roots don't just wither and die like carrots do; amazingly, the roots grow bigger still, while (get this) remaining tender-crisp and palatable. More of a perennial really, though no one describes them that way. It makes them a potentially more practical crop, though it complicates their use in regular garden companion arrangements. What I have yet to figure out is whether they could be late-sown after an earlier crop, yielding a small root that then overwinters and grows large the second year. Ideally they would not try to flower, but merely bulk up more, sort of analogous to growing onion sets. Speaking of flowering, I've tried steaming the young immature blossom-buds with scallions and liked them very much. The idea was inspired by dandelion blossom-buds, which I've always enjoyed. I'm not sure if this all applies equally to salsify and scorzonera, but it certainly works for the latter.

The Tomato Family

The Solanaceae are generally a pretty sketchy crowd, all of them toxic in some form or other. Members range from the sensible pillar of the family's reputation, the potato, to that nefarious assassin tobacco; from the alluring moonflower to the butt-ugly jimsonweed. The family is better represented in the Americas, the eggplant being, to my knowledge, the only Old World food crop in this group.

People tend to refer to members of this group as solanums, though that is the name of the genus including only potatoes and eggplants, whereas the larger group to which these all belong is Solanaceae. I shall unscrupulously propagate the error, as it is too inconvenient to correct it. Here, I'll offer my advice for the members of the family that I actually grow. That doesn't include eggplant, which does not mature well in my Maine garden, at least not without black plastic, which I'm unwilling to use.

Tomatoes

I'm guessing the most widely grown of these, certainly in the United States, is the tomato. I'm speaking of the veggies here now, so I ignore the filthy tobacco—filthy meaning not only the human consumption, but also the fact that its plants are laden with a mosaic virus that readily infects all of its kin. Being a New World plant, tomato, like pepper, was unknown in Europe before 1492. Columbus didn't discover America, but he did discover something every bit as important: the ingredients for pizza! You'd think Europeans would have been ecstatic at this new find, but at first they were very cool to it—well, you see, it wasn't mentioned in the Bible. Although the scriptures didn't actually forbid tomatoes, their silence regarding the species was taken as condemnation, especially since it was suspected of having aphrodisiac properties (aka "love apple"). Tomatoes were widely reputed to be poisonous, probably based on their similar-looking nightshade relatives. In order to fully discredit this myth, in the early 1800s some gentleman took out a newspaper ad stating that at a certain date and time he would stand on the Camden County courthouse in New Jersey and eat (drumroll) six, count them, six raw unpeeled tomatoes. At the appointed hour crowds gathered around to watch the man, tended by a physician, in his last extremity. After he ate them, and the doctor declared him quite fit after all, many folks began to take the threat less seriously, to the point that Camden is the headquarters for Campbell's Soup and tomatoes are a major New Jersey crop.

Tomato is by far the most popular garden veggie in the United States today, which means you probably already know a thing or two about growing them, so I'll try to limit my comments to what may be new or useful.

Tomatoes are not presumptuous; they grow happily in garbage—witness your typical compost heap. They do prefer a mulch, even if it is rubbish. I usually have a bin of half-cured compost at about that time (or at any time), which I could turn yet again for finished compost, suitable for the more finicky crops. Indeed, that I do, but I also take out a generous amount of the coarse stuff and mulch the young tomato plants with it. I usually put a light mulch of shredded leaves or spoil-hay atop that, to keep the compost nutrients from drying out and evaporating. Whatever crop follows this will need no added soil amendment.

By the way a common trick is to plant tomatoes sideways in a deep furrow to encourage them to form roots all along the buried stem. It's an excellent idea, but I have a better one. I set the plants *deeper* to bury the stem (I usually broadfork that bed anyway); not only does this have the same rooting effect, but the deeper roots get more steady moisture and minimize transplant shock.

I've read somewhere that tomatoes can be grown in the same place for up to six consecutive years without disease buildup, which seems suspiciously precise to me, but also surprising because tomatoes have a number of disease problems (septoria leaf spot, alternaria blight, early and late blight, verticillium wilt, anthracnose, and more) that are usually easy to avoid by rotation. While it is true that tomatoes grow blithely in the half-decayed remains of *other* plant species, their own residues are apt to work them woe, due to the persistence of those disease organisms just mentioned. In years when we have had severe

Figure 8.18. A heavy grass mulch keeps these young tomato plants free of early blight; though a determinate variety, they will soon be staked.

regional outbreaks of late blight, other gardeners have insisted that it is critical to collect all the tomato debris and remove it from the garden. I'll buy that, it only makes sense. I've also been told that one absolutely must burn all diseased plants, that composting will not destroy the fungus but only spread it around. That I do not buy, as it flies in the face of my own experience. I have to suspect that they really mean *improper* composting will not destroy the spores, which I'm sure is true. What I do know is that when I've shredded the diseased vines and composted them so they generate the proper heat (162°F, or 72.2°C), I have not had problems the following year. Although it is entirely possible that those particular batches of compost did not get used on tomatoes, but only on other crops that are unaffected by it, I still assume that if the spores survived the composting

they would have lingered on in the soil of the other beds until tomatoes were rotated there again, or else spores would have drifted. At least with many diseases an outbreak hinges less on the presence of the pathogen than on the conditions (like weather) that allow it to proliferate. I do know that one of the worst outbreaks was introduced on Florida-grown plants sold by Walmart. Since I grow my own plants from my own seed, that might explain why I got hit later than others (though it still wiped me out).

Tomato varieties are either determinate or indeterminate; when determinate vines reach a certain point, they stop growing longer and axial buds take over, forming a bushier plant. Indeterminate vines are apically dominant, meaning that they never know when to quit. The tip of the vine will keep stretching out indefinitely, regardless of what any side branches

Figure 8.19. Hanging the Royal Chico vines from the overhead joists lets their fruit ripen indoors long after frost.

on that vine do. Indeterminate varieties need stakes or rather poles (and sometimes radio towers) to support them. Letting the vines sprawl, whether on dirt or mulch, seems to encourage blossom-end rot and other fungal diseases. I always stake my tomatoes, even the shorter determinate varieties. Free airflow around plants and a maximum sun exposure solve a lot of problems before they happen.

Also, determinates are generally earlier, so I prefer those for my short cool summers. Some of my favourites are Bonnie Best (aka John Baer) and Siletz. For my main crop, however, I grow Royal Chico, a paste or canning tomato. Royal Chico is a solid heavy yielder, but a bit late to ripen, which is exactly why I prefer it. Around the equinox we typically get our first frost, and there is oh-so-much to be taken care of right away; do you think I have time to stop and can

tomatoes? If they were all ripe I'd have no choice, but only about a third of the Royal Chicos usually are. So I uproot the entire vines and shake off the dirt plus those tomatoes that drop off easily. I gently haul the plants into my high-ceilinged cellar, where I hang them from nails in overhead beams. I used to use string, but the lower crotch of each vine is always strong enough to hook over a nail.

Most of the fruit is still green, but full-grown and shiny. In the cool, dark cellar they will ripen slowly. As I find time I pick them off, beginning with the redder ones, which get canned quickly. I spread a tarp on the floor beneath to catch any that drop off on their own—it helps that my cellar floor is dirt, not concrete. If I wish to hurry things along I also pick the lighter-green ones to put in shallow trays on a sunny closed-in porch, where they can ripen faster.

Figure 8.20. We store tomatoes of many varieties on the sunporch. We'll sort and blend the fruits as they ripen.

Figure 8.21. These paste tomatoes and cherry tomatoes are ready to be sauced ASAP.

The non-paste tomatoes plants are harder to hang without losing fruit, because the fruits are bigger and juicier. I strip those off the plants at harvest and sort them into trays according to ripeness. Those also sit in a sunroom until I can process them. They have a richer, sweeter flavour than paste tomatoes; I usually can the two types together, although if I use them for spaghetti sauce, I'll need to cook them down more. The cellar-hanging trick works well enough that many don't have to be canned at all: For a few weeks we can just use them fresh off the vine. I know that having the roots attached helps them store longer, as the fruit draws some life force from them, but is it worth it? As they dry some dirt drops onto the fruit and onto me as I'm picking them. It's probably no big deal, but next year I'll try cutting some off and hanging the rootless vines.

For fresh use our favourite is an old *in*determinate variety from England called Gardener's Delight. It's a cherry tomato, packed with sweetness and over-the-top tomato flavour. I'm impressed at the English showing such good taste in tomatoes of all things. It's slightly late for us, but never fails to crop heavily. I always have lots to can (as strained puree), but it is so juicy I don't mix it with the others, instead bottling it separately for highly flavoured soups. I can add it to sauce for its rich taste, but then I need to cook it down more. Sometimes a small amount of terrasol meal or potato flour will go a long way toward thickening it while further enriching the flavour (no, you haven't read that somewhere else before). In fact pureed Gardener's Delight is an exquisite ingredient in our New World soup.

I have a new favourite variety of my own making, derived from a cross of Royal Chico on Gardener's Delight, which I call Gardener's Sweetheart. It is a conical heart-shaped cherry type. It is almost as sweet as Gardener's Delight, with the same robust flavour and more solid texture, and completely free of cracking, which is a flaw in its female parent.

Gardener's Delight and Gardener's Sweetheart are both rampant growers. I train them on a 6-foot (1.8 m) trellis down the centre of the bed, and I sometimes plant bush beans along the outside edges of the bed.

Peppers

I could easily grow hot peppers here in Maine, but I do not. I like a little cayenne—a very little cayenne—so little that the amount I use in a year would not fill a little spice bottle. Therefore I just buy a little spice bottle, knowing that when the *Titanic* lifts its stern, I can grow hot peppers at the drop of a hat. Sweet or bell peppers are another matter: I eat bushels of them, both green and ripe. I love them far too much to let others control my supply, unlike pineapples and macadamias.

When I first went to gardening, most available pepper varieties were too late for me to get reliable crops. I could only grow them here by using black plastic, but since I couldn't grow black plastic here, either, where was the gain? The few "early" varieties might give a few green fruits just before frost. That was all before Lipstick, which I consider the crowning achievement of Rob Johnston's breeding career. Lipstick was not merely an improvement, it was *the* difference, the key to abundant field-ripe thick-walled peppers, at least in August, without the reliance on ticky-tacky. Since then many other early peppers have been developed, possibly as good or better than Lipstick (although those I've trialed are not); however I continue to rely on those flat-tapering cerise-red . . . well, the name says it all. Mind you, I'm not even comparing it with Ace, which is an F1 hybrid and therefore not part of my world. In recent years I have developed a variety I call Mountaineer. It is basically an elongated Lipstick, sired by a semi-hot that seems to have passed on none of the heat. I'm still far from insisting it's an improvement on Lipstick. I also grow a stable open-pollinated variety called Deuce that was bred by my late friend Bob Langevin, who derived it from the hybrid Ace. Deuce is very early but not as uniform in shape as Lipstick.

I have had to learn to crowd peppers, after reading that they set fruit better that way (although for me peppers are largely self-pollinating). I set them a foot (0.3 m) apart in the row, with at most 2 feet (0.6 m) between rows. I find that they have moderate fertility needs, though abundant humus and phosphorus are helpful for heavy fruiting. I rely on fall-tilled shredded leaves and/or a light application of compost (two barrow loads per bed).

After mulching I pretty much ignore them; however when heavy-laden they tend to lie over, and I'm concluding that some light support is warranted. Not like tomato stakes, just short lightly branched twigs between them, stuck in deep enough to do what I call "starkening up" the plants.

In recent years we've chopped and frozen most of the pepper crop, but they also dehydrate very well—not so handy for stir-fries perhaps, but a reassuring backup in the event of powerlessness.

Ground Cherry

The ground cherry (*Physalis pruinosa* or *P. peruviana*) is delightful as an occasional snack. These tropical fruits love plenty of heat but will *not* tolerate drought at all. I have to start them very early indoors and set out fairly advanced plants in order to harvest a large crop of ripe yellow fruits. All too often the first frost finds me with oodles of fruit, most of it green.

Figure 8.23. Protected by their papery husks, these fruits will continue to ripen in a sunny window.

Like tomatoes, ground cherries will after-ripen in a bowl on a sunny windowsill, especially since their papery husks hold them loosely protected. Unfortunately they will have increasingly little of the sweet pineapple tang that characterizes the vine-ripened fruits and more of a soapy solanum taste. Tomatillo (*P. ixocarpa*) is a close cousin of the ground cherry,

Figure 8.22. Ground cherries, or husk tomatoes, in the days before I started using moisture-retaining mulch.

and it's a classic ingredient for making salsa verde. But I find green tomatoes are a good substitute for tomatillo in salsa, and tomatoes are easier for me to grow, so that's what I do.

Potatoes

Potatoes are almost a category to themselves. The USDA considers them a vegetable, reasonably enough (although for the school lunch programme they also consider ketchup a vegetable serving), yet in their Andean homeland they are a grain substitute, since grains do not grow well at higher elevations. Far denser in starch and protein than most vegetables, potatoes have been a staple support for whole civilizations. The Quechua, Aymara, and Chibcha peoples prized the potato so much that they made gold and ceramic likenesses of the earthy tubers. As with tomatoes the Europeans were slow to appreciate the value of potatoes, despite government efforts to promote this cheap easy-to-grow foodstuff. A French aristocrat came up with an idea to make them popular with the peasants. He planted them in his palace garden and conspicuously posted a guard over the place during the day. Just as conspicuously he removed the guard at night, and sure enough, potatoes began disappearing from the palace garden and reappearing in peasants' fields throughout the region. Anything needing that kind of security must be truly precious.

During the 16th, 17th, and 18th centuries there was a dramatic power shift from the Mediterranean basin to the countries of Northern Europe. Italy and Greece fell into decline as Britain, Holland, France, and Germany became the main power brokers of the era. At least one historian has suggested that the lowly potato was a big factor in this rearrangement. As long as grain, particularly wheat, was the energy substrate for the population, Southern Europe held the upper hand. But potatoes, which greatly outyielded wheat in the cold northern soils, could support a much larger population of labourers, taxpayers, soldiers, and so on, fuelling a population explosion that turned nations as diverse as Sweden, Poland, Russia, and Lithuania into important players on the world stage.

This reliance on potatoes had one important drawback. Being asexually propagated potatoes lacked the genetic diversity within any variety to adapt to environmental threats. That is achieved by breeding for resistance; however the potatoes grown in Europe all derived from a very few clones collected in South America, which apparently lacked any resistance to late blight (*Phytophthora infestans*). The variety most people were growing, Lumpers, was very susceptible to the disease. When in the 1840s a particularly cold, wet autumn saw an outbreak of the fungal disease in Ireland, it spread through potato fields like Sherman's march through Georgia, leaving a wake of devastation. Irish peasants grew potatoes on their own subsistence plots, unable to afford the grain their wage labours produced. The Irish potato famine is always portrayed, rightly enough, as the result of *P. infestans*, but it's not so simple. You see, while the Irish peasants were losing their main subsistence base, the landowners continued to ship their wheat and beef to wherever the price was highest—that would be England. The government eventually halted the export of grain from Ireland, but not in time to avert massive suffering. Whenever I am narrating the story of the Irish potato famine to classes, I hold up Lumpers as the reason why *O* is the biggest section of the South Boston phone directory.

The Irish potato famine was more than an ecological collapse. It is an indictment of imperialism and of the marketplace, particularly when unregulated. I digress into this topic because it is a classic example of the need to recognize that our gardens have no borders, that most of our global problems arise from a lack of sustainability, indeed a refusal to even consider how we might shift to more sustainable approaches. The checkout aisle, the pulpit, Wall Street, and Pennsylvania Avenue all run right through your garden-without-borders.

My Scatterseed Project maintains several hundred varieties of potatoes. People often ask me which are my favourites, assuming I must have tried cooking and eating each of them. I can only answer the question by asking them: Favourite for what? Baking? Boiled and mashed? In salad? For early use or winter storage? And each of those categories might still only narrow it down to a couple dozen varieties. I would say that for baking, two of my favourites would be Butte Russet and Acadia Russet. For earliness there is an old Pennsylvania heirloom variety called July, which is indeed very early and fine eating quality. For high production, great flavour, good keeping, and general use, I am particularly fond of Daisy Gold and Alby's Gold (which may be the same).

Potatoes with coloured flesh are all the rage nowadays. Personally I find the blue- or purple-fleshed varieties less appealing, as the pigment seems to compromise the flavour, despite its antioxidant properties. Yellow-fleshed varieties, on the other hand, are pigmented by carotene, which not only is very health-promoting but also gives a flavour some describe as buttery. The yellow pigment seems to be particularly characteristic of the subspecies *andigena*, although it is also common in the closely related species *Solanum acaule* and *S. stenotomum*. That these "species" can all be cross-pollinated with relative ease shows that speciation has not proceeded so far as to completely isolate them reproductively. By the way I have noticed that those varieties having the brightest-yellow waxy flesh and richest flavour tend to be poorer yielders and poorer keepers, the so-called fingerlings being a case in point. However, a few varieties, like Daisy Gold, combine intense-yellow colour and rich buttery flavour with very large tuber size and superior keeping quality. Another of my all-around favourites has been Granola, a German variety very much like Carola, but with better keeping quality.

Spuds are stolons, which grow from the underground stem portion of the plant. That means that the more of the stem is buried, the more stolons will grow from it and the more spuds you will get. This is the main reason why most people hill up their potatoes, but here's another one: Exposure of the potato skin to daylight triggers the formation of the glycoalkaloid solanine, which tastes bad and is pretty toxic as well. For spuds to be good for you they must be kept dark.

All of the solanums have glycoalkaloids in some parts under some conditions. Of course, all plants

Bitter Potatoes for Chuño

Of course certain potato varieties are best for baking, or boiling and mashing, or for deep-frying, but there are a couple of uses you never hear about, each having its own particular varieties. Where my sister-in-law grew up in the Andes, a particular sort of potatoes—bitter potatoes—are used to make chuño, a starchy staple of the Altiplano region. The newly dug tubers are piled in outdoor heaps on frosty nights (which at that altitude come very early). They freeze and the cells burst; they are tread upon to squeeze out any free liquid, and the process is repeated often until the tubers are freeze-dried. You see cholos trudging down the hill to market with a backpack thrice as high as the bearer, yet weighing half as much. Ground into flour, chuño is used to thicken soups and stews. I have not made it myself, but I have eaten it: a slight tangy cheesiness, but not too exciting.

have some chemical components that are more or less toxic to something, to protect themselves from their foes. By "foes" I mean other organisms that like them, perhaps even love them to death. That may include us, although the substance that is toxic to some other creature may be the very flavouring agent that makes it attractive to us.

Nobody I hang around with loves the taste of glycoalkaloids, but some people think you can get rid of it by cutting away the green part. No way; you see, the green part isn't the poison. That's just the chlorophyll indicating the presence of invisible glycoalkaloids, in this case solanine, which are pretty much spread throughout the tuber, though it may be diffuse enough to make the taste less noticeable. No, the way to prevent the bitter taste is to keep sunlight

Figure 8.24. These potatoes were planted on fallow ground under hay mulch, with more hay added later and shredded leaves to exclude light. Though some seed pieces rotted, those that emerged are more vigourous and healthy than those planted in soil.

off the forming potatoes and those in storage, and since some spuds insist on forming on the surface, the most effective way to avoid that is by hilling up or hoeing up soil around the stems—which, if you recall, is also the best way of getting more spuds, period.

Of course hilling up is also a great strategy for controlling weeds, but what about my "cover-the-earth" strategy? How can I use my universal mulch policy while hilling up the potato plants? I can simply use hay mulch for hilling instead of soil, although it takes a very great amount to bury the stems to any depth. It's a widely known method, but I first heard of it from my elderly neighbour Lucian, whose grandfather lived in Harpswell on the Maine coast. He, the grandfather, never planted his potatoes before the Fourth of July, and even then he technically did not

plant them. Rather he laid the cut seed pieces on top of the soil and covered them with many loads of rockweed (the marine plant *Fucus vesiculosis*), which he, living by the seashore, had in unstinting quantity. According to Lucian, he always had robust healthy plants and a great crop of spuds with no dirt to wash off.

Well, I live too far inland to do that, but hay works almost as well, although it doesn't have that wealth of trace minerals that seaweeds contain. Others have told me of the advantages of growing potatoes under hay mulch: You needn't dig the spuds, merely pick them off the ground after pulling back the mulch; they were right about that. No dirt on the spuds, just rinse them before cooking; right they were. No need to fertilize the following crop, that hay mulch will add plenty of everything; sho 'nuff. Flea beetles will be no problem, since they like heat-stressed plants growing in hot, dry, bare soil. That was a slight exaggeration, although it certainly made a noticeable improvement. Oh, and Colorado potato beetles would be a mere memory. Those that hatched out in the soil would be so hampered and discouraged from climbing up through all that dense hay that the few who made it would take off and never return. Not so; in my experience planting in mulch is a slight deterrent to PCBs, but no panacea. That's okay, I'm fine with partial solutions, but I don't assume that I can ignore the beetles, especially since new ones are always wandering in.

However, there is one huge disadvantage to growing potatoes in heavy hay mulch that everyone seems to downplay: rodents, especially voles. They will fill the patch with a network of tunnels and destroy a significant part of the crop. Few things are as exasperating is reaching for a big beautiful spud and finding it hollowed out from the other side. Even moles can do a lot of damage, although they are carnivores. As they tunnel down the rows seeking grubs and larvae, they leave the spuds pest-free; their tunneling leaves the plant roots damaged and drying out, however, and the plants stunted.

There are several approaches to the rodent problem, and some of them are applicable to all crops;

therefore I shall discuss them in more detail in chapter 15, *Pests and Diseases*. However one remedy is so closely related to potatoes in particular that I shall describe it here. The trick—which I learned from Eliot Coleman—is to put a shallow wooden box upside down under the mulch, preferably where a run or tunnel already exists. Notch out an entrance on each of the four sides and place a mousetrap just inside each door. No bait is needed; the critter's curiosity will compel it to enter and step on the spring. I have a few of these scattered in my heavy grass mulch and if checked and emptied regularly, they seem to have a big impact on the rodent population. I have not yet tried them in my leaf shred mulch areas, where rodent damage is much less intense though still present.

It is partly because of rodent problems that I still grow some of my potatoes the conventional way, hilled with soil (although the problem is not completely solved by that, either), but if I continue to find Eliot's method to be sufficiently thorough and reliable (if I am diligent enough to check and empty the traps regularly), I will probably grow all my potatoes in deep hay mulch, as everything else is in its favour.

There is another advantage to the mulch-hilling method: I like to intercrop potatoes and dry beans in alternate rows; supposedly they repel each other's pests (see *Companion Planting* in chapter 14), something of which I'm still skeptical. It seems to help somewhat, but again I welcome partial solutions, as long as they don't create a worse nuisance. In this case the beans interfere with hilling up the potatoes, as it is difficult to scrape up enough soil around the potatoes without damaging the shallow roots of the beans. Moreover, whether the potatoes are grown alone or in combination, hilling forces me to space the rows farther apart than either crop requires for itself. Either crop can be planted as close as 2 feet (0.6 m) between rows rather than 3 feet (0.9 m), if only hilling were not required. The mulch method removes the need to move soil around and thus allows for the closer spacing. By the way it also prevents rust in the beans.

I often find that however thickly I apply the hay mulch I don't get it thick enough close around the plants, the very place where spuds may be exposed to sunlight. My solution to that is to wait until the mulch has settled down a bit, then apply a further mulch of shredded leaves right on the row itself. This will serve to block any interstices in the hay where sunlight might penetrate.

In food plants, as in life, I've had to humble my assumptions about what is good or better; at the very least I must add the question: "Good or better for what?"

Cucurbits

The cucurbits are not a huge group, at least not the cultivated species; nor do they loom large in importance among vegetables. Perhaps just as well, since they are among the least eco-efficient of crop species—you put in a lot in proportion to what you get back, whether in mass or in macronutrients. The only major contributors of food energy are the winter squashes, with abundant starches, sugars, and carotene. Another is the protein- and fat-rich pepitas, but I group them with the oilseed crops, and discuss them in chapter 11. Here I'll talk about growing squash (including pumpkins), melons, and cucumbers.

All that being said cucurbits' main contribution is not in protein, carbohydrates, or fats, but in their watery parts. Water itself is a very important part of our intake—in fact the most important, as everything else requires water for its metabolism. Of course we can just drink water out of a cup with much less ado, but the water that is bound up, physically and chemically, in our food goes much farther toward making that food digestible and assimilable, so don't short-sell the succulent veggies.

The water in cucurbits is especially rich in emollients, which we associate with skin care products, but which also contributes a lot to a healthy digestive system. They are some of the same substances that are found in aloes, comfrey, borage, purslane, sedums, and cacti. The succulent plants seem to be the highest in electrolytic alkaline minerals (K, Ca, Mg, Na, and others) as well as soluble silica.

All members of the cucurbit group are highly frost-tender, although a couple of cucumber varieties have exhibited a slight cold hardiness, presumably due to an exceptionally high electrolyte content. Most cucurbits are quite sprawling of habit—even zucchinis sprawl, though not very far—and most require a long warm season to mature fruits. Some of them show their resentment of drought by dramatically wilting, but this strikes me as an adaptation to compensate for their large transpiring leaves by shutting down, because they will perk up suddenly when water arrives.

Members of the subgroup Cucumis (cucumbers and muskmelons) are somewhat atypical of Old World crop species, both in their cold-tenderness and in their sprawling habit, whereas the Cucurbitas (squash and pumpkins) are wholly typical New World crops in their big-fruited tropical luxuriance.

Cucurbita pepo

The species *Cucurbita pepo* is best known for pumpkins and some squashes, including zucchinis, scallop, and yellow, but it also includes spaghetti squash, acorn squash, and the delicata types (Tennessee Sweetpotato, Sweet Dumpling), plus certain gourds.

Zucchini is infamous for its heavy summerlong production. My friend Jack Kertesz once suggested that breeders should develop a *modest*-producing variety, which he would name Underwhelm. I would prefer that breeders turn their attention to developing a more flavourful variety, one you'd eat because you're hungry, not because you're thirsty.

While zucchini and other summer squashes may overproduce at certain times, in late season production may wane as the plant decides it has done enough. You can boost late-season production by applying urine to the soil around (but not on) the plants. If you're unable to use your own urine, there is a good substitute: green pee. I make that by filling a trash can with fresh grass clippings and covering with water. After a few warm days I strain the rank-smelling "tea" and apply it, preferably before rain or irrigation: The neighbours may not appreciate the smell of green pee any more than the real

Figure 8.25. Some zucchinis outyield Costata Romanesca, but few, if any, surpass it in flavour.

thing. Of course mulch will soak it right in and neutralize the odour almost immediately. This works for most crops for which extended production is more important than ripening.

At risk of seeming opinionated, heaven forfend, if you're growing any other variety than Costata Romanesca (Ribbed Roman), you're not really serious about this. I mean, this is the main variety grown in Italy. It has flavour, it has texture. People have told me that their kids wouldn't touch zucchini until they tried Costata.

Sure, there are those cute little yellow crooknecks and those spooky-looking patty pans; all very nice, but I'll take Costata. Moreover it is more forgiving: If the fruits get a little overgrown they're still fine.

Figure 8.26. If it's flavour you're after Buttercup is the one.

In fact when they get *really* overgrown (as for seed saving), they dwarf all the others. Ever since the first year I entered a 28-inch (71.1 cm) specimen in the Big Zucchini Contest at Common Ground Fair, all of the winning entries have been Costata Romanesca.

You can make ridiculous stuff like zucchini cake and zucchini pickles, but I enjoy large quantities of zucchini in what we call Summer Stew, which also includes green beans, green shell favas, bell peppers, rocambole garlic, scallions, tomatoes, and fresh basil. Don't ask me for the recipe; I just gave it to you. Serve it over pasta or slabs of fresh-baked whole wheat bread. And don't try to make this in the winter; even Birdseye can't make summer squash tasty out of season. That's why we have winter squash.

By "winter" squash we usually mean *C. maxima*, the large warty thick-stemmed varieties like Buttercup and . . . well, there is little point in mentioning any others after Buttercup, although Sweet Meat and a few others are fairly respectable. I admit to being a "pepo bigot" when it comes to winter eating; some delicatas have high sugar content, but they lack the carotene coupled with a mealiness that gives some maximas, especially Buttercup, their rich comforting quality.

I have an additional reason for favouring Buttercup over the rest: Several other squash varieties have nice dry, sweet flesh when fully ripened, but when slightly immature they are bland and watery. Buttercup develops its eating qualities relatively early; if an untimely frost curtails the growth, I'll have smaller squash, but all of them good eating compared with other varieties at the same stage. I discovered this one long-ago year when I was unable to plant them until June 20; the crop was poor in size, but what I did get was fine. In fact, I always set aside a pail of the small fruits that fail to size up. For a couple of weeks after harvest they have a special quality that I really fancy; cooked up whole (the seeds are still tender), they are a bit like summer squash but much more solid and flavoursome. Squashes other than Buttercup are not half as good this way.

I have recently discovered some amazing challenges to my pepo cynicism in a most unsuspected place. While trialing a number of hull-less pumpkin varieties (see chapter 11), I noticed one or two varieties from Eastern Europe that had coarse bright-yellow flesh. I never expect pepitas to have tasty flesh, but my curiosity led me to steam and mash and eat some. Not too shabby at all. I found one of them such good eating that I froze 10 pounds (4.5 kg) of it. Knowing how open-minded I am *not* on the subject, you may draw your own conclusions. Having an oilseed pumpkin with tasty flesh has always been an elusive fantasy. It still is, until I can further increase the seed for distribution, but I am very encouraged at the prospects.

C. moschata is a very popular species, though I don't usually grow it myself. It includes the various strains of butter*nut* squash, Waltham, Ponca, et al., with thick-necked, thin-skinned fruits. They are fairly good keepers, though not as good as Buttercup, but

Figure 8.27. Curing is the secret to long storage.

more important they are a southern species, preferring a longer, hotter season than most maximas, which generally do better in my short, cool Maine summers.

I have discovered a radical way of increasing the fertility of the squash plot *while* the crop is growing and controlling striped cucumber beetles at the same time; I described that in chapter 2.

Fully half of handling a good squash crop is in the curing. Market gardeners often send their crop to market with little or no curing, which is okay if it isn't expected to keep beyond a few weeks. Long-term storage demands that the squash be left to sit in a warm, dry place after picking, when it no longer receives sap from the plant (which presumably has been frost-killed anyway). The idea is for this skin to toughen and the flesh to dry slightly, which fully develops the carbohydrates. Field-curing in piles is the usual practice, which has mixed benefits. The direct sunlight will destroy fungal spores on the skin and the airflow is great; however being piled means that some are shaded and still in contact with the damp soil. Some growers throw some of the vines over the piles to protect them from subsequent frosty nights, but those are too wet for me; I prefer tarps for temporary protection. Eventually I spread mine on a sunny porch or other wooden platform, placed on their sides or upside down when possible to expose the bottoms to UV and dry air. Fresh-picked squash should be handled very gently; an imperceptible bruise can cut months off the storage time. After a couple of weeks, I place the fruits on rough wooden shelves at the end of an upstairs hallway, cool and draughty. Properly cured squash of the right variety should keep in good condition until March.

Muskmelons and Cucumbers

I dearly love muskmelons (*Cucumis melo*), but getting enough heat-units to mature them is extremely challenging where I live. The usual remedy for that is black plastic, which I am unwilling to use, as petrochemicals seem inconsistent with organic methods. Not that I don't make little compromises here and there—irrigation hose, barrels, tarps—but I balk at something that must be routinely replaced.

Aside from black plastic I know of two other strategies to succeed with melons: Start with advanced greenhouse-grown plants, and grow extra-early varieties. The first I do, but not as well as I might, in my attached solar greenhouse. As with the other cucurbits I use 4-inch (10.2 cm) Jiffy-Pots, thinning to three plants each.

I do not use the living mulch strategy mentioned above with cukes or melons, although I presume it would deter beetles in those crops as well. Instead I use a ticky-tacky row cover that excludes pests, but also excludes pollinators if you leave it on too long. I have used some 3 × 3-foot (0.9 × 0.9 m) screen frames, too, which would soon be too small, but I only need them until the vines begin to run, by which time the beetle population has been greatly reduced or eliminated by starvation.

The reason I don't use a concurrent green manure, as with squash and pumpkins, is that I have a better use for the space between rows and hills, and that is to grow companion food crops. Cucumbers (*Cucumis sativus*) and melons are Old World cucurbits and more compact-growing in form, and so between snow-go and the Fourth of July I can grow quick cold-hardy crops like rape greens, leaf lettuce, spinach, or bunching onions. With a row 2 feet (0.6 m) away on either side of the melons/cukes, even when the latter begin to run it will be another week or more past July 4 before the companions are wholly overrun, by which time I either use them up or pull them out. Another reason for not sowing oats and peas around the melons is the need for heat. The one thing that can be said for bare soil is that it heats up faster; when I do finally mulch the piece, I usually use fresh grass clippings as opposed to dry hay or shredded leaves. The grass clippings may heat up the soil slightly as the bottom decomposes, plus release a dose of nitrate-forming urea each time it rains or is watered.

As for varietal selection we encounter the same dilemma that confronts breeders of heat-loving species: When you develop a variety that will mature in our cool summers, it usually lacks sweetness and flavour, which are, after all, the products of intense photosynthesis. For example, my neighbouring town of New Sharon, Maine, is home to an exceptionally early-maturing melon—the Dyer melon—which has little of the sugary redolence of a Charentais, yet is more reliable than most others. Our favourite variety in recent years it is an excellent compromise, having the uninspired name of Delicious 51. It is truly delicious, yet fairly early. However, there are many others I've not yet trialed, even among non-hybrids.

Obviously any crop that contains as much water as melons do must take in that much and more; moreover it must come regularly—a small amount every two or three days is more useful than a gully washer every fortnight. Of course mulch helps keep moisture levels steady, but I've found to my own grief that natural rainfall is rarely adequate. Most summers supplemental watering from my irrigation tank is crucial to success.

Cucumbers may be in the same genus as muskmelons, but they are wholly separate species and wouldn't even dream of cross-pollinating. Cukes are easy to grow and even easier to eat; I rarely eat fewer than six a day in season, and that's just walking past the patch, not counting what comes to the table. I can imagine making a potato salad without diced cucumbers, but only if they are out of season; otherwise there is just no excuse.

Everything I've said about growing muskmelons applies to cucumbers, although cukes are much earlier and more reliable. I was intrigued to learn that cucumbers are rich in soluble silica, which is important for healthy teeth among other things. The key word is "soluble"; otherwise you might as well eat the dirt they grow in, which is mainly silica, though *not* particularly good for healthy teeth. I'm always super-impressed by plants that can actually dissolve silica—that is, eat rocks. Yet there are some plants that can do it in a heartbeat, most of them evolutionarily primitive plants like horsetails, some ferns and lichens and sedges. To find a highly evolved angiosperm dicot that hasn't lost the knack is exciting, doubly so because cucumbers are much more fun to eat than lichens, and because they build up protective enamel armour on your teeth; what's not cool about that?

Of all the fine cucumber varieties available, there's only one that I bother to give space to. Many years ago on a camping trip, my parents collected a cucumber from a member of the Boothby family of Livermore, Maine, an old family heirloom for at least five generations, never offered in the marketplace. I introduced it via Pine Tree Seeds and now it is all over the place. It is an ivory-skinned pickling type, short and blocky, very early and heavy yielding. Unlike modern varieties, which are bred for thick flesh and small seed cavity, Boothby has lots of "guts," the slushy centre surrounding the seeds. In this case that is not undesirable, as that centre is very sweet and tasty and the seeds are slow to fill out, perfect for succulent fresh eating on a sweltering July afternoon. For pickling this is more of a liability, but pickles are low on my priority list. And I'd much prefer to pickle kohlrabi, terrasols, or green nasturtium seeds than cucumbers.

Watermelons (Citrullus lanatus)

Considering that watermelons are native to South Africa, it is hardly surprising that they are challenging to grow where I live in the Western Maine Mountains. To grow watermelons here it is very important to start plants early indoors, and you absolutely must choose the right early varieties, which are all too few. Several F1 hybrids are quite early, but choosing those only makes you a captive of the seed companies. My friend Glenn Drowns has done as much as anyone I know to help northern gardeners grow their own watermelons. As a teenager in Idaho, Glenn crossed and selected early strains to develop the open-pollinated variety Blacktail Mountain, a roundish little melon with rusty-red (not pretty pink) flesh, lots of southern watermelon flavour, and small crunchy-delicious seeds. It is my favourite variety by far. By the way "studies have shown" that watermelon, especially the darker-red varieties, contain considerably more lycopenes (assuming you believe in such stuff) than tomatoes. Another variety well adapted for the North is Cream of Saskatchewan, fine flavour if you can get around the colour, or lack of it; bye-bye lycopenes. I've tasted these varieties grown at Johnny's Selected Seeds in Albion, Maine, and found them superb, but

Albion has a lot more heat-units than my place, plus the folks at Johnny's use black plastic.

Chenopods

The chenopods (or goosefoot family) include useful food crops, such as grain amaranth and quinoa (which are discussed in chapter 9), but only a few common vegetables, such as beets, chard, and spinach. I don't plant any spinach in the spring anymore because it bolts so quickly. I sometimes grow fall spinach, when the increasing cold weather favours production of succulent leaves. However, beets and chard are much more important goosefoot crops in my garden than spinach.

Beets and Chard

I see little merit in the yellow or white novelty varieties of beets, and I find the cylinder types lacking in flavour. The various Long Keeper types may indeed store better than others, but all beets keep so well in the proper conditions (damp!) that the contrast seems less than striking, whereas their course and irregular shape is a bit wasteful. In general I prefer the Detroit types with their smooth round shape and superb flavour. The biodynamic variety Feuer Kugel is very well bred and selected.

Beets are native to the Balkan region and are more tolerant of hot, dry alkaline soils than many other Old World vegetables. The viable lifetime of beet seed is exceptionally long, or at least it appears to be. In fact a beet "seed" is actually a dried fruit with numerous tiny seeds embedded in its surface, sort of like a strawberry. Nevertheless you are actually sowing a cluster of seeds, no matter how carefully you space them, so obviously the rate of germination will be high. Moreover, being still attached to the placenta must surely create a nurturing environment, even if the true seed is not actually drawing nutrients from it. Many gardeners don't like to spend time thinning beets and would rather risk a few gaps here and there. For them the breeders have developed a type of beet called monogerm, meaning that only one seed per "seed" will germinate, after which the others

Figure 8.28. Sugar beets are easy enough to grow, but then what? I'm working on a practical method for making tasty beet molasses.

will shut down and give up. This is important for commercial production, since farmworkers typically refuse to be paid in beet greens. I'd rather thin my beets and eat some beet greens.

I have little difficulty growing good beets. As with carrots early thinning is important, and I usually add at least 2 gallons (7.6 l) of wood ash per bed. Indeed beets are especially tolerant of hot alkaline soil, which may be why sugar beets are widely raised in Utah and Colorado. My main pest problem is deer and woodchucks, which love to browse the leaves of young plants. But I have seen beet plants come back after the young beet tops were completely nipped off and yield a quite decent crop (as long as I solved the grazing problem). A greater concern is voles eating the mature beets. The crop often looks perfectly fine until you decide to harvest it and find a lot of hollowed-out or bottomless beets. The remedy is simple: Pay attention, and as soon as the beets are full-sized, pull them. Don't leave them sitting there just because there are a few weeks of mild weather left.

A note on storing beets: Cold is very good, but damp is essential. Beet crowns are prone to withering in open-air containers. Under proper conditions I have known people to throw away perfectly usable 12-month-old beets in order to make room for the new crop.

Beets are fine pickled, roasted, or shredded in salad, but we prefer a kind of borscht that is the centrepiece of our Winter Solstice feast. It is rich, redolent, vegan, and completely homegrown. You can find the recipe in chapter 17.

Chard, whether Swiss or Elbonian, is merely another form of *Beta vulgaris*—did you notice its seeds?—but instead of forming a nice neat root-ball chard puts its energy into lush foliage. More particularly the foliage has very thick fleshy stems, like celery, hence its name. *Chard*, or *carde* in Italian and Spanish, means . . . well, those kind of stems. Hint: The root word pops up again in a Flemish vegetable cousin of the artichoke, cardoon. To my personal taste I prefer the leaf blades of chard to the stem. Fortunately for people like me, there are some heritage varieties that grow that way, such as Perpetual Spinach and Cut and Come Again. The only tip I have for chard that's different from growing beets is to thin them further, at least 6 to 8 inches (15.2–20.3 cm) apart, to stimulate growth of bigger tops. Also, it may help to give a late-season side-dressing of green pee (described in the cucurbits section of this chapter) to boost languishing production—green pee is a no-no for beets, though.

Sweet Corn

I grow a few veggies that don't fit well into other categories. For example, corn is really a grass, but sweet corn is really something else again. It is one of the least eco-efficient plant foods I can imagine (which still puts it ahead of the most eco-efficient animal foods). About 3 square feet (0.3 sq m) per plant, most of which is stalks and leaves and cob,

Figure 8.29. Twinned ears are not uncommon with Baxter.

with a little pawful of immature kernels to justify all that space and fertility. Corn of any kind is a gross feeder, demanding all the nitrogen you can give it. It is the one crop on which we regularly use composted animal manure—our own. Sweet corn might be more wasteful of earth resources than, say, carrots or chard, but the taste alone would justify it all in my eyes, and anyway, I do a couple of things to mitigate the waste. First I never grow just corn by itself; I always interplant it with edamame—vegetable soybeans—which are highly eco-efficient (see chapter 14). Second I undersow ladino clover between the rows after hilling the corn. This gives me a great deal of high-nitrogen cover crop to replace what the corn is removing, so

Figure 8.30. This typically heavy crop of Baxter sweet corn will provide oodles of good old-fashioned corn flavour.

that I can follow it with other heavy-feeding crops, like potatoes, without additional fertilizer.

In the early years I usually grew Ashworth or Golden Bantam, which are excellent, but ever since I discovered Baxter I've had no interest in anything else. The Baxter canning company once had several "corn shops," or canneries, in northern New England, and this was the variety their growers used.

It is early and very productive, especially for a northern type. Many of the ears have a double tip, almost a double cob, sort of like a Popsicle. The heavy yield might almost compensate for less-than-stellar flavour, but it doesn't need to—Baxter has the richest, corniest flavour of any I've tasted. The modern corns with the sugar-enhanced gene are bred for those who mistake sugariness for flavour. Baxter is for those who remember and crave the taste of real corn.

Other than the issues of fertility and space, there is always the challenge of getting a sweet corn crop past the raccoons, which is really rather simple, as I'll explain in chapter 15, *Pests and Diseases*. Otherwise there is no reason why sweet corn should be a luxury. As a youth I remember visiting my grandparents in the country and seeing huge platters piled with fresh-picked corn-on-the-cob, a rare luxury for us townies who ate off-the-shelf. It was my first powerful impression of the gracious living available to those whose paths lead away from the marketplace.

Asparagus

Asparagus is another gourmet indulgence for those who live away from the land. Although it does not require much space, it requires forethought; you don't just stick in some seeds (as with radishes) and a few weeks later start bringing food to the table. Asparagus is a perennial; you need to buy plants, although you can grow your own from seed, which is less predictable than the original clone. That's how all the feral asparagus along railroad tracks and irrigation ditches came to be: from birds eating the fruits and then "sowing" the seed. It is quite acceptable eating. Once established an asparagus bed will produce more crowns, which you can use to expand the patch.

A peculiarity of asparagus is its tolerance, indeed its preference, for salt. It is native to the banks of the Danube, Don, and other Black Sea estuaries where it is occasionally flooded by brackish water, thus it has a big edge on the competition by utilizing what is poison for them. I know a man who every spring swept the winter's road sand out of his ditch and put it on his asparagus bed, where the mixture of sand and salt kept it vigourous for many years. What I do not know is whether or to what extent rock salt could be used as an herbicide, especially for quack grass, which is my main weed in asparagus. I never added enough to have that effect. I believe quack grass itself has some salt tolerance, which may explain its presence. Curiously, asparagus itself is actually very low in sodium. It also craves the other alkaline minerals, so wood ashes and lime are beneficial. When I planted my present asparagus patch several years ago, I made a trench and set the crowns at least 10 inches (25.4 cm) deep with lots of rubbishy material like twigs and wood chips, as recommended. I knew that when you bury woody stuff too deeply the anaerobic environment creates phenols and other slow-to-decay substances; however the asparagus seems to have no problem with that. After planting the crowns I only filled the trench half full, until the shoots came up through that, and then I filled it level. That's what the directions called for, but I'm not certain how important it is; in subsequent years those same plants sent their shoots straight to the surface from the full depth with no help from me.

My late friend Orlando Small used deep planting as part of his weed control strategy: Early in the spring he would run his rototiller right over the patch, well before the new spears had reached the surface. I've tried it with success, but timing is crucial, so as to do no damage. In addition to the original deep-set crowns, your crop depends on satellite crowns that spread out from there, and I have found that those may form at a somewhat shallower depth and thus be more vulnerable to damage. By the way I have been terrible about controlling weeds in my asparagus beds, and they are not mulched, yet they seem to yield plenty despite that. Still, the weed competition, especially quack grass, doubtless shortens the life of the beds. While I don't recommend my method or lack of it, it does suggest that asparagus is a resilient crop.

Aside from fertility a key to growing good-sized spears is harvest management. Remember that asparagus isn't producing those spears for you, at least not intentionally. Its purpose is to strengthen and propagate itself, but it can do both, as long as you're not too heavy-handed. It's all in the management: For the first two or three years don't pick it at all, let it grow strong. When it does begin to send up fatter spears, harvest them sparingly and leave the slender ones to grow tall and branch and send photosynthates back into the crowns, further empowering them for another year. Eventually you'll get mostly fat spears, but even then, as the season progresses, you'll notice a tapering off of size. This tells you it's time to let them take care of their own needs now.

Weeds as Vegetables

In addition to pigweed and lamb's-quarters at least two other weeds have domestic counterparts in my garden. Purslane and sorrel, the pests, share a cultivated space with purslane and sorrel, the gourmet vegetables, and the main distinction is that the latter have achieved greater size and upright habit; regarding palatability there is little difference.

Purslane

In the case of purslane I'm just as glad that it still has a tendency to sprawl and form mats, as that makes it work well as a companion crop. It can creep among its neighbours without competing for nutrients, because its own root zone is several inches away from there. As for its top growth tending to smother others, that's easily dealt with by harvesting: Cut it back to its own space. Purslane is a succulent, having some of those emollients that aid digestion and cool the body, much like cucumbers, borage, and watermelon. It is also a rich source of some essential minor nutrients, like omega-3 oils. A professional nutritional adviser once assured me that those had to come from fish oil, so vegans are in big trouble. Of course I never actually saw an omega-3 oil, so I had to take his word for it. I did wonder, however, how the Tuareg or the Kickapoo, who rarely if ever saw a fish, managed to get their omega-3 oils. Now I know: They obviously ate massive amounts of purslane.

What I'm still experimenting with is just which crops companion well with purslane. If hints from nature count for anything, then beets may be a special candidate, seeing how wild purslane seems to prosper among them and vice versa. I suspect it may have something to do with how they manage their water needs, perhaps to mutual benefit or at least with minimal interference.

Most people like purslane raw in salads, and I don't mind it, either, although so far I prefer it lightly steamed or added to stir-fries; it is a shame to destroy that mild crunchiness by overcooking. It seems to blend well with other stronger-flavoured greens like (well, whadyaknow!) beet greens or chard.

Sorrel

Cultivated, or French, sorrel has a different growth habit from its low-class cousin, wild or sheep sorrel. Instead of the shallow rubbery roots that make the weed such a nuisance, the cultivated form has an almost woody, more compact root structure. I recently transplanted some old sorrel plants, expecting failure, yet was completely successful. Sorrel propagates easily from teeny triangular black seeds, shaped like buckwheat, to which it is kin. Both wild and cultivated sorrel are classed as *Rumex acetosa*, yet I've seen no sign of mine crossing, which I assume would be obvious. The varieties I know are Belleville and Blonde de Lyon, both excellent.

I'm not really big on eating lots of sorrel by itself, but it adds a nice sour accent to blends with other steamed greens. Particularly in Near Eastern dishes, spinach is often served with lemon juice (my hometown has a large Lebanese population, so I ate lots of that stuff even before I was into natural foods). I cannot grow lemons, so if I wish to be ultra-self-reliant, then sorrel is really appreciated. Also, I usually prefer chard over spinach (because it's much easier to grow, and because chard and sorrel mature together long after spinach has gone to seed). A very filling and satisfying one-piece meal can be made by mixing the sorrel-chard blend with mashed chickpeas (also homegrown—see chapter 10) and stuffing a pita.

Cultivated Burdock

There are certain plants in our food system which are technically wild, yet we semi-cultivate them. Now I am no Euell Gibbons, but a few of them are just too important to ignore. Burdock, also called gobo, grows wild in the neighbourhood, some of it gone wild from a Japanese variety, Ouro, which I planted many years ago. I also grow the commonly available variety Takinogawa, but the fact that the heavy taproots can reach down 2 to 4 feet (0.6–1.2 m) isn't very helpful in my shallow, rocky soil. If I plant it in my very deepest-dug soil, perhaps 18 inches (45.7 cm) deep, I still must waste most of the crop. The reason I originally planted Ouro is that it's much shorter but much thicker; I hope to get it back sometime. I'm not a fan of foods just because they happen to be weird or exotic, but the long, delving root tells me that it must be pretty rich in various minerals, and anyway I like it sliced up thin and diagonally and added to stir-fries. Aside from the need for very deep, friable soil, gobo is very easy to grow, drought-tolerant, and moderate in its fertility requirements. I plant early for large roots.

Condiments and Spices

Condiments and spices loom rather low on the list of priorities for self-reliance. When the first-century rabbi Yeshua ben-Yosef spoke of "tithing mint and dill and cumin," he wasn't disparaging those condiments so much as pointing out their peripheral role in the diet, indeed in life. But they do add an "accent," a point of interest to staple food items that may bore our palates with starch and protein, and as such, I find it worthwhile to grow some. They're not vegetable crops, but I tend some of them in my vegetable garden, so this seems the best place in the book to discuss them.

Some spices and seasonings have been given much more importance than they seem to merit. Wars have been waged over access to salt and pepper and cinnamon and cloves, right along with whiskey and opium. But before we allow such condiments to be taken so seriously, we should rather consider whether we can provide ourselves some satisfactory alternatives. My farm has never produced a decent crop of nutmeg or cardamom, but there are some homegrown substitutes that I find quite acceptable. They do not necessarily taste like the tropical spices do, but they work in similar kinds of dishes. For example, coriander is often used in savoury dishes like curries, but I find it even more appropriate in applesauce and stewed fruit compotes. Sweet puddings that usually call for vanilla get an aromatic sweetness from dried and crushed rose petals. We have found that ground shepherd's purse seed reminds us of black pepper. I've read that shepherd's purse seed is bad for the kidneys if used in large quantities, yet who would want to use it in large quantities, any more than black pepper? By the way black pepper itself is not exactly a wholesome item. It is what some people consider a harsh stimulant (like ginger, turmeric, cinnamon, et al.), which is after all why we use them: to stimulate our taste buds to trigger an increased flow of saliva, helping us digest bland foods. Some of these spices really overdo it, and wear out the sensitive organs that are constantly exposed to the irritation. For example, turmeric contains colchicine,

a chromosome-distorting substance used to make interspecific crosses. Black pepper contains compounds contributing to pipestem arteries (whereas *red* pepper—cayenne—is said to be a preventive). The northern spices I have in mind are not as strongly stimulating, but neither are they harsh on our systems. I don't know that any of them make any great nutritional contribution to our diet, any more than a winter bouquet makes the house warmer.

I find it interesting that so many of these flavourful species are members of the Umbelliferae (parsley family), including dill, coriander, caraway, cumin, anise, fennel, and sweet cicely. These are all characterized by various terpenes, yet each particular group of terpenes is so distinct that you can recognize each flavour at first taste. One of the benefits of all these terpenes is that they act as a kind of antifreeze, enabling us to sow these herbs very early, which indeed is when they do their best.

The annuals dill and coriander are extremely easy to grow; earlier in this chapter, I described how I companion them with other crops, especially beets. A major way I use ground coriander is in a mixture we call baloney seasoning. When I first became vegetarian, I didn't miss steak in the slightest (we rarely ate it anyway), but sausage was another matter. I realized that what I missed wasn't the taste of meat but rather the greasy, salty, spicy taste of the various sausages, particularly Bologna sausage, or baloney. When staying in Amish country I had discovered that the signature taste of summer sausage (basically baloney) is a combination of coriander, paprika, garlic powder, and sugar (mustard and black pepper are optional and less essential). Any meat analogue I made with those ingredients (say with tempeh or soy burgers) completely satisfied any "sausage" cravings I thought I had.

Caraway is also easy to grow, though as a biennial it must be overwintered; it will make seed in the second year. That's easily done, given its great hardiness, although I wonder whether it could, like parsley, be sown rather late in the first year and just get itself established before winter. That might allow an early crop (like spinach?) to precede it and perhaps a late crop to succeed it in the second year, instead of having it occupy garden space for two full seasons (the seed ripens and is harvested fairly early).

Cumin, anise, and fennel seed are annuals, yet I've had some difficulty with them, I think because they need to be started very early in the Far North. I remember in our neighbourhood in San Francisco, you could tell where Italians had lived by the gone-weedy fennel that crowded up through cracks in the sidewalk. I use fennel seed (as opposed to the bulb type) only in spaghetti sauce and occasionally with its old garden friend, beets. Anise is delightful in anything sweet. Son Kindle and I are partial to it with applesauce and sunflower meal on toast.

There is one umbel species that I never bring into the house or combine with anything else, and that is sweet cicely. I have this hardy perennial growing near a ditch beside the nursery, and whenever I pass it on a hot day, I pull off a few green seeds and chew them for a licorice-like rush of cool sweetness. They say you can preserve the green seeds in sugar for a special treat; I can well believe it, though I've never bothered. When gone by the seeds become very fibrous and lose most of that exciting flavour.

Regarding common culinary herbs I have little to add that isn't obvious. Some of the herbs we use in greatest quantities—basil, summer savoury, parsley, and dill—are annuals that fit in well with our regular beds, companioned with other food crops. However, several of the herbs we use are perennials that don't incorporate with the constantly rotating crops, they require a permanent place of their own. The perennial herb garden is a small terraced area close to the perennial flower garden. Its narrow rock-walled terraces slope steeply southward and catch plenty of sun. Those herbs include sage, thyme, oregano, winter savoury, chives, and a few medicinals such as lavender and lady's mantle. Though they are mostly of Mediterranean origin, they all seem fully hardy here in the Western Maine Mountains. We mulch them with shredded leaves, but that is for weed control, not winter protection. The tiny space (less than 300 square feet, or 27.9 sq m) produces all we can use, and we use them quite freely. We make a point of not fertilizing them heavily, especially with nitrogen, since lush growth is at the expense of aromatic oils, which is mainly why we want them.

Three of my "alternative" spices are crucifers, a family known for its pungency. I've already mentioned shepherd's purse, which is a common weed, but a few times I have sown it deliberately for its peppery seeds. In fact the only difficulty in growing it is that the seed ripens very fast and shatters easily, sometimes before you get around to harvest; it bears watching. Mustard is a well-known condiment; it is as easy to grow as rape, and fully as vulnerable to flea beetles; be prepared. I grow the French variety Burgonde.

Horseradish is a hardy perennial that I grow in a plot of its own; a small plot, as I'm the only one who eats it and my needs are very moderate. I was told that a common way of adulterating commercial horseradish was to mix it with a larger quantity of shredded turnip. I've tried that, plus salt and vinegar, and I would say that "adulteration" could be considered more of a value-added than a fraud. I use it mostly on soy burgers, often further thinned with ketchup.

Ideally, I grow two short rows of horseradish and, when I am systematic, let each one grow for two years before harvesting, alternate rows in alternate years. In reality I may let plants grow several years before digging and grating them, which results in a coarser, more fibrous sauce.

My first year on my land, neighbour Lucian shared some cooked spring horseradish greens from the patch behind his house. They were delicious, and I have tried in vain to duplicate them, but mine are always excessively pungent. I suppose Lucian either managed to get them earlier or else poured off the cooking water at least once.

My wife has a small patch of Canada snakeroot (*Asarum canadense*), or wild ginger, growing in a shade garden in the hedgerow. Occasionally I pull up one of the small runner-roots to nibble, and they are indeed strongly suggestive of ginger, though sometimes bitter. I have tried to chop some finely into stir-fries, but the ginger taste doesn't come through as well when cooked, or my timing may have been off. I should try them in spring. Snakeroot is reported to be mildly toxic to the kidneys, but the real ginger is pretty harsh, too. I advise against using either in any great quantity.

One of the spices I grow—fenugreek—is actually a legume, though that's not obvious until you grow it. In fact southern farmers sometimes use fenugreek as a green manure. I haven't grown it for many years, but I believe I planted it after the last frost, like dry beans. Apparently some varieties don't mature here; I guess I lucked out with generic seed off the spice rack. Ground and lightly roasted fenugreek seeds give a very distinctive flavour to curries, which is how I've used it. Commercially it is used to produce artificial maple syrup. I've heard of sprouting fenugreek, but never tried it. Since dry raw fenugreek seed is very hard to grind, I can only assume that it softens up considerably in the sprouting process. As it matures it must be watched very carefully; the many sabre-shaped pods are often well hidden in the foliage and may shatter before you realize they're ready.

CHAPTER NINE

Grains

It's impossible even to contemplate a self-sustaining garden that does not include grain. It is the main nutritional staple that has sustained humanity for many millennia. Eating grains coupled with pulses (beans, peas, and the like) completely removes any need for meat in the diet. I emphasize the combination because each food type (grain and legume) is rich in its own set of amino acids, which together supply a complete balance of protein. But grain is also especially rich in energy-giving starches; the only thing remotely like it is potatoes, and they are an imperfect substitute.

It's almost impossible to discuss the role of grains in the diet without mentioning macrobiotics. I must admit that I'm not much of a traditionalist. I consider myself a rationalist, a radical even, happy to retain and embrace any tradition that makes sense, not just because it's a tradition. The focus of the macrobiotic system is on eating whole grains that have not been milled or broken in any way. And what doesn't make sense about that? What's wrong with chewing your food carefully instead of bolting it down? And this stuff they tell me about eating mainly foods that are grown in your region and in season: if that isn't preaching to the choir! I'm less convinced that rice is the ideal perfectly balanced (male/female) food that should be at the centre of your diet no matter where you live. Don't get me wrong, I love rice. I'm only suggesting that we should accept new ways because

they are reasonable, not because someone else has been doing them for a long time.

All that being said, cooking and eating really whole grains is kind of a no-brainer, whether instead of breadlike things or in addition to them. This idea clearly arose from the fact that rice is low-gluten and doesn't work as bread, but is light and tender and delicious in its whole form. The other grains are not quite like that, but the same principles can be applied, especially the part about long, slow cooking. For years I tried to develop a liking for whole-boiled barley, wheat, oats, and rye, and while I love the taste of them all, the chewiness was rather off-putting. A macro friend, who was more open-minded than I, suggested I cook them macro-style, that is, very long and slowly. I have done that, and while I cannot say they're quite as tender as rice, I must admit that he was much more right than he was wrong. Incidentally cracking grains very coarsely is sometimes a reasonable compromise, still with the long slow cooking. Treating all the grains like rice offers a much greater scope for using them in a new (our own) culinary tradition, as opposed to their conventional uses. No longer is wheat only all about bread, oats are not merely a breakfast cereal, rye is more than crackers, and so forth. Even in regions where rice cannot realistically be grown (I grow it on my farm, but not realistically), it points to new possibilities for the other grains. Three cheers for the macros!

The Wrong Way to Grow Grain

There's no denying that as a nation we grow too much grain. Modern industrial agriculture has gotten so good at producing grain surpluses, especially wheat and corn, that we cannot begin to eat it all ourselves, so we feed it to livestock, animals that are not evolved to eat grain and whose digestive tracts are harmed by it, making them fatter but sicker (and those who eat that meat . . . ?). For that matter our grain is increasingly harmful to ourselves and to our land. Aside from the heavy burden corn puts on the land (and the sea: much of the pollution in the Gulf of Mexico is chemical runoff from the midwestern Corn Belt), it has become the main ingredient in our store-bought diet. While modern Americans eat rather little corn in its recognizable state, they consume vast amounts in the form of high-fructose sweeteners, modified starch, and alcohol, to say nothing of meat and exports to feed foreign livestock.

Likewise most commercially produced grain is used as livestock feed, but even that which we eat directly seems to be harming us. The great increase in wheat allergy incidence seems to be telling us something, though no one seems clear exactly what. Gluten intolerance, as in celiac or sprue, is one identifiable problem, but there is also an alarming increase in allergies to other wheat proteins or something else altogether. But why now? What's new in the way that wheat is grown and processed? We know that prior to the 1870s wheat was harvested and milled in completely different ways. The replacement of stook-curing by the combine, and of stone-ground by roller-milled, have made radical changes in the nature of our bread (not to mention bromating and various additives). Modern wheats have been bred

Figure 9.1. Having a variety of homegrown grains is pivotal for a self-reliant gardener.

with very different priorities from older types—bred for traits that may be important to the grower, the miller, the baker, and the distributor, but perhaps problematic for whoever finally eats the loaf. Is that part of the problem? Or is it just that we are eating too darned much wheat at the expense of everything else? Like corn, it is in practically every processed food we buy and always in some highly fractionated form: malt syrup, modified starch, gluten.

Americans today eat much less oats, barley, rye, and buckwheat, than our grandparents did; are our tummies rebelling against the nutritional monotony of it all? Are any or all of these things factors in what I call The Great Wheat Malaise? I certainly do not know, and I may never find out: Homegrown wheat and other grains used to be a bigger part of my diet, and I'm struggling to move back in that direction. Said wheat is mostly of older, traditional varieties that perform more reliably in my system with fewer inputs; it is hand-mown and stook-cured (more on that later), ground fresh in a hand mill as I need it, and fresh-baked with only water, yeast, and salt added. Preservatives? Quelle question!

There is also the little matter of independence, of which self-reliance is the cornerstone. No matter how much of my veggies I grow, without staples I'm still wholly at the mercy of the marketplace (which, if you haven't noticed, is not very merciful).

Grains may currently be the cheapest items in my diet to purchase, but they're also the most basic. Do I really want to be completely dependent on strangers for something so indispensable to my life and happiness?

I should emphasize that *any* milled grain product is far tastier and more wholesome when freshly cracked or ground on demand, as the oil in the germ starts to go rancid very quickly. Granted that processing small amounts of grain right when we want to cook it can be a nuisance, the difference in quality may offset the inconvenience. After all, it was the removal of the nutrient-rich germ that made industrial-scale milling, shipping, and storage of flour feasible.

There's been lots of discussion in the media about future global shortages of all grains and the spectre of Third World famine and political chaos washing up on our shores (North America is where much of the grain is grown, after all). And so forth: global warming and all the *Titanic* scenarios mentioned earlier; the sky *is* falling in, just ask the polar bears. Isn't it about time we took a closer look beyond our gardens' supposed borders?

There are other reasons, less dramatic perhaps but compelling nonetheless, for incorporating grain in your garden: For one thing it is aesthetically attractive. Even a few square feet of the various grains tucked among your perennial beds can offer a stunning contrast with the surrounding broadleaf flowers. The muted grey-green tones and close vertical stems help show off their bright-blooming neighbours to great advantage. Even after harvest, the grain heads add enormous textural variety to dried arrangements with long black awns (whiskers), reddish oats, bluish wheats, and more. I have often seen bouquets of these ornamental grains featured all by themselves. Eli Rogosa, director of the Heritage Grain Conservancy, has a great flair for making a mere grain sample look like a prom corsage.

Specialty grains are also useful for strawcraft, including the British folk tradition of corn dollies. "Corn" of course means any grain, and wheat straw is favoured for braiding fancy mats, doilies, hangings, et cetera, the heads sometimes left on for effect. The Germans used oat straw to craft stars and snowflakes for Christmas tree ornaments.

There are actually grain varieties especially suited for strawcraft: Squarehead Masters wheat comes to mind. If that sounds absurd keep in mind that straw thatch was an essential construction material (along with slate, tile, and cedar) in the days before asphalt roofing. According to my neighbour Lucian, rye straw used to be preferred for stuffing horse collars because it resisted compression.

Speaking of straw, one of the goals of the Green Revolution was to breed dwarf varieties of grain that would resist lodging (laying over), and not waste photosynthates (the material products of soil alchemy) on making straw. It assumed that the grain was the only desired product and the straw was mere waste, a bulky nuisance. However, for most of humanity's farming history, straw has been a significant part of livestock fodder and bedding and an all-important factor in maintaining soil fertility (unlike most crops, grains return to the soil a significant part of what it took to make them). In *Farmers of Forty Centuries*, F. H. King points out that the Chinese always expressed their rice yields as two numbers: the yield of grain per acre and the yield of straw, the latter being nearly as important. In contrast Green Revolution breeders took it for granted that soil tilth was poured from a fertilizer bag, and that draft animals would be replaced by American-made tractors and imported herbicides.

To pursue that critique a bit further, I would point out that the Green Revolution was a programme initiated by the Ford and Rockefeller Foundations to counter socialist trends in the Third World. Even assuming that those board members were motivated by a sincere desire to improve the lot of Third World farmers (and I am not cynical enough to assume otherwise), they were nevertheless guided by a left-brained notion of what is needed, an unsustainable vision. Likewise Heifer Project: When people are poor and hungry because their land has been degraded by the rooting and grazing of livestock, let's solve their problems for them by sending them *more* livestock; and not the scrawny scrub local breeds that can cope with that resource-poor landscape, but improved American breeds that will demand yet more from those barren hillsides. Okay, so much for the discussion of straw.

The methods I'm about to describe are *not* appropriate for large-scale production but may contribute a lot toward your personal security. An important aspect of grain growing is the interplanting with other crops, either for food or for green manure. I've already dwelt on those in the appropriate sections.

Planting a Grain Crop

The first step toward planting grains is to procure seeds. I would point out some issues of varietal adaptability that can apply to any grain. When you buy grain seed from your seed company, the variety is not necessarily the best adapted for your conditions;

indeed it almost certainly is *not*. Seed companies do not produce the seed themselves; they buy it in from some grain dealer in the Midwest. It's whatever variety farmers out there are growing. For me that's a problem, because prairie soils have a very different character from my Western Maine Mountain soil, which was formed by forests. The prairie soils are warmer, drier, more alkaline soils based largely on bacterial decomposition of grasses and herbaceous perennials, whereas our northeastern soils are cooler, wetter, and more acidic, having been originally built up by fungal decay of woody residues. Two centuries of cultivation have radically altered them, but they are still not prairie soils; the imprint of the forest remains upon them, and any grain variety that's uniquely adapted to one ecotype will not fare as well in the other. Generations of US grain breeders have focused completely on varieties for the Grain Belt, so what has happened to those varieties that were once grown in places like Maine? Mostly extinct, I'm sorry to say; after all, who has been growing wheat here since the Civil War? A few dabblers like me, but with no continuity to sustain those old land races; they are largely gone forever. Later in this chapter I discuss variety choices for each of the grain crops that I grow. Some are available from commercial suppliers, some from me, and some from the National Germplasm System.

I do not have special areas for growing grain and others for veggies and pulses. I rotate all of these crops with one another, with several benefits. Vegetables, grains, and pulses are very different *types* of plants: The "true" grains are grasses, monocots, with a very ancient evolutionary background, as opposed to the veggies, most of which are dicots. The grains produce significant amounts of soil-building biomass in proportion to the seed, whereas most of the succulent vegetables give back little for what they take out. Soil can get "sick" of either crop type, I suppose because they each keep taking out the same blend of nutrients and building up a soil biota that may become imbalanced. Rotation gives all players (grains, veggies, soil) a break from one another.

Rotation properly done also breaks weed cycles, which is important when growing grains. If you wait

Too Many Farmers

Many occupations have systems of certification, licensing, regulation, and so on, whose real purpose (ostensibly to guarantee quality and safety) is to limit the field of practitioners to less than what the market demands. It's an ingenious policy harking back to the medieval guilds, and it guarantees a fair (or perhaps better-than-fair) remuneration to the elite group of professionals. Today this includes not only lawyers and chiropractors, but pet groomers and hairstylists. Now, the farmer is surely a professional. Yet he does not have a closed guild, and there certainly are too many of them. What! Too many farmers; how could that be possible? Just imagine if for every farmer in the United States, there were several dozen movie critics, scores of neurosurgeons, whole villages of futures traders. If that were so, a bunch of radishes would cost more than having a will prepared, a jar of pickles would buy a kidney transplant or face-lift, a loaf of bread would be enough to hire a personal money manager—and you'd need one.

Too many farmers at the marketplace is bad for all of them. For all of the inputs—seed, energy, fertilizer, pest control (even organic growers have that expense)—they rely on the marketplace. No point in haggling there, when every other farmer is standing in line with the same needs. Then to sell their crop they go back to the same marketplace and queue up with all the others selling the same stuff.

They cannot expect to demand a fair reward for the surplus products they have so zealously produced; there are too many others in the same boat. The only farmers who have the faintest hope of winning at that game are those who are not selling corn or wheat or soybeans. Try growing organic squash blossoms (for stuffing) or dried herbs (preferably as tinctures or other value-added products), or maybe grow crops for a niche-filled CSA (lots of fresh basil and arugula). For the time being anyway you may be kept afloat by the growing interest in organic and local. It's a very healthy, positive trend, but it's still too niche-y and I don't wish to depend on it. I'm truly glad to see folks finding better roads to the marketplace, but I'm still seeking paths that lead *away* from there.

Who then will grow the staple foods needed by the masses? I say let the marketplace worry about that. When there are fewer farmers selling grain in the market, those who remain will be more fairly compensated. Bottom line: I'm very skeptical of taking grain to market; for most of us it makes more sense to stay home and eat it there. Wheat is a natural choice, since wheat and bread are so important in Western culture. But there are plenty of other grains that can add variety to our diet, and it is absurd not to grow and use them, especially since many people find them more tummy-friendly.

to deal with the weeds *after* they're in the crop, you have an expensive proposition: expensive in labour if you remove them or expensive in lost yield and quality if you don't. So let me describe some pre-planting strategies I use to manage weed populations while building fertility.

I could use my veganic compost to fertilize grain plots, and at times I have done so, though I prefer not to. I need all my compost for my more demanding vegetable crops; in addition, the one weed that

seems to survive the compost process unscathed—lamb's-quarters—is also one of the most problematic to have in the grain crop, especially at harvest. If I am to use compost I prefer to apply it in autumn for a spring-sown crop, giving the weeds time to sprout and get killed, either by winter or by my spring cultivation. Generally, though, I prefer to rely on rotation for grain bed fertility, and here's one of my favourites.

Where any garden crops have been harvested early (field peas, rape, flax, early potatoes, what have you), I could immediately sow a crop of winter wheat (or rye) to make maximum use of the season. However, when I anticipate a weed problem, I immediately after-sow a catch-crop of oats with field peas. They grow into late fall, but die and decay in time to offer fertility and weedlessness for next spring's early-sown barley, wheat, or oats.

Sowing the Crop

So let's talk about planting. The method I use is the same for most of the grains I grow: wheat, oats, barley, and rye, as well as spelt, durum, and emmer. Early planting is good; earlier is better; here's why. Grains are very photoperiodic, or day-length-sensitive. When the days begin to shorten (after June 22), they feel an irresistible urge to bolt, or go to seed. That of course is exactly what we want, but not until the plant has been able to bulk up a bit by making a certain number of leaves. If I plant a wheat or barley seed on June 1, it will have plenty of season in which to grow and mature, but only three weeks to develop enough plant mass to sustain a crop; it will be a very meagre crop. Boost the fertility to your heart's content, water it daily, chant mantras over it, all to no avail; the plant has its eye on the calendar and so must you. No worries; there's no such thing as *too* early. Those species are all extremely hardy and will not self-activate until *their* instincts (much better than yours or mine) tell them it is safe to get sprouting.

The most ancient low-tech way of planting is by broadcasting; it is also the most wasteful and inefficient way. I have lots of experience sowing green manures that way, but for food grains you want the right density and consistency. Thinly seeded spots will grow weeds; sowing too thickly wastes seed and gives poor quality and yield due to overcrowding. The best way to precision-plant is by drilling. That means seeding in close rows like carrots, which is usually done with a tractor attachment (a drill box), but you and I can do a splendid job on our small plots using a hand-seeder such as an Earthway or Planet Jr. I use the former and it does a fine job except for one problem: It comes with several feed plates for different-sized seeds, but none is intended for small

grain; clearly the manufacturers don't expect gardeners to use them for that (you and I must prove them wrong). The closest size plate is the one marked for beets, chard, spinach, and okra, although the number of feed buckets per revolution needs tweaking. I used to feel that the spacing was a bit too sparse, sometimes leaving gaps of a foot or more. My solution to that was to go back and forth twice, figuring that it would average out, but I'm coming to the conclusion that a thinner spacing would be better, to encourage tillering. Tillers are analogous to strawberry runners but shorter; they grow out from the parent seedling and root themselves while sending up yet more shoots. Thus each seedling becomes a clump of several stalks, each bearing a head of grain. The idea is that with thinner spacings (some say as little as one seed per square foot, or 0.09 sq m!), you get much more tillering and thus more yield per plant, but also more overall yield per acre. I'm not quite sure I understand that last part and indeed some researchers refute it, claiming that, in our climate at least, the tillered heads do not mature as well or as full. They advocate the opposite extreme, with drills as close as 4 inches (10.2 cm). That's all hearsay to me, as I've never tried the extreme spacings, but I have found that 8 inches (20.3 cm) between drills and 3 to 6 inches (7.6–15.2 cm) between seeds seems to work very well for me. I imagine that much greater spacings would risk weed problems; however I also know a possible remedy for that, which I describe below.

I typically grow four to eight beds of each grain type, depending mainly on how much we use. I've never grown even half of the wheat we consume in a year; for the other grains we are more fully self-sufficient. Each bed is 6 × 40 feet (1.8 × 12.2 m), so each bed is 240 square feet (22.3 sq m), or approximately 1/180 of an acre (0.002 ha), handy for extrapolating yields. Here is how I lay out the area for drilling: At each end of the plot I put in marker sticks 8 inches (20.3 cm) apart. (I use old label stakes, but have even used twigs.) I use a string of corresponding length and tie each end to a pointed peg stick. I stretch out the string and insert the sticks at each end of the plot, not at the #1 stakes, but rather at

Figure 9.2. When I prepare a grain plot for seeding, I mark both ends of every row with stakes and then stretch a length of string between every third set of stakes, starting with row #2. The string serves as a visual guide while I walk rows #1 and #3, sowing seed with my Earthway seeder.

the #2 stakes, as you'll soon understand. With stakes and string in place I'm ready to plant.

I don't need to make a furrow; the seeder has a little projecting trough that slices the soil, drops the seed, and buries it all in one neat operation. I hardly get to see the seed, so I am grateful for the little *tick-tick* sound that assures me it's dropping as desired. Or not; if the soil is at all sticky, it tends to plug up the trough end, and then the seed drops in irregular clumps or not at all. Now, if I were to push the seeder along the marker string, I would be constantly shoving the string aside and wouldn't be making a straight furrow. Rather I plant row #1 by eye, just rolling along parallel to and 8 inches (20.3 cm) away from the string that's stretched between the stakes of row #2. The flat wheel print shows pretty clearly where I have been. Next I can just as easily sow the #3 row on the other side of the string. I can then move the string from row #2 to #5, to guide me while I plant rows #4 and #6. But first I must go back and plant #2, now unmarked by a string, but obvious enough in relation to the wheel prints of #1 and #3. And so on across the piece. I like this method because it would be tedious to move the string every row. This seeding method works quickly and easily. It is certainly not as fast as a tractor-mounted drill that can seed many rows at each pass, but it is vastly cheaper and more precise.

How deep? Grain can be sown from ½ inch to 2 inches (1.3–5.1 cm) deep, probably even more or less, but that's not the point. It's more important to be as consistent as possible, so the seedlings emerge uniformly. Planting too shallowly leaves the seed at risk of drying out or being stolen by birds; too deep may delay emergence and result in uneven ripening. Once planted it's helpful to roll a barrel or lawn roller over the plot to firm up the seedbed, unless your soil is naturally heavy and prone to compaction. Another reason for rolling is to smooth the surface of the bed.

The seeder wheel leaves a slight depression that can be annoying some weeks later when I spread mulch (as I'll explain below). Until the shredded leaves have settled into place (a couple of showers makes them mat nicely), they're easily blown around. If the seed rows are a little sunken, they can fill up with mulch, leaving bald spots in between rows, which defeats the purpose of the mulch.

Mulching After Planting

I've always had some puritanical resistance to naked soil; after all, if our croplands are to mimic any natural ecosystem, should it be desert? But how do you mulch crops that are as close-spaced as our intensive vegetables are, some as close as 8 inches (20.3 cm)? We long ago devised the system of using shredded hardwood leaves, not carefully placed but strewn like a blizzard of organic confetti from a trash can atop my shoulder. It worked splendidly, creating an efficient and attractive blanket overall. Well, one day my wife Molly asked, "What if you were to mulch your grains—like oats—that same way?" With a benign smile and a patronizing chuckle, I assured her that it wouldn't work. "But why not?" "Because it's just not right, nobody mulches grain." "What about that guy Fukuoka with the teeny mud balls?" "Well, that's in Japan and this is America, and that's rice and this

Figure 9.3. Thanks to a mulch of shredded leaves, this crop is weed-free at every stage.

is oats, and he's trippy and I'm sensible and . . ." She, being Molly, persisted, and I, being me, stubbornly resisted, but eventually I tried it just to prove her wrong. Treated the whole patch like it was carrots or onions, practically buried the little blades of grass, but not quite. They kept poking up through, grateful to have their toes buried in cool, moist mulch. Oat heaven it was to them; they shot up tall and stout and made the most fulsome crop of plump, heavy seed I ever saw, shade-smothering any weeds that dared show. I've mulched my grains ever since, save a couple of occasions when I neglected to get to it early and regretted so afterward. Yep, I'm totally convinced that's one of the best ideas I've ever had.

Quantities? I am not so good at keeping track, but here are some rough numbers: I fill a large (30-gallon, or 113.6 l) galvanized trash can with as much leaf confetti as I can easily stuff in there. About three of these will cover a 6 × 40-foot (1.8 × 12.2 m) bed (remember, I plant the paths, too); adding still more mulch a week or two later would probably be a clever thing to do, though I rarely get back to it.

When the leaves contain some twigs the woody bits help hold everything in place. The delicate spears should be at least 2 inches (5.1 cm) tall when mulched so you won't risk smothering them. They can poke up through a thin layer, but it's much better if you can see them when you're done. Even if you feel you cannot put on as thick a layer as you might wish, it's better not to delay for the sake of weed suppression; you can always add some more next week.

Leaf mulch may be a radically improved way of producing high-quality grain, but don't look for it to catch on with the big growers; I just can't picture a tractor farmer with 40 acres (16.2 ha), or even 4 (1.6 ha), doing the mulch thing, even if his farm lies amid forested country. But that's no reason why we can't do our own that way.

After you've mulched a plot of grains there's not much to do but relax or work on the more demanding vegetables. If the mulching was well done there should be no weeding, none. Mulch won't suppress milkweed or quack grass, but if those are there you shouldn't be sowing grain until they're under

control. However, if the occasional pigweed or dandelion should erupt through a skimpy spot, there's no reason you can't walk in there and yank it out. The 8-inch (20.3 cm) row spacing should be enough to allow you to step between the rows of young plants, but don't fret too much if you trample some; after all, it's grass, right? It might not abide trampling as well as your lawn, but neither will it lay over and croak at one treading. The point is you shouldn't *need* to waste much or any time on hand-weeding; pre-management and mulch should have taken care of that.

Now that you know how to plant grains, let's look in more detail at specific grain crops.

Wheat

Whenever I am asked to speak at conferences on grain growing (which is often, as the topic is wicked trendy), what is tacitly assumed is that "grain" means wheat. In fact there are 15 or 20 other grains that we could as well be addressing, but no, it's all about wheat. Understandably, perhaps, since wheat is the main grain most people in the West eat, and has been for many millennia, whenever there was a choice. That, too, is understandable; since the invention of bread, wheat has been the most convenient (even before *sliced* bread) grain to eat. Wherever rice will

Figure 9.4. Wheat: the staff of life, or just one of them?

grow, it has always been very popular for its light, tender texture and simple preparation, but in the cooler climates where Western agriculture has prevailed, rice is less well adapted; hence wheat.

I would submit that bread was the original meat analogue. Even among pre-agricultural hunter-gatherers, it was surely noticed that certain glutenous grass seeds could be ground, shaped, and roasted to make a chewy filling staple, especially in times of scant hunting. It is to gluten that wheat owes much of its popularity. With the discovery of fermentation, which aerated the bread dough while converting some starches to more digestible forms, the gluten level became especially crucial because gluten imparted a viscosity that trapped the gas bubbles in the dough and resulted in a lighter loaf.

Types of Wheat

Speaking of gluten, that's the main reason we must speak of two wheat types, often described as hard and soft. The "hard" refers to high gluten level, desirable for yeast-bread making. The "soft" refers to high starch content (in proportion to gluten), desirable for cookies and cakes, and therefore called pastry wheat. We also generally associate hard with red and soft with white, although that's where the complication arises. The reddish-brown colour is from pigments in the bran, compared with the pale amber of the "white." The gluten and pigment are not necessarily linked; indeed there are some hard whites, marketed as golden flour, I suppose for people who want their whole wheat loaves to look more like puff bread. Still, you mostly hear about hard red and soft white wheat.

Another distinction among wheats (as with barley) is whether they are bearded (having awns or whiskers). This seems to be a negligible consideration, but I find it amusing that the iconic sheaf of wheat portrayed on so many bread wrappers invariably has awns, which are more picturesque than the actual grain contained in that loaf, which more than likely was awnless and was certainly never in a sheaf. The only advantage I can see in awned types is that the heads are held more separate when curing in the shock, or stook, which may be conducive to better drying.

The other main distinction, and the far more important one, between types is winter and spring wheats. Spring wheats act like most other crops: You sow them in early spring (the earlier the better) and harvest in early autumn. Winter wheats, however, are sown in late summer or early autumn. They make a modest growth before gathering winter cold shuts them down. They are very hardy and simply go dormant under the snow—hopefully plenty of snow, to protect from alternate freezing and thawing, the bane of winter wheat. I should mention here that although most winter wheats are hardwired for this behaviour, some are not. For example, if you were to sow a real winter wheat in early spring, it would not bolt but simply make grassy crowns with tillers, just as if you had sown it in August. However there are also "facultative" wheats that can go either way, depending on when you sow them. An example is the traditional hard red wheat Red Fife.

Winter wheats are potentially better yielding, since they get established in late fall, then make most of their massive growth in very early spring, when the cool, wet weather is most to their liking. That said, they suffer badly when winter snows are delayed, and they must endure frost heaving that exposes and damages their young roots. The cold they don't mind as much as the extreme fluctuations. That's why spring wheats have been traditionally favoured in northern New England. Surmount that difficulty, and the preference shifts dramatically to winter wheat.

The leaf mulch method described earlier in this chapter is particularly useful with winter wheat, which as I mentioned is susceptible to early-winter damage. Given the mulch, it is superior to spring-sown grains; moreover both the sowing and the reaping happen at times when I'm less preoccupied with other farmwork. By the way, with winter wheat and rye, the mulch may well have weathered enough under the snow so that it needs replenishing in the spring; easily done, provided I have stockpiled an ample supply in a dry place. Now tend to your other work; your wheat should need little or no need for attention for several weeks. Do stop and appreciate it

Figure 9.5. This winter wheat will be safe from frost-heave damage until snow comes.

when you go by, though; I think it likes that. More to the point familiarize yourself with its growth, in order to anticipate your next operation, which is harvest. That's not altogether obvious, as I'll explain soon.

Back in the late 1820s, with the Erie Canal nearing completion, the Maine state legislature feared an influx of cheap western wheat ("western" meaning western Pennsylvania and New York and Ohio—I once had a neighbour in San Francisco who spoke of growing up "back East," meaning Wyoming). Concerned about the effect on Maine farmers they enacted a "bounty" (today we would call it a subsidy) on Maine-grown wheat. To implement the new policy, every town had to compile and publish a bounty book listing every farmer, how much acreage and yield of wheat, and—all-important to us today—what variety. Some of these books are still around, and in town after town the same name pops up: Banner. It's a hard red spring wheat, originally imported from Latvia to improve our local crops. Latvia, of all places, not Western Europe (too mild), not the Russian steppes whence came Turkey wheat, the ancestor of all our modern prairie wheat varieties. Latvia, a forest-rich Baltic nation with the climate much like my own. Likewise, we know that Banner was well adapted here, if we could only find it. There is a bogus version of Banner in circulation, which may be every bit as good as the historic variety, but it is not the real thing. More recently an apprentice with Estonian roots (next door to Latvia) returned there to collect stuff for me, and he brought back a sample of a winter wheat called Sirvinta, not a traditional variety, not even native Estonian (it's a river in Lithuania), but highly adapted there and—guess what?—better adapted for my Western Maine Mountain farm than anything in US catalogues.

We may have lost much of our northeastern grain heritage, but that isn't to say that well-adapted varieties are not to be found. Marquis is a hard red spring wheat bred for northeastern forest soils. Also, Red Fife works very well as a spring *or* winter wheat. No doubt there are many others, but I have focused my search on the Baltic region and northwestern Russia rather than the American Midwest.

Here's a strategy to pre-fertilize the winter wheat plots while reducing weed pressure: When I plant corn (any type), not only do I plant soybeans in the same rows, but after tilling (early July) I undersow the area to ladino clover. This barely gets established until the corn is removed (late September to early October), when it gets a brief flush of growth before snow sets in. Early next spring it bursts into life and gives one or two heavy cuttings, at least the first of which is removed for compost. Then I chop in the rich residue in time to thoroughly die and commence decay before I plant winter wheat (by the way everything I'm saying here also applies to winter rye). The tilling of the corn (which was fertilized by weed-free night soil), plus the repeated mowing of the clover, pretty well guarantees the demise of most annual weeds and prepares for a clean, labour-free wheat crop.

Harvesting Wheat

What I'm about to describe refers specifically to wheat, but it can apply to the other grains as well. As the crop bolts to seed and forms the soft grey-green heads, take note. In fact, take a sample: Pull off the head and peel away a soft hull from an immature kernel (aka berry). If you squish that between your thumbnail and forefinger, a milky juice will squeeze out. Amazingly this is referred to as the "milk stage," and it is quite useless for food, but take note nevertheless.

Actually, oats in the milk stage have cosmetic uses as an ingredient in specialty skin care products, due partly to their emollient properties, somewhat like aloe or cucumber juice. In Peru, moreover, immature wheat is sometimes harvested and dried, later boiled as a specialty treat called *trigo verde*—green wheat. I assume we could do this with any grain; in fact we routinely eat (sweet) corn in the milk stage and call it corn-on-the-cob.

If we leave the immature grain and return some days later, we will find that the kernels have dried up somewhat and now resemble soft dough, hence the name for this stage. Some days later still, a kernel can be readily dented by a thumbnail, but not actually squished. More like hard dough, in fact, and so we call that stage. Now hold that thought, please; we'll

Figure 9.6. With blossom remnants still clinging to it this wheat is in the milk stage.

be coming back to it. If we wait a few more days, all of the green will be out of the straw and the kernel will be undentable, as hard as flint. When you store grain or *grind* it into flour, it must be at the flint stage—but that is *not* the best stage at which to *harvest* it. If you were to cut the wheat back at the hard dough stage and let it finish drying in the shock, the grain would be of much higher quality, both for baking and for digestibility. That is partly because of the balance of proteins and starches when the grain was severed from its roots before ripening, but it's also because the bran is more attenuated—thin and flexible, rather than coarse and thick.

In the old days sheaves of wheat were stacked in the field to cure in circular arrangements called

Figure 9.7. At the hard dough stage this wheat is within a few days of being ready to cut.

stooks, and thus the grain was of better quality. The invention of the grain combine in the 1800s made the harvest much less labour-intensive (and thus cheaper; grain began to be a US export). But farmers had to cut all the grain in the flint stage, which resulted in coarser bran. No problem, because around the same time an Austrian miller was developing the process of roller-milling, which instead of grinding the kernels passed them through a series of rollers and sifters. This crushed the grain into a fine flour while separating the coarse bran and germ, which was then fed to livestock. Unfortunately it also removed several key vitamins and trace minerals, but once that fact was discovered, the milling industry learned to "enhance" the depleted flour with some (not all) of those nutrients. That's how much respect we've shown for the staff of life, and that's what happens when you rely on strangers in the marketplace to decide what you eat.

So now we have this natural foods movement putting whole wheat flour back in the marketplace, but it is still combine-harvested, it still has coarse crudely ground bran, and it is difficult to bake with. Don't suppose that commercial bakeries rely on that stuff; they insist, correctly enough, that the whole wheat flour they can obtain will not make a decent loaf the consumer will buy. Therefore they either blend it with white flour (hey, the label says "wheat

Figure 9.8. Harvesting wheat with an antique grain sickle.

flour"—that means whole wheat, right?) or rely on additives like refined gluten to give the dough good "falling numbers," the measure of elasticity that produces a light loaf. So if your family does not prefer your home-baked whole wheat bread to that from the store, don't blame yourself; you must either add that refined stuff or use stook-cured grain. And where can you buy that? Nowhere, gotta grow your own. So let's get back to the harvest, so you can learn how.

When the crop is in the hard dough stage, I take my grain sickle and set to. A grain sickle is not like a common grass or weed sickle such as you find in the hardware store (if they have those anymore in this age of string trimmers). One of those would not do at all. A grain sickle has a long, slender sweeping blade—think Soviet flag, if you're old enough—and requires a particular style. (See appendix B for sources of sickles.)

You don't hack at the grain; that would risk snapping a valuable tool and possibly slashing your knuckles in the bargain. No, you slide the well-honed blade behind a section of grain row, perhaps a foot (30.5 cm) or so. As you grasp the tops of the straws with one hand, use the other to draw the sickle up towards your waist while sliding it slightly sideways, neatly slicing the straw as low as possible. The longer the straw, the better. If the sickle is dull or you hack at the straw, it may well uproot it; you now have a mess of dirt mixed with the crop.

Figure 9.9. Cutting wheat and bundling the sheaves: Any job goes faster with friends. Photograph courtesy of John Paul Rietz

I lay that fistful down on the ground and continue down the row or step forward into the next row. Either way I pile the handfuls until I have a bundle I can barely wrap both hands around, perhaps 4 to 5 inches (10.2–12.7 cm) in diameter when squeezed together. I lay those altogether in their own neat little pile and move along until the plot is covered with those little bunches. If they're held off the ground a bit by the stubble, that's nice, although I don't intend to leave them lie for long. Next I bind the bundles into sheaves using the straw itself as a convenient and organic binding. At this hard dough stage the straw is still partly green and not yet brittle. I choose a nice long straw and draw it between my pinched fingers, making a tough ribbon of it. I wrap this twice around the waist of the bundle and draw it snug before tying a double knot. Don't pull it too tight or the straw will snap, usually at the head. With a little practice you'll catch on. This way of binding works for wheat, oats, and rye. Barley straw is short and weak and has all the tensile strength of wet toilet paper, so I use baling twine for that.

Once I've bound all the sheaves I stook them, or shape them into a self-supporting shock. It's awfully handy to have a helper at the start, until the shock is self-supporting (though I often don't). I lean two sheaves against each other, then hold them thus while I lean two more from the other direction. I keep leaning sheaves into this pile—at some point it holds itself up without my help. I stagger them so as to leave lots of passageways for air to circulate and dry the stack. I put something like 20 or 40 sheaves into a single stook, standing in the middle of the patch on the stubble. Some folks lay a couple more sheaves over the top of these in case of rain. Those sheaves will not repel the water, but rather let it run along the straw and drip off the ends before soaking in. Mind you, a shower or two will do little harm if followed by dry weather; that's why we build it in that shape. If I am cursed with a wet spell (which earlier I might've called a blessing), the stooks may need some protection to prevent mould or sprouting. Then I may favour a smaller number of larger stooks so I can cover them with tarps.

Figure 9.10. Binding wheat sheaves using itself for the tie.

Figure 9.11. This half-built stook of wheat will have at least 30 sheaves when complete.

Figure 9.12. Flail at rest.

Figure 9.13. Son Fairfield takes a turn at threshing out the wheat.

Threshing

You kind of hate to cover them, though, they are so satisfying just to look at. You feel like you're living in a Millet painting, or—as we Mainers say—"they're some wicked cunnin'." However, they're not meant for decor; they're food, and furthermore birds and mice are also a threat to the grain, so as soon as it is cured, usually within three to five days in fair weather, I haul the wheat to the threshing floor. I have a special threshing room in the barn, handy when weather bodes ill, plus I can thresh and winnow beans and seed crops there in the winter. But I tend to thresh my grain outdoors on a tarp-covered bit of hard-packed ground, situated where I can drag the tarp into the barn if necessary. It's nice to work outdoors because it's easier to avoid breathing the dust and chaff. Also, basking in a sunny breeze makes the grain thresh out more easily and completely. As I toss each sheaf onto the tarp or threshing floor, I give it a flick of the sickle; the dried binding straw snaps and the sheaf bursts open. I soon have a huge pile of loose grain—heads and straw altogether and awaiting the next step: threshing.

To thresh you need a flail, and since I cannot tell you where to purchase one (how the marketplace fails us at critical moments!), I'll have to tell you how to make your own, and so I've done so—you'll find those detailed instructions in appendix A.

So now that you have a flail (or two, I like to have more than one size of dasher, or swingle, for different crops) and a heap of wheat or oats or whatever, you're ready to do something your and my ancestors have been doing for at least 9,000 years. It might seem intuitive to just swing that thing over your head and wallop that stuff something fierce. Go ahead, but you will wear out quicker than a skunk's welcome and not make much headway for it. Instead I roll the thing low and sideways so I can wrist-flip that dasher down with a moderate but firm rebuke. Allowing the handle to rotate in my grip, I can follow each blow by another and another without undue exertion, sidling slowly around the pile to address a new spot every time. Eventually the braided twine thong that connects the dasher to the handle wears and breaks. I've never had one fly off anywhere; due to my rotating style of swinging, the released dasher always just flops down in the grain and patiently waits there to be repaired. Still, there's no guarantee that it won't someday fly off like a slingshot, so I always insist that no one I love is standing right across from me. Indeed, I sometimes do have a helper with another flail, but I'm careful that we are no more than 120 degrees apart, lest one of us should miss and mar the other's good humour.

As I beat on the pile I can hear the rattle of grain shattering from the heads and sifting down to the bottom. Nevertheless much of the lower stuff doesn't get fazed much, buried as if under a springy mattress of straw, so from time to time I take a hay fork and turn the pile, exposing ever more grain to the flail's

abuse. There is always a little more to be got by persistence, but sooner or later you must decide that you've gotten all that's worth the trouble and fork all the straw away, sifting out any trapped grain. The remaining heap is a mixture of grain, chaff, broken heads, and bits of straw, all ready for the next step.

Winnowing

Winnowing (I assume that's cognate with *wind*) is about as simple as it gets. The grain is the densest part of that pile (hoping we've included no dirt or pebbles); I sweep everything into 5-gallon (18.9 l) buckets and slowly pour from one bucket into another while a mild breeze (or draft from a floor fan) is sorting out any chaffy stuff. I don't try to get it perfectly clean all at once, but repeated winnowings

Figure 9.14. Cleaning wheat with a hand-crank winnower. Photograph from Will Bonsall's collection

give purer grain each time. Sometimes I may see what looks like good grain blowing over the edge of the bucket. Waste? Possibly, but I gather it up and re-winnow by itself, perhaps rubbing it in my hands to free any attached hulls. More often than not I find that what looked like good grain is just empty hulls. If not, I either adjust the wind or lower the upper bucket to gain closer control.

Of course there are machines for winnowing grain, and I'm not just talking about powered combines, but rather the elegantly funky (another oxymoron?) wooden devices that used to be found on any farm in northern New England and probably elsewhere. Around here they were used pretty exclusively for winnowing dry beans for the canneries; however the large assortment of screens that came with them show that they were designed for a much wider assortment of crops. I know for sure that the machine pictured here was used for grain as well. On the bottom of one side the owner apparently knelt on the chill barn floor to autograph his machine and to inform posterity that "I am threshing oats today." It is doubtful that the farmer (in Bingham, Maine, as he informs us) ever went beyond grade school, yet his handwriting, done in lead pencil on planed pine on a February morning in 1874, is far more neat and legible than my college-trained scribble. Humbling.

No matter how carefully I winnow my grain, it seems there are always a few kernels that still retain the innermost hull, even in the so-called hull-less grains (more on that later). No biggie; after all, there are plenty in the commercial stuff, too. If you are milling into flour, the grinding and sifting will either remove that or render it unnoticeable, but if you're using it whole (like rice) or rolled like oats or coarsely cracked like bulgur, it's nice to not have that stuff sticking in your teeth. For small batches (less than two bushels) as most of mine is, I find I can separate out the hulls by flotation, thanks to the tiny air pocket that's trapped underneath. I half fill a large (4-gallon, or 15.1 l) canner with cold water and dump in a peck or two of winnowed grain. Stirring it around, the not-quite-hull-less kernels float to the surface and are easily skimmed off. Finally, all the clean grain must now be spread on screens to re-dry for storage. I usually re-dry the floaters, mini thresh them in my palms, and blow them clean. Any that still retain a hull go in the bird feeder. It's a method that only works on our small scale, and may not be worth it even then if it's clean enough to serve your purpose.

The best place to store grain is in an outbuilding—extreme cold is nothing but good for it. However, it should be kept in tight metal barrels, *not* plastic, to deter rodents and to exclude dampness.

Other Wheat Species

The kind of wheat we've been discussing so far is all *Triticum aestivum*, which is sometimes referred to generically as bread wheat, even including pastry wheat. That is to distinguish it from other wheat species like durum, spelt, emmer, and einkorn. I have only a little experience with most of these, but will share what I think I know, much of it from hearsay (the Internet).

Durum wheat (*T. durum*) is a high-gluten type used exclusively for pasta and couscous. When people speak of semolina flour, they're talking about durum, although it is usually refined, with the germ and bran removed. Essentially it is white flour, although it's actually yellow in colour. We have made our own pasta from whole-grain durum, but it is difficult to get the smooth, glossy surface texture of the refined semolina. I have heard people say they don't like the flavour of whole-grain pasta; that it tastes like "livestock feed." I just figured they had over-refined taste buds that didn't appreciate the robust flavour of real food. More recently I myself have had some batches of store-bought pasta that were disgustingly stale and rancid. They usually come from some discount retail place, so it makes sense. If you buy white flour products, like cookies or crackers, from those same places, they taste okay—or at least as good as white stuff tastes—because refined flour, being dead, lasts forever. But whole foods are perishable; they're not meant to be entombed. If you try to store them for years at warehouse temperatures, of course they will taste like mummy-wrapping. Most health food stores do not keep their pasta products properly refrigerated, but at least they have higher turnover so the flavour is generally acceptable. However, to have really vital, full-flavoured pasta, you probably need to grow your own durum and make your own pasta fresh, as I describe in chapter 16.

Having said all that, I've never grown large enough quantities of durum to make my own pasta; I've always used store-bought flour. I have grown many durum varieties experimentally and am satisfied that they can be grown here successfully, although they do better in hotter, drier climates, like Nebraska. All of the many durum varieties I have trialed were spring-sown.

Durum and common, or bread, wheat are the only two *Triticum* species that are naturally naked or loose-hulled. The others, such as spelt, emmer, and einkorn (one-grain), all have tight hulls that are very difficult to remove, and they have remained minor crops. However, they all have unique flavours and nutritional merits that have secured them a small but enthusiastic following.

Spelt (*T. spelta*), aka dinkel wheat, is a specialty grain that is growing in popularity—and for good reason. Like most wild or semi-wild species, it is exceptionally robust in growth, taste, and nutrition. All kinds of livestock (who know a good thing even if we don't) relish it as grain or even as hay. One woman who received seed from me complained that she couldn't propagate hers because her geese kept grazing it to the ground and tearing it up, although there was other forage nearby.

The big problem with spelt for human consumption is the same as with common oats and barley. It seems to be an analogy across all grains: The more primitive types have tenacious hulls, or glumes, which render them very difficult to process for eating, whereas some types have been developed with loose hulls that can be easily cleaned by ordinary threshing and winnowing. However, while naked oats and barley are viewed as special types, with wheat it is the naked types—common wheat and durum—that are the norm, while the tight-hulled forms like spelt, emmer, and einkorn are considered the special types. Spelt in particular has an additional quirky feature of a tough rache, or

central stalk, with fragile spikelets—the tiny stems uniting the kernels to the main head. This causes the head to shatter easily into smaller clusters, which are then very resistant to further separation. That may be a helpful trait for survival and dispersal in the wild, but it certainly is a limiting factor for domestication. The solution has been to simply continue pounding the grain until these clusters break apart and the hulls separate enough to be winnowed. It is difficult, which is probably why spelt continues to be a specialty product, while its naked cousin is cultivated in the millions of acres. Fortunately the pounding process results in an uncompromised whole-grain product, unlike pearl barley or white rice.

I used to be under the impression that all spelt varieties were spring-sown; those few I had trialed had acted that way. However it's possible those were facultative, capable of going either way depending on when I sowed them. I've since learned that winter types are far more common. I recently trialed a variety called Braveheart, which clearly is set in its ways. Although I sowed it in early spring along with my other spring grains, it just sat there all season long, determined not to bolt. Come snow-go it finally did its thing. Next year I'll know to not sow it until September and let a green-manure crop occupy the space in the meantime.

I have only grown spelt experimentally and haven't produced enough to practise cooking with it. It is somewhat lower in gluten than standard wheat. Moreover it yields significantly less than regular wheat, so it can only compete with wheat in the areas of taste and nutrition. Indeed the flavour of spelt is often described as sweeter and nuttier than wheat, plus it is higher in several minerals including zinc and selenium. Some people with wheat intolerance find they can eat spelt with little or no trouble, but not those with celiac. Celiac sufferers are specifically bothered by gluten, which spelt has, whereas other wheat allergies may be reactions to other proteins or other components altogether—substances less abundant in spelt.

One disadvantage that is always cited for these primitive wheats is that they yield less than wheat;

after all, they haven't been carefully bred and selected for thousands of years. Yet it is also maintained that they require less fertility, particularly nitrogen, to produce that yield. Now, more grain is better than less grain, but in our garden-without-borders shouldn't we weigh *all* the costs required for that improved yield? I'm not suggesting that we should be content with poor yields from weak soils. Far from that I'm suggesting we should maintain optimal levels of all tilth factors, including nitrogen, but in a manner that's more eco-efficient and thus more sustainable.

When it is mentioned that spelt requires 25 to 30 percent less nitrogen than wheat, it doesn't just mean that it doesn't *need* as much, but that it shouldn't be given as much as wheat, else it will fare poorly. Sort of like a horse foundering in clover, its system cannot deal with the surfeit of richness. One result is that the straw will be weaker, especially the spreen, or crown, and be prone to lodging or laying over. To some extent it is a problem with any grain that is overloaded with nitrogen; it's aggravated by a relative deficiency of minerals, particularly the alkaline minerals calcium, potassium, magnesium, and even sodium. I'll tell you an analogous story.

A friend of mine once grew a small crop of grain—rye, in this case—in his vegetable garden. Always wanting to grow the biggest and best of everything, he loaded the garden with chicken manure from somewhere. The rye predictably grew outrageously tall and heavy and threatened to lodge. He remedied that by oversowing a small amount of salt (NaCl), which toughened the straw and helped it stay upright. I thought it was terribly clever of him to know about that, yet not so clever to have created the problem in the first place. It may be a paradigm for our civilization: We will probably discover a cure for cancer long before we address those things that we already know cause it.

It all comes back to The Big Footprint, which is central to our garden-without-borders perspective. The concept of a crop's footprint is inherent in our use of the words *per acre*; if we always insist on looking at only The Little Footprint, the actual space physically occupied by the crop itself, while ignoring all the externalities—if we continue to look at the world through a magnifying glass rather than a telescope—we will continue such quixotic quests as finding a cure for cancer, ending world hunger, and creating a lasting peace. We will feel better about ourselves as we accomplish nothing very profound. What is it about straw that makes me keep digressing into philosophy, anyway? Dunno, but perhaps it's because straw is a major source of humus, and humus is a near-panacea for so many things. Indeed, if my friend had built up his garden's humus level proportionate to the huge glug of nitrogen he was dumping on it, the lodging might have been less of a problem, since balance is itself a remedy for much. However, the balance must go beyond The Little Footprint; concentrating that much of *anything anywhere* is of dubious sustainability.

Triticale

I'm not going to try to tell you anything about emmer and einkorn, as I have little experience with them, and much of what has been said about spelt applies to them as well. I do have some experience with triticale.

This trendy grain has something in common with rutabagas and beefalo, and for that matter with wheat and corn: None of them exists in the wild. These crops are all intraspecific hybrids that have occurred spontaneously (without deliberate human intervention)—except in the case of beefalo—and that cannot maintain themselves outside of cultivation. There are wild carrots and wild turnips, but no wild rutabagas and no wild triticale. Triticale was originally discovered in wheat fields adjacent to where rye was growing. The two species, *Triticum aestivum* and *Secale cereale*, are not even in the same genus, so such a cross should be totally impossible. Considering that the original criterion for categorizing species was reproductive isolation, perhaps our taxonomy is flawed here. In general the biblical criterion for *kind* ("each according to its kind") seems to serve us better, though it is not so simple. Since all the diversity of life was not created in a single event but evolved gradually by a process of speciation, it should not surprise us that some species are not yet fully separate, but still retain enough genes in common with others of their "kind" that the occasional cross may occur. This is hard to swallow when we observe such variations as occur between wild turnips and wild kale (the progenitors of the rutabaga) or wheat and rye, or cattle and bison, yet the fact that they can get it on successfully is a powerful statement of something.

This crossing between allegedly different species doesn't always occur without human interference. For example, treatment with colchicine, a naturally occurring substance from a species of crocus, causes chromosomes to break and rejoin in less controlled fashion and thus create new genetic entities that would be unlikely to occur with normal sex. This is sometimes presented by GMO advocates as evidence that bioengineering is really quite natural, but there is a world of difference between crossing plants that may occasionally cross on their own anyway and splicing organisms that are as unrelated as corn and fireflies. Now back to triticale.

Whether by the hand of man or with the help of serendipity, triticale exists, and it has some very useful traits. The plant is as hardy as winter rye, yet more productive. It has less gluten than wheat, and the kernels are much larger than either wheat or rye. Many find the taste and texture of triticale superior to both parents, although for bread baking triticale by itself produces a rather heavy loaf.

The only triticale variety I have grown is Heritage, and it acted like a spring wheat, although it is possible that it is facultative and could just as well have worked as a winter wheat. It performed well enough, though the planting was too small to be conclusive.

Barley

Barley has been cultivated for as long as or longer than wheat, but has always been relegated to second place because it makes a coarser, heavier bread, even though it tolerates poorer, drier soils than wheat. Indeed, on good soils I find that wheat consistently outyields barley, so it's easy to see why barley would've been pushed to the more marginal lands. In the biblical story of Gideon, barley bread is portrayed as a hardier loaf, while fetching half the price. That said, barley is a flavoursome grain, rich in minerals and more easily digested. I find it works especially well as a rice substitute here in the North, either whole or very coarsely cracked. Here's the catch: Common barley has an inner hull, or glume, which unlike its outer hulls does not come off easily in regular threshing and winnowing. The main way people have made it edible for humans is by grinding off the last glume by a process known as pearling. That also removes much of the nutritious bran and leaves a gooey product with a bland taste. A friend who attended convent school remembered with chagrin the pearled barley soup they were frequently served for the sake of frugality. It always struck me as curious that whole-foods stores that would rightly disdain to sell white rice are

Figure 9.15. A six-rowed hull-less awned barley in the soft dough stage.

proud to stock pearled barley. Barley doesn't have to be that way, but how else to render it useful for more than hog feed or beer brewing?

Fortunately there is a type of barley, called naked or hull-less, which is in fact not hull-less at all, but is rather loose-hulled, which is to say that *all* the hulls thresh off, leaving the grain naked like wheat or rye. There are numerous varieties of naked barley, but few are available in the garden seed trade, and those not necessarily the best adapted for my conditions, and you need varieties adapted for where *you* live and garden. A great assortment are accessible through the National Germplasm System (see appendix B), but those, like everything else they offer, are distributed as packet-sized samples only to those who intend to further propagate their own. So once again: Learn how to grow and save your own seed and do so.

Although only naked barleys are practical for home-scale processing as human food, the common barleys are not useless. They are suitable for feeding poultry and brewing beer, both of which I consider wasteful, but they're also good for green manuring. A blend of barley and oats (and field peas) is particularly resilient when rainfall is uncertain. With abundant moisture the oats will tend to predominate, but in a drought the barley will give a welcome back up. The two grown together will provide much of the dovetail effect inherent in companion cropping, much more so if peas or vetch are added to the mix. By the way research (hearsay to me) has shown that two or more barley varieties sown together (whether for grain or green manure) yield better than either variety alone. Is this an example of the companion principle, even within a species? If so I wonder how that works with other grains. Of course, when you plant a mixture of varieties, it's important that all varieties used have the same maturity date, and you must maintain seed of both or all varieties separately so you can control the mixture over time.

The barley varieties I've offered over the years are nearly all naked. Someone asked me which of those would be best for malting/brewing. I'm not certain, but I believe the *non*-naked types are better suited for that purpose, so if you want to grow barley for

Figure 9.16. Barley is stooked like wheat, although it must be binded with twine because the straw is too weak to use for that purpose.

eating *and* drinking, it's best to grow both types. Helpfully, barleys, like wheat and oats, are completely self-pollinating, so there's no risk of genetic crossing between two patches. There are some fall-sown varieties of naked barley; I've tried two French varieties that were promising, though perhaps not as hardy as winter wheat, but as yet I haven't enough experience to compare them with the spring types.

Barleys of any type have either two rows or six rows of seeds per head. The two-rowed types are reported to be more productive, which seems counterintuitive; indeed my own experience doesn't seem to support it. I mean, the two-rowed types are noticeably longer of head, but three times longer? Maybe there are other factors, like tillering, that I'm overlooking.

Still another distinction among barleys is between awned and awnless. Unlike wheat, the awned types are far more common and the awns stick out more sideways and are very stiff and serrated. The serrations are actually barbs going in one direction. Ethnobotanists have suggested that this was an evolutionary adaptation for self-planting, as the barbs, given their tendency to swell with nighttime moisture and shrink with daytime heat, would push the heads deeper into the soil. Quite possible, though I've seen no tendency for missed heads to plant themselves; rather, they lie on the surface trying to sprout and root before a bird or mouse discovers them. My own theory is based on the experience of getting an awn stuck in the throat: they will only go one direction, irritating and choking every inch of the way. I

Figure 9.17. This amount of hull-less barley wouldn't go far as livestock feed, but it's more than enough to last my family a year. Photograph courtesy of Debbie Burd

can imagine a large mammal being put off by them. Moreover those awns, while not Velcro-armed like burdock, must easily catch in the woolly hair of wild or domestic sheep and goats, thereby hitching a ride to a potential new habitat. All that matters to us in all this is that those awns are a nuisance, yet I generally prefer them to the awnless varieties for threshing. In fact, threshing barley is one of the few things that can induce me to keep my mouth shut, although I've often had greater difficulty in swallowing my words. A really mean trick (which I would never contemplate) is to slyly insert a barley head into the shirt cuff of an unwary bystander. Within a very few moments, that head would be jabbing his armpit, causing great annoyance. Of course I wouldn't know that; I only presume that's what would happen.

Regarding varieties my region has no real tradition of barley raising, especially naked barley, so I have sought varieties wherever I could find them, mainly from the NGS or European germplasm sources. Many of those are bred for commercial production in places like California or the Palouse region of eastern Washington. Therefore they may be imperfectly adapted to climate and soils like mine; however, some of them have done quite well for me. Some standouts among US varieties include Burbank, Lompoc, Mesa, Junior, Excelsior, and Thual, none of which is from my region, but all are good performers here nonetheless. A couple of northern Italian varieties have caught and held my attention: Milan and Leonessa. Arabian Blue is an attractive and productive variety. The hull-less type seems to have originated in Central and East Asia, and even today is mainly grown in small terraced garden plots far from the nearest commodity market. Come to think of it, that sounds a lot like my place, so I'm not surprised when some of my best performers hail from Korea or Nepal. I'm still forming opinions of those, but my eye is on a particular Mongolian variety with a unique long, narrow shape and pale grey-green colour, which seems especially well suited to cooking whole like rice.

There's a great deal to be said about how to use homegrown barley, and I discuss that in chapter 16.

Oats

Raising oats in many ways parallels the cultivation of barley. Most oats, or common oats, have an inner hull that can only be removed by an elaborate process used by the folks at Quaker Oats, et al., but not very helpful for you and me. Fortunately there are naked oat varieties that we can easily process in our own backyards with funky equipment we can either find or make ourselves.

Among naked oat varieties my hands-down favourite is Terra, but a Swedish variety called Rhiannon is very impressive. Of those available in the marketplace, Pennuda is fine indeed. Several others are close contenders, including Chinese Hull-less, Shadeland, Spokane, Brighton, and Torch River.

Figure 9.18. These Terra hull-less oats are maturing for seed. Photograph courtesy of Mary Doyle

A woman once requested a naked oat variety from me to grow feed for her goats. I told her that didn't make sense, because I've "heard say" that common oats are better yielding than naked oats, so if the goats can deal with the roughage, indeed may well need the extra fiber to avoid acidosis, what's the advantage? On the other hand, from my own experience I couldn't see that the yields were so noticeably different. Still I assume there's some good reason why most oat farmers grow common oats, even when the crop is to be processed for human consumption. In fact I've heard there are some ways of home processing common oats for people fodder, but I haven't found them yet. Meanwhile I believe common oats have their place in veganic systems, if only for green manure. But for those of us looking for something we can grow and process for food on our own homesteads, naked oats seem to be the best option, at least for now.

To most people, oats means rolled oats, eaten as a hot breakfast cereal or as granola. But oats can also be cracked or flaked. We have a tiny clamp-on table device that makes oat flakes in small quantities (but fresher). You can buy much larger, more efficient models, but for cracked oats we simply use our Corona hand mill at a looser setting (ditto with barley), and the product is much like steel-cut oats.

The main way my family consumes oats is as flour and that flour in the form of waffles. Our version of waffles is quite different from the wheaten version, which calls for lecithin-laden eggs to make them light but rich. Ours uses sunflower meal (up to a third) for a rich oily texture and flavour, and baking powder for leavening. There are serious objections to the use of baking powder and baking soda (they are discussed in part 5), so when we are behaving ourselves, we use an alternative method. We use a high-speed blender to incorporate lots of air bubbles into the batter, which after all is what both baking powder and yeast are all about, but you must not let the batter sit around too long unless you reblend it. Those waffles also incorporate lots of homegrown sunflowers and amaranth, and occasionally other grains; no one yet has been disappointed with them. They are traditional Saturday-morning fare at our house, although my kids are convinced they don't like homegrown oats or amaranth any more than some of the other things Daddy grows out there in the dirt; I'd appreciate it if you didn't enlighten them.

Rye

Rye is such an easy crop, most of the people I know who have grown it have done so by accident. That is to say, they planted it as a fall cover crop, fully intending to turn it in come springtime. Then came springtime and they were too busy, and then it got too high, and then it looked so pretty, and then, oh, what the heck. Rarely have those people reaped and eaten the crop they so carelessly grew. A real pity, since rye is so tasty—and muscle-building, as opposed to fattening like corn. Eating rye gives you a sense of purpose: Did you ever meet a wimp from Finland? Of course not, that's where all those rye crackers come from. But I would like to put this thought deep in your heads: Rye is not just about crisp breads and dark sourdough loaves (much as I adore both of those). Coarsely cracked, it is a wonderful change from oatmeal for breakfast, especially with a palmful of dried blueberries or cranberries. Whole rye, steamed long and slow (macrobiotic-style) is a hearty alternative to rice, especially served with kale and leeks. In short, rye is too good to be relegated to the crackerbox. And it doesn't hurt that it is such a resilient and undemanding crop. Although I occasionally grow rye as a cover crop, when *I* grow rye for grain, it's not a mistake; it's good food.

Any rye I've ever encountered in the United States is strictly winter rye, although in Europe they also have spring-sown types (sommerkorn). I prefer the former for all the same reasons I prefer winter wheat: It dovetails into my crop calendar, giving me more flexible use of time and space.

The only rye I've ever seen sold by variety name is Aroostook. I presume there are others, but I don't know of them.

Millet

Millet is a whole different kind of grain, different in needs, structure, taste, texture, and nutrients. First, unlike wheat, barley, or rye, millet doesn't form a central head or rache. Instead its seeds are held in a loose panicle (the very word comes from the Latin form for "millet": *panicum*), which is a broomlike cluster of straws or stems, one seed per stem. Indeed the "broom" reference is totally apropos: Natural straw brooms are made from broom corn, which is actually a form of millet. Millet belongs to a large family of crop plants including broom corn, grain, and cane sorghum, plus our pesky weed foxtail millet. Unfortunately I cannot ripen most of those where I live, but for those who can, sorghum molasses is a splendid alternative to the tropical cane sugar molasses. These all require much longer, hotter summers than I have; even the Japanese millet that folks here sometimes use for a cover crop will not reliably mature seed in the North. It has no frost resistance whatever; folks who use it for a forage crop must be careful, because frost makes the leaves form cyanide, which is poisonous to livestock. Even if I could mature Japanese millet, the grain is not suitable for human consumption because the tight hulls will not thresh off readily. There is a loose-hulled or free-threshing version of millet called proso millet, which has the added attraction of requiring a very short season, much like buckwheat. Although it is frost-tender it tolerates, indeed prefers, hot, dry weather to perform well. One reason proso millet is so early maturing is that it doesn't waste time making lots of biomass; I've rarely harvested a crop that exceeded 2 feet (0.6 m) in height. For this reason proso millet is no good for cover cropping, but I have plenty of other options for that; I want the stuff for food. Food for me that is; I'm not keen on sharing it, and therein lies my main problem. You may have noticed that millet is a main ingredient in commercial bird feeder mixes. That's for a good reason: Millet has an abundant supply of calcium in a highly assimilable form (hearsay: "studies have shown"), which is especially important if you're laying eggs. I am not at the moment, but like everyone else I supposedly need calcium, and I believe that dairy foods are a poor source of dietary calcium. I mean, they certainly contain lots of it, but in a context that is acid-forming to the human system and causes it to precipitate in inconvenient places, like kidneys and joints. Dairy products are typically part of high-protein, high-fat diets, so is it surprising that osteoporosis is an ailment of Western societies, whereas in poor countries where most of the calcium comes from grains, pulses, and leafy greens, osteoporosis is unknown (they may have plenty of other nutritional problems, but not bone loss).

Anyway, birds totally get it, and they seek calcium where it is most easily gotten: grains and seeds, especially millet. That is problematic for me: It's easy enough to grow, but hard to beat out the birds. With other crops I can usually be diligent and harvest as soon as birds show an interest, but I have known birds to start eating millet when the seeds are still quite green. A strategy that works just fine is covering the plot with some kind of hoop-supported mesh, like berry netting. That severely limits the quantity of millet we can grow practically, but it may be worth it, at least for small quantities. The only variety I've grown so far is Dawn, which yielded well enough—just ask the birds. I am currently trialing other varieties from the NGS, but no conclusions yet. One concern is that most of those are from the Dakotas, Nebraska, and Colorado, good hot, dry millet country, so they will just have to get adapted here if we like each other enough.

In culinary terms millet is perhaps the most rice-like of all the grassy grains in that it is easily cooked whole into a light, fluffy dish—words that aren't so applicable to the others. In fact I've sometimes used whole boiled millet with coarsely cracked wheat, and oats cooked separately before mixing. Baking in a casserole, perhaps with herbs, accentuates the ricelike quality of the combo. A little millet flour is a welcome addition to my oat-sunflower waffle recipe. Also a corn bread recipe with millet replacing some or all of the corn is nothing to sniff at, unless with pleasure. Corn bread recipes usually include wheat for the gluten, so you can easily use yeast instead

Figure 9.19. Duborskian rice plants started in seedling trays are more likely to succeed. By mid-August my Duborskian rice is just heading out, still time to mature.

of baking powder. I never sow millet before June 10, which gives ample time to ripen a crop. The few times I've grown it, I've harvested it fully ripe by stripping the panicles by hand, but again, the birds didn't leave much for me to harvest.

Rice

We occasionally mention rice here in Maine, but usually with a dismissive chuckle, as in "The way my bottomland is flooded, I may as well have planted

rice." Well, the ironic sarcasm is generally justified, but perhaps not wholly. I have grown upland rice right here in Maine, as have several other Mainers, so gardeners farther south should certainly take heart. Nor is it necessary to construct floodable paddies; the variety most of us have had various degrees of success with, Duborskian, is a short-grained upland rice that can be planted on well-drained soil just like wheat or oats. And it will mature here in a good season, although direct-seeding is highly dubious; seeds should be started a few weeks early in Speedling trays (yes, made out of ticky-tacky, but fortunately reusable again and again) and young plants set out after the last frost. That, I believe, is what will severely restrict the production of rice this far north: There are limits to how much greenhouse space you can afford to dedicate to a marginal grain when other crops are sure of much greater productivity. Still, it is through persistence and adaptation, both by plants and by their cultivators, that we have a variety like Duborskian to work with; who can say what further advances will be made in breeding?

Another obstacle to home rice production in the North is the difficulty of removing the innermost hull, or glume. Traditionally this was done by various pounding methods, where a large mallet or log was driven into a barrel of grain (which was already threshed and winnowed). It is not so much the direct blow from the beater that separates the hull as the hard rubbing of grain against grain that eventually sloughs it off. There are small hand-powered machines for hulling rice, but the cost is impractical for a home gardener. Perhaps a group of gardeners could pool resources to co-own one (and other specialized or hard-to-find equipment).

Wild rice is not rice at all but a native American grain that grows only in certain riparian habitats, not in your garden.

Field Corn

Corn (*Zea mays*) is an incredibly diverse and plastic species. Sweet corn is often grown in the vegetable garden (see chapter 8). The other types of corn—popcorn and flint, flour, and dent corn—are grain crops. (Broom corn, however, is actually a type of sorghum.) Corn's obedient cooperation with human cultivators has caused it to be the most widely cultivated of all plants, so much so that it has been suggested that it is not we who have domesticated corn, but rather the other way around. To a very high degree both species have come to depend on the other for their very existence. In fact corn is nowhere to be found in the wild, its nearest living relative being teosinte, a cobless Mexican perennial grass that is itself rare.

While I agree that we grow and consume (in myriad forms) corn to great excess, I insist that we must grow our own nonetheless. The manner in which corn is grown commercially is so wasteful and destructive, there's no way you can help but improve upon that, and while we should remove King Corn from its central throne, we should keep it as a dutiful citizen of our crop-world. Stop feeding it to livestock until they're sick; stop making processed sweeteners for junk food; stop putting it into our gas tanks; just eat it, as simply and directly as possible.

Dent corn is named for the dimple that forms in the end of the kernel as it dries. It is the main type grown in the Corn Belt for fattening livestock, generally my least favourite for eating. *Flour* corn has a soft, starchy white interior which is easily ground for corn bread, although I find it very bland-flavoured. Some flour corns have multicoloured kernels and are used in decorative arrangements; so do some popcorns. *Flint* corn has a hard amber-coloured (sometimes red) almost translucent kernel; it tends to be earlier maturing in the North. *Popcorn* is classed by some as a flint corn, due to its glossy impervious skin, yet its soft starchy interior is more like a flour corn, which is more obvious if you grind it instead of popping it. It is the combination of the soft interior with the hard seed coat that enables popcorn to pop.

I prefer flint corn for my own use, for several reasons. I've found varieties that tend to mature more reliably in the North and have more robust flavour than the others. I grow two or three heritage varieties of flint corn (in different years for purity's sake),

but my favourite is a very old regional variety called Byron Yellow. It was originally grown by the Abnaki in the mountain valleys of western Maine and narrowly rescued by me from the last grower. It is an eight-rowed type often with a distinctive swelling at the fore-end, from trying to crowd in a couple more rows. The cob is exceptionally slender and long. There are corn varieties that yield much better than Byron, yet those don't mature as reliably; likewise there are even earlier corns, but they don't produce as well. Byron is very well adapted here; its roots in the neighbourhood go much deeper than my own, so it's staying. Hopefully you can find a field corn that fits so well with your place and your needs.

Because flint corn lacks the soft starch of the dent and flour types, it makes a heavier bread, or pone. However, corn bread is usually made with at least two-thirds wheat flour, and baking powder for leavening, in which case the wheat part should be pastry wheat. I prefer a corn bread mixed with hard red (bread) wheat, which is more glutenous but allows me to use yeast. To my taste it helps that the flint corn has a richer, cornier flavour, so it is not dominated by either the wheat or the yeast. If I want to use less or no wheat, I make a polenta instead of pone bread. The yeast will not raise it that much, but better than not at all. I bake it long and slow in a casserole, which

Figure 9.20. Byron Yellow flint corn is well adapted here in Maine because it originated here.

Figure 9.21. This Neil popcorn was left on the cob until near ready to pop, an ideal way to store it if you have the space.

I can cover if it gets too dry. To this I sometimes add terrasol or chufa meal (see chapter 11 for more about these crops).

For popcorn I grow one of two old Maine varieties, Andrew Wilson or Neil. Both mature well here, and their stalks are relatively short and skimpy. I grow yellow soybeans among them, as with the flint corns, but if I were to grow a common (*Phaseolus*) bean among corn, I would use a semi-runner (such as most black beans) and I would use popcorn, because of its more open growth.

As I mentioned in chapter 1 (under *Composting Humanure*), corn and squash are the only crops under which I put composted night soil (humanure). Of course those areas will be rotated into other crops over time, but by then the soil biota will have resolved

Figure 9.22. I spread Neil popcorn and assorted dry bean vines on tarps in early fall to dry outdoors.

any hygienic concerns, even if the composting did not. Also, I have a very particular way of growing both soybeans and clover together with corn, which you can read about in *Companion Planting* (chapter 14) and *Living Mulch* (chapter 3). I make an extra-deep furrow and add about 2 gallons (7.6 l) of composted humanure per 40-foot (12.2 m) row. On top of that I put a dusting of wood ashes; directly on top of that I drop the two kinds of seeds. I plant both by the middle of May, and they're ready to harvest right around the fall equinox.

Broadleaf Grains

There are a few grains, notably buckwheat, amaranth, and quinoa, that agronomists classify as pseudo-grains.

This seems annoyingly dismissive to me; of course they are grains, we call them grains, we prepare and use them as grains. What scientists mean is that these grains are not *grasses*; they do not belong to the family Poaceae. Well, quack grass *is* a member of the Poaceae, and so is bamboo, but who calls those grains? Still, there is an important distinction: The genuine honest-to-goodness, right-out-front dyed-in-the-wool card-carrying real-McCoy grains (which I'll henceforth call "grassy grains" for clarity) are all monocots, which have a very different worldview from the fake, fraudulent, bogus, ersatz, sham pseudo-grains, which I'll henceforth refer to as "broadleaf grains."

All the grassy grains make their new growth from the base of the plant. That new growth emerges directly from the crown, whereas the broadleaf grains

Figure 9.23. This is the way we used to grow field corn, without soybeans or clover growing as a living mulch; however well this stand of corn yielded, we missed out on two other crops!

just keep adding new growth onto older growth. Consequently if you, or a grazing animal, snip off the top of a wheat plant, it will just keep growing, as in a lawn; do that to a field of buckwheat and it will stop dead, as in a parking lot, or at most branch out and grow bushy.

And then there's the ability of grassy grains to "eat the rocks" and convert the dissolved silica into a protective coating. It's a trick that most dicots aren't so good at; they rely more on taking up the already soluble silica in the soil humus. Moreover grazing animals then use that soluble silica to make their otherwise chalky teeth strong and sharp so they can chomp at that grass; it doesn't seem quite fair.

One of the helpful things about the broadleaf grains is that they lack gluten, which apparently is strictly a grass thing (and not all of those). By "helpful," I mean of course to people who are gluten-sensitive, like celiacs; to the rest of us it's a flaw, in that we cannot make bread from them. But isn't that seeing the glass half empty? I mean, might as well say that wheat is flawed because it's too sticky to make a nice light cereal like kasha.

The seeds of broadleaf grains contain complete proteins, whereas all grassy grains are low in the amino acid lysine (which is why it is good to combine them with legumes to make a balanced diet).

Buckwheat

No, buckwheat is not related to wheat, any more than a seahorse is related to a horse. Rather it is in the Polygonaceae, along with rhubarb, sorrel, knotweed, and smartweed. The *buck* part is believed to come from an old Dutch translation *boec-weite*, which actually means "beech-wheat," referring to buckwheat's strong resemblance to beechnuts. In fact the words for "buckwheat" in different languages tell much about its story.

You see, buckwheat originated in East and Central Asia. The common type is called Japanese buckwheat. The other is called tartary buckwheat, but it actually hails from the Himalayan kingdom of Bhutan. According to one story buckwheat arrived in Europe in the wake of the Ottoman Turk expansion in the

13th century (unless it was brought back still earlier by returning Crusaders). In any case it was associated with Islam, which Western Europeans equated with the Saracens. Therefore, the French named it *ble de Sarazin*, Saracen wheat, which name persists to this day. In German-speaking regions it was called either *buchweizen* or *heidenweizen*, meaning "heathen wheat" (an epithet the Muslims likewise bestowed on Christians). The humble grain is a tactless reminder that Eastern Europe has often been a fracture zone for conflicting views. Likewise the name *Tartary buckwheat* reflects its hostile Asiatic heritage.

Another story is that it was introduced into western Russia by Byzantine Greeks in the 7th century, which would comport nicely with reported archaeological finds in the Balkans of buckwheat being cultivated at least 4000 BCE.

At risk of repeating some of what I wrote about buckwheat in chapter 2, I want to point out that it's very frost-tender but matures in a very short season. Unlike grass grains, which ripen all at once, buckwheat plants bear seed in all stages of development—from flowering to overripe and shattering—simultaneously. This raises the question of when to harvest it. The Acadian buckwheat farmers of the St. Johns Valley between Maine and New Brunswick sow buckwheat in early July. It is still making late blossoms when the first frost shuts it down in September. The freeze-killed blossoms quickly turn dry and crumbly; when the crop is reaped the dried stems and leaves are simply winnowed off, with a minimum loss of ripe seed. The Acadians use the roller-milled flour (it is the tartary type) exclusively to make traditional all-buckwheat griddle cakes called "ployes." This is a curious example of an heirloom variety being grown on thousands of acres and sold in area supermarkets by the 10-pound (4.5 kg) bag.

I don't know how buckwheat was reaped before the days of grain combines. It would be difficult to put into sheaves and stooks without shattering, plus there would be little point to it, as the seed is pretty much dried when it's cut. Lacking machinery I simply cut mine with a grain sickle and toss it onto a 9 × 12-foot (2.7–3.7 m) tarp, which I then drag to the threshing floor. (I never grow more than a bushel or two of seed.) It is incredibly easy to thresh; indeed if you cut it in the cool morning or evening, it will shatter less and dry on the tarp in the waxing sun. If you look at it cross-eyed it will be half done, a few blows of the flail will finish it off nicely. I often spread my clean grain on screen racks under cover for a few days, in case of any residual moisture that might cause mould.

Of course the hulls, or seed coats, are still on (there are no hull-less buckwheats); removing those is the main obstacle to using homegrown buckwheat. See chapter 16 for a discussion of removing buckwheat hulls.

Buckwheat is not just outrageously tasty; "they say" it is also extremely nutritious, although I have doubts that anything so flavoursome can be good for you. It does not contain gluten but has plenty of selenium, an essential trace mineral that doesn't turn up just anywhere. And it has a full complement of amino acids, making for a balanced protein source.

No less in its favour in my garden-without-borders perspective is how it makes all those goodies and at what cost to the immediate and wider environment. For example, the phosphorus with which buckwheat is so replete is not necessarily from added (imported) sources; buckwheat has the exceptional talent of extracting phosphorus from inorganic minerals in the soil, such as apatite (yes, you might truly say said it has "an appetite for apatite"). In other words it can "eat the rocks." This mineral-dissolving trick is something we expect from lichens, but higher plants are generally much less adept at it.

In *Facts for Farmers*, Solon Robinson cites one farm where buckwheat was grown continuously for 17 years, with no fertilization of any kind being added. Indeed a crop of buckwheat grain was taken off and sold every year, only the straw and stubble being returned to the soil that produced it. Instead of wearing out the soil that residue alone built up the tilth year after year until eventually the farm was sold and the new owners immediately planted it to oats, corn, and potatoes (the latter two very heavy feeders), harvesting fair crops without adding any other form of fertilizer. Now, there's an example of eco-efficiency.

Figure 9.24. Here's Japanese buckwheat being grown for food, not green manure.

When sowing buckwheat I make a distinction between buckwheat sown for green manure and that sown for food crop. If I want to smother weeds and build a maximum of humus, then I will broadcast it at a rate calibrated to leave no bare or thin spots. That rate could result in reduced seed yield, but I don't intend to have any seed yield anyway, since I will turn it all under in the flower stage, growing the seed for it on a separate plot. For a food crop I drill the seed at a density comparable to wheat, encouraging each plant to grow robustly and form branches. I never dream of companioning buckwheat with any other crop for the same reason it makes such a good cover crop: It just doesn't get along with anything else.

Amaranth

I first tried experimenting with amaranth back in the 1970s when Rodale Press (publishers of *Organic Gardening* magazine) ran a big readership-based project to trial and promote various varieties of the then-novelty grain. Knowing of its Central American origin I had some doubts, which seemed justified when the variety they selected for me failed to mature. I realized I had probably not done enough follow-through, but I also knew that its close cousin, red-root pigweed, had no trouble whatever here. So many years later, when Cousin Tom showed me some big beautiful plants he had matured, named Opopeo, my interest was piqued again, although the fact that Opopeo is a town in Michoacan, Mexico, wasn't too confidence inspiring. I tried it and was quite successful; since then I've tried other varieties with varying success, but I can always count on Opopeo.

Direct-seeding would probably be reliable, but I no longer do that, for three reasons: In its seedling stage amaranth is nigh impossible to distinguish from the pigweed, which I value for greens but want to pull long before it goes to seed. Also, by planting earlier and thus lengthening the season, I believe I harvest an even greater quantity of ripe seed. Remember, the broadleaf grains don't ripen all at once like the grasses, nor are mine usually senescing (calling it quits) by the first killing frost, so why not just keep cranking along?

The most important reason for pre-starting amaranth has to do with my intercropping system. You see, Opopeo has bodacious stalks over 7 feet (2.1 m) high; I do believe some would grade out for veneer logs, but the very skimpiest would serve nicely as support for pole beans, much as sunflowers do. Indeed the sunflower analogy goes further: Both sunflowers and amaranth are fairly frost-hardy,

Figure 9.25. Opopeo amaranth grows over 7 feet (2.1 m) tall and fast—the crop reliably matures even here in Maine.

amaranth not so much, but far more than the pole beans. Therefore I can plant both support crops well in advance. Rather I *must* plant it earlier; if I do not give them a two- or three-week head start, the leggy beans will quickly outgrow them and then have no place to go. Amaranth in particular is such a pathetic threadlike wimp of a seedling, it seems to take forever to get established. But when it finally does take off it makes explosive growth, so much so that if the beans get *too* late a start, they'll suffer from the amaranth's shade. Amaranth tends to branch much more than sunflowers, which is fine with me, as the bean vines have more to cling to. Branching is encouraged by wider spacing, so rather than plant them one per foot (0.3 m)—or three plants per 3 feet (0.9 m) like sunflowers—I often give them 2 feet (0.6 m) in the row (and 3 feet between rows). I typically grow four to six 40-foot (12.2 m) rows.

Figure 9.26. Amaranth curing inside.

I've trialed some shorter varieties that are somewhat earlier, like Plainsman, but they are not as well suited for my particular system. At about 3 feet (0.9 m) in height they're bred for mechanical harvesting by combine. I must harvest my giant Opopeo by hand, which I'm going to do anyway, so why shouldn't I also get a second crop, the pole beans? I can't do that with the short industrial varieties.

I should point out that there are two types of amaranth bred for two separate uses: grain and greens. Although you could use either type for either purpose, it's not worth it. The greens varieties have black seed (like wild pigweed) and a strong flavour and tend to be late; the leaves of the grain types are totally edible, but I'm loath to reduce the plants' leaf area that is feeding the grain.

I don't pay undue attention to soil nitrogen for amaranth, as I must with corn. I rather strive for a high humus level and good drainage, which they really appreciate.

The first frost seems to check any further growth of the plant—it usually has stopped its upward growth by then to focus on ripening its seed—but doesn't destroy the plant. The foliage keeps its composure through several heavy frosts before it begins to wilt in resignation. The seed is loaded with cold-resistant cell salts and continues to cure. When I feel it has suffered enough of such treatment, I harvest with a machete and hang it indoors. I used to lop off the whole plant, but since that massive stalk takes forever to dry, I rather cut it where it branches out, usually 2 feet (0.6 m) above the ground; if there are grain-laden branches below that, I lop them off separately. In the overhead beams of my shed I have nails every foot (0.3 m) or so, where I hang the amaranth stalks upside down, hooking onto the lowest crotch. There I leave them for as long as I can, sometimes until there is snow on the ground and I am forced indoors. After a few days of drying I can consolidate them on fewer nails without their moulding. That frees up some space, which is always at a premium at that season. When the plants are very dry I align them carefully on a tarp, even if I'm using the indoor threshing floor, because the tiny seeds get caught in

Figure 9.27. Opopeo amaranth is easier to thresh by treading than by flail. Photograph courtesy of Yaicha Cowell-Sarofeen

Figure 9.28. Winnowing amaranth in a light breeze. Photograph courtesy of Yaicha Cowell-Sarofeen

cracks. I thresh the grain with my rough-shod feet, scuffing crosswise like a chicken scratching, turning the stalks occasionally.

Amaranth plants are said to be cross-fertile, which I do not doubt, so when I save several varieties in my Scatterseed Project, I isolate them by at least 220 yards (200 metres) to keep them pure. I'm not certain that is enough, because all the Chenopods (amaranth, quinoa, lamb's-quarters, spinach, beets, et cetera) are wind-pollinated, and therefore have very fine pollen that is easily wafted; so far, however, I've seen no signs of mixing between Opopeo or Plainsman, which are very different in height. It is also claimed by some that the grain amaranth and red-root pigweed are cross-fertile; if so, I've seen no indication of it, no intermediate types or mixed variants, although they have ample opportunity to cross.

Looking at my first amaranth crop I was somewhat disappointed to see that, although the yield of biomass was impressive, the grain yield seemed much less than, say, wheat, although they are said to yield comparably. When I hefted the bucket of grain, however, I was far more impressed. It was like lead, and I had to conclude that such a dense grain must really be as nutrient-packed as they say. The rich flavour also does not belie that. Even without the accompanying pole bean crop, I consider the space very well used.

Amaranth can be cooked whole like rice or millet, yet is it is extremely unlike those in taste and texture. Indeed the gelatinous (read "gloppy") texture of amaranth makes it rather unappealing by itself. Cooked separately and added to those lighter grains, the texture and flavour of both are improved.

We use amaranth mainly as flour, and that poses another problem. The seed is so tiny and hard, I must put it through the hand mill as much as four times to get it fine enough for me. Fortunately it doesn't take long to do that. Then it can be used in corn bread recipes, although I prefer it in pancakes, waffles, or corn porridge/polenta.

The ancient Aztecs and others have made a special confection by popping amaranth in a hot clay pot and mixing it with honey or other sweetener. I've done that, using a cast-iron pot, and it certainly works, although it's a bit of a trick to get it all popped before it scorches.

Quinoa

The first time I ever heard of quinoa (*KEEN-wah*) was from my sister-in-law Yolanda, who grew up in the mountains of Peru. There it is a staple food among the poor people (which is to say, everyone). At the higher elevations quinoa is one of the few crops that will mature, whereas rice, corn, and wheat must be imported from the lowlands at prohibitive expense. It is something of a paradox that the increased popularity of quinoa in the developed world as a trendy health food is threatening the food security of people in the Altiplano region who depend on it. So far 99 percent of the quinoa in the marketplace is imported from Peru and Bolivia, and the very people who produce it cannot compete with the purchasing power of the US dollar; they cannot afford to eat the very food they grow. Although US production in places such as Colorado is slowly ramping up, it is not keeping pace with the demand, which makes it all the more compelling that we gardeners should grow our own.

I loved the first bag of quinoa Yola sent me but assumed I could not grow it here; after all, Maine is a lot different from Peru. But it also occurred to me that the quinoa-producing areas were at high elevation where seasons were cool and relatively short; after all, they invented the potato. Yola's hometown of Tarma is renowned for its carrots and even rutabagas, not exactly tropical crops. The main difference between our growing seasons is day length. At our latitude we have exceptionally short days in winter,

Figure 9.29. Homegrown quinoa awaiting threshing.

but also exceptionally *long* days in summer, a phenomenon that folks living near the equator do not experience, nor their crops. Well, some quinoas (and some amaranths) are affected by day length and some are not, so it is all-important to plant a suitable variety, never mind the days-to-maturity. I've mainly grown Faro Red, which does all right, though I suspect I need to do a lot more trialing. My friend Mark Hutton is researching quinoa as a possible rotation crop for the potato growers of Maine's Aroostook County. It seems the main obstacles are diamondback moths and leaf miners, the latter of which I have observed as a minor problem in my own crop. The wild lamb's-quarters is a reservoir species for those pests, although the weed itself is not half as vulnerable to their damage as is the cultivated quinoa, surprise, surprise.

Quinoa doesn't require high fertility; indeed it does best with fairly low nitrogen levels, although I don't believe it objects to ample humus, and it does resent poor drainage and high acidity.

A not-so-helpful feature of quinoa is the coating of saponin that protects the grain from birds (that of course is helpful) but also makes it unpalatable to humans. If you insist on eating it unprocessed, the soapy stuff (what *saponin* means) will be very laxative. We've had a couple of generations now who have never tasted soap—presumably because their speech is more decorous than mine was—so take my word for it that you should first remove the saponin before cooking and eating quinoa. It's not difficult; simply line a large strainer or colander with cloth and rinse-wash the grain well, rubbing it by hand and changing water until any sudsiness is gone. Since the coating may help protect the quinoa in storage, it may be wise not to wash the entire crop all at once, but in occasional batches as needed, and of course re-dry any that's not for immediate use.

Quinoa can be direct-sown, but I prefer to start plants in two-inch cell trays and set them out when they're at least 2 to 3 inches (5.1–7.6 cm) high. They're very hardy, so you needn't wait until last frost to set them out. I set out quinoa plants for the same reason as amaranth: to avoid confusion with wild weed seedlings, in this case lamb's-quarters (nearly identical at this stage). As with amaranth I don't rush to harvest them at first frost; they may be done growing but they continue to cure, perhaps even ripen a bit more on the plant. Birds are not a problem as they sometimes are with amaranth, partly because of the saponin coating. The quinoa I grow doesn't exceed 3½ feet (1.1 m) in height, so I don't interplant it with pole beans as I do my amaranth. Semi-runner bean varieties might grow well among them, but I'm not sure quinoa is rugged enough to support other plants.

However, because they're so late to take off, I have grown quick early crops of lettuce and kohlrabi (both from transplants) in between them and they were out before the quinoa created much shade. (For more about this refer to *Companion Planting* in chapter 14.)

Quinoa is supposed to be "perfectly balanced" in everything, although some other skeptical killjoy naysayers claim there is a dearth of another amino acid: leucine. I am skeptical of the whole superfood concept, preferring to partake of as great a variety as possible in case some crucial nutrient (like maybe the element pandemonium) is missing from most of them. I'm often asked where I get protein in my diet, and I cannot give an authoritative answer, because as far as I know, I mostly eat stuff like cabbage and corn and beans and onions and wheat. If some of those have protein in them, well, I don't mind. Curiously no one asks about the protein in the food I grew up eating, probably because they know it was perfectly balanced Maine milltown cuisine, based on the four major food groups: Velveeta, Spam, Miracle Whip, and Marshmallow Fluff. With such a solid nutritional background behind me, I figure it won't hurt me to eat a little leucine-deficient quinoa once in a while.

One of the most popular ways of using quinoa in natural foods cuisine is in salads and for good reason. Its very rich flavour (I'm reminded of eggs, although I haven't tasted those in 40-plus years) makes a bowl of crispy, crunchy raw veggies feel like an all-sufficient well-rounded meal. Just to give you some idea of how excellent quinoa is, the United Nations named 2013 the International Year of Quinoa, and they never agree on anything! Maybe you think Peru has such clout in world affairs as to pull this off, but I gotta believe that quinoa sold itself.

By the way quinoa is said to have 14 percent protein, second only to legumes, if you want to believe in that kind of stuff.

CHAPTER TEN

Growing Pulses

The various pulses, or dried legumes, are pointed out as a critical source of protein for vegetarians. In a way that's correct, but the same could be said for grains. In fact neither group supplies in itself the complete set of amino acids that our bodies need to be healthy. Well, what do you know? My hay field reflects that same dovetailing of needs: grasses with legumes, as in timothy and clover, each keeping the system sustainably healthy. Long before Frances Moore Lappé wrote *Diet for a Small Planet*, my tummy told me that baked beans go with corn bread, rice with tempeh, rye bread with split pea soup, lentils with barley; no lab tests needed.

It is fortunate that the great variety of food legumes do well in my so-called temperate climate. In particular I can reliably grow dried beans, runner beans, field peas, favas, chickpeas, and soybeans, and so these are the crops I discuss in this chapter. You can also find information about growing the vegetable forms (such as snow peas) of these crops in *Vegetable Legumes* in chapter 8. My apologies for any redundancy.

The true beans belong to the New World genus *Phaseolus*. However, the word *bean* is also applied to several other pulses that belong to different species and even genera, such as soybeans, garbanzo beans (also called chickpeas), fava beans, even mesquite beans. The New World pulses, kidney beans and runner beans, are frost-tender, having evolved in Central America, whereas the Old World pulses are generally quite cold-hardy.

Growing and Harvesting Pulses

While it may be a challenge to grow all my own grain, producing all my own pulses is much more realistic, even though my vegan diet puts a lot of emphasis on these high-protein meat analogues. For one thing pulses don't require as much space: Of bush dry beans alone I can usually count on 10 pounds (4.5 kg) per 40-foot (12.2 m) row, perhaps thrice that with pole beans (even when sharing space with sunflowers or amaranth). My family and I eat three to four bushels a year of all pulses combined, and that doesn't require an unreasonable amount of space, especially since most of them are grown in combination with other crops.

Pulses are generally easy to grow. I don't bother applying high-nitrogen compost, as long as the humus levels in the soil are adequate—remember, the rhizobacteria on legume roots fix nitrogen from the air, but only if the soil structure gives them access to that air. What pulses do need are sufficient amounts of calcium, potassium, and magnesium, all of which are abundant in wood ash. They also need full sunlight and good drainage.

The basic processes of harvesting are shared by all the pulse crops. They should stand until nearly

Figure 10.1. Hand-threshing a crop of Jacob's Cattle beans. Photograph courtesy of Yaicha Cowell-Sarofeen

dry-ripe, but not much longer. The simplest, fastest method of hand-harvesting is to simply pull up whole plants, roots and all, knocking off the dirt if necessary. Although it takes slightly longer, if you clip off the plants near the ground and leave the roots behind, the crop will be cleaner and what is saved for seed will be less likely to carry soil pathogens, which may get into the seed coat. Moreover, the root nodules, those rhizobacteria colonies that are now charged with nitrate, will remain in the soil to break down and release their boon of fertility for the ensuing crop, which may not be as self-reliant as the legumes. The part of the nitrate that has been converted to the plant will soon go to the compost heap, and the part that's stored in the crop as protein will go into your tummy to strengthen you for further life. By the way those beloved cooties, the rhizobacteria, got there on their own in my gardens. I never inoculated my seed or my soil. All I did was create the conditions to let them proliferate, and they have.

Pulses are all processed the same way, by threshing and winnowing. I will describe the process as I do it for beans. Use the same flails as for grain; I have one with a 1¼-inch (3.2 cm) swingle, or dasher, and another of 1½ inches (3.8 cm). I begin with the lighter one so as not to send beans flying off in every direction; as the remaining pods resist shattering I switch to the larger one, which packs more of a wallop. For small seed lot samples I sometimes stick whole plants into a woven-fibre feed bag and beat them with the flail handle (or any other stick). Then I pull out most of the stalky debris; the beans and empty pods remain in the bottom.

Some people actually pick the pods off the plants, which is fine, but then you would need to either tread them out (as in a galvanized washtub) or shell them out one at a time, which is impractical with a large crop.

Once the coarser debris has been removed I dump the mixture of beans, pods, and chaffy dust into a 5-gallon (18.9 l) bucket and pour it into another bucket from a height of 2 to 4 feet (0.6–1.2 m) in a steady breeze. For more nuanced control or on calm days, I sometimes winnow indoors with a large floor fan (I don't mean in the house, silly; that would make a dusty mess overall). Repeated pouring back and forth will quickly remove all the trash. There may be a few whole unshattered pods; these I pick out and open by hand. Often they have resisted shattering because the seeds are not wholly dry-ripe, so it's just as well if they aren't mixed in with the drier ones. If they are really unripe I cook and eat them soon as shell beans.

After winnowing I pick over the beans, removing any broken ones or tiny stones (if you clipped the plants instead of pulling them, there shouldn't be any). However dry they may appear, I prefer to leave them in an open bucket or, better yet, on screen drying racks for a few days before I bag them up and put them in tight-lidded galvanized barrels. Beans are safe out in the shed, but peas, chickpeas, and soybeans need to be protected from rodents (see chapter 15 for suggestions). If they are full of chaff I winnow them enough to reduce the volume and finish them later in the barn, on a day when I need to work under cover. A few pod fragments mixed with the beans actually help them store better, as the drying air can penetrate them better. Properly stored in a dry, cool place pulses can be kept for a few years in fine edible condition, but when they get very old it becomes difficult to cook them tender, no matter how long they are soaked and boiled.

Beans (*Phaseolus vulgarus*)

For many folks, the hallmark of field beans is flatulence. Beans are the butt of many jokes (or the joke of many butts?) because of the paradox of their being "good for your heart," yet hard on the digestion. The embarrassing phenomenon is mainly the result of the formation of methane in stomachs unaccustomed to digesting certain amino acids from plant foods. I have always loved the flavour of beans, even when I was a meat eater, and presumably made as much noise as the next person. When I became vegetarian I partook heavily of all food legumes, partly because I like them, but also because others convinced me that getting enough protein was a major challenge. For months I had considerable bloating, which wasn't helped by my being such a hearty eater. I was also moving to whole foods in general and my system wasn't used to so much bulk, including a different balance of protein. Other than the discomfort of bloating I felt better than ever, shedding some health nuisances I had previously endured, and so I persisted. As months passed I ceased to be as affected and gradually I found I could eat any amount of beans with little or no "repercussions," as my gut became accustomed to providing more of certain enzymes and less of others.

From the beginning beans were the easiest food legume for me to grow, demanding little in fertility or attention. If I had a piece of newly turned ground, with all the fertility locked up in the just-ploughed sod and a legion of weeds poised for resurgence, I would plant bush beans, and if I was late getting onto the land, I would plant Jacob's Cattle in particular. I would not mulch, but instead would spend the first year hoeing and hilling, until the sod had succumbed and released its considerable fertility for ensuing crops (especially phosphorus, which is abundant in old sod and much needed by legumes). The ability of beans to thrive in lean conditions is why they were such a central part of the shifting agriculture (slash-and-burn) system upon which New World civilizations were based.

Figure 10.2. A thrifty crop of Jacob's Cattle beans.

Dried beans have likewise been an important cash crop for northern New England farmers—easy to grow, easy to store, easy to ship. Canneries that sprang up across the region were known as "corn-shops" because their primary product was sweet corn, but many of them also canned beans—baked beans or shell beans—which gave the canneries a longer operating season. The Burnham and Morrill Company was an industry leader, and the alliterative advertising mantra "Mmm, boy, don't them B&M Brick Oven Baked Beans from Boston taste good!" became a challenge we schoolboys struggled to master (although I somehow assumed *B&M* stood for "beans and molasses"). Indeed Boston was often nicknamed Beantown. Nearly every small farm in the region grew a few acres; and the hand-crank wooden winnowing machines can still be found in many falling-in barns. Baked beans are as much a hallmark of New England cuisine as chili and refritos are in the Southwest.

Dried beans are familiar to most gardeners, and I have few insights that are innovative. Basics of cultivation are the same as for green beans, especially the use of mulch, as described in chapter 8. I have gotten into the habit of companioning bush beans with potatoes. They are reputed to repel each other's pests, which I find partly true, but not as conclusively as reported. As I experiment with growing potatoes under hay, this has obvious implications for the beans. I cannot hill them up, but so what? The main reason for hilling potatoes is to bury the stem, which encourages formation of stolons (spuds), not a factor for the beans. The reason for hilling beans is to quickly and efficiently destroy small weeds and to support the plants from laying over when pod-heavy, and if the hay mulch accomplishes that, then fine.

There are a gazillion varieties of bush dried beans, all of them good, some better. My favourites are those that mature early, are easy to thresh, and have some resistance to anthracnose (a fungus that causes sunken pits in the pods) and mildew, as well as good beanie flavour and texture when cooked. Jacob's Cattle are among the earliest and most versatile. I grow a bright yellow bean from the Azores, which I believe is a Canario, surprisingly early with a lovely texture.

Pintos are not a variety so much as a type, and several varieties grow fine in wet New England as long as they are early and upright; my current favourites are Agate and Agassiz. In my garden semi-runners tend to get mould. I like Ternier Kidney and Ukrainian Sugar and similar meaty beans, which make a solid stew (superb in borscht), and Black Coco, with a quality somewhere between a black turtle bean and a horticultural bean.

Horticultural beans are a distinctly different type, typically but not always having a beige ground with maroon streaks. They are usually used in the "green shell" stage and as such are described under *Vegetable Legumes* in chapter 8. However many of them are also superior as dried beans, such as the Vermont Cranberry, True Cranberry, and Taylor Horticultural, and of course my favourite, Orlando's Horticultural. The very best green shell varieties, such as Tongue of Fire (which usually has wrinkled skin due to the soft starches not filling the skin tightly), are in my opinion the poorest for dried beans as they lack the denser qualities of a good stewing or baking bean; just heft them.

Pole beans are much more productive than bush types, but require more labour in setting poles and handpicking the pods, or stripping off the vines. Bush beans work better as a potato companion (so far, anyway), and pole beans don't have to climb on poles (as described elsewhere). For still greater space efficiency I mainly plant them with sunflowers and amaranth as natural support, as described in chapter 14. Therefore I grow plenty of both types. I've long thought it would be neat to have pole forms of my already favourite bush varieties—pole Jacob's Cattle and so on. Well, I have found a pole pinto bean, called Roark, which I am experimenting with while on the lookout for others. The vining trait is dominant and the pole type exhibits great variability, including largest and smallest seeds, unique robust flavour, and flamboyant seed coat colours. It is a generally swashbuckling crew of food plants that bear more investigation. They can't be machine-harvested, which means you're unlikely to encounter them in the marketplace.

Figure 10.3. Hefty white runner beans: It won't take too many of these to fill the pot!

Runner Beans

Runner beans (*Phaseolus coccineus*) are another sadly neglected food crop, especially as a dried bean. They are more commonly grown as an ornamental (especially Scarlet Runners) and a lure for assorted butterflies, moths, and hummingbirds. People who do use them for food seem to view them as a green bean or green shell bean, sort of a northern gardener's lima.

That's all as it should be, but to neglect them as a dried bean is just silly. I greatly prefer the taste of the not-so-pretty kind, the white runner bean, which are in almost every way like lima or butter beans, except for two things: They are much fatter and they reliably mature here, which no lima will. For more about growing runner beans see chapter 8.

Field Peas

Pea seeds can be wrinkled or smooth. The smooth types are hasty to develop starch, whereas the wrinkled ones are bred for "arrested development." Because of their denser structure the smoothies are often preferred for canning and freezing. Most people think that all garden or table peas are wrinkle-seeded and field peas are smooth. However, there are numerous table peas (and snow peas, too) that are smooth-seeded and can (to my knowledge) be used

Lentil Substitutes

I know that lentils can be grown here in Maine because I have grown them, but I have never gotten practical yields, and so I don't take them very seriously so far. I do love the taste of lentils, however. To my palate yellow split peas and black turtle beans, when mashed together, make an extremely similar and satisfactory substitute.

for field or soup peas. (But most soup peas cannot be used as table peas, although some of the older table pea varieties like the Alaska types were originally used as field peas and later for canning.)

As with the other Old World pulses the trick to growing peas of any type is early sowing. In fact most of the Old World legumes don't just tolerate cold but actually require cool, moist conditions for good germination and early growth. Most of their diseases result from drought stress and untimely heat. When you learn that peas originated in places like Ethiopia and Iraq, it seems intuitive that they must require long, hot summers. In fact it is just the opposite, which makes more sense when you realize that in those regions, such crops are grown during the short, cool winters (which are more like our short, cool *summers*), when rain is more abundant and the return of hot, dry weather is well timed to ripen the crop without mould or other complications. That is why those pea varieties recommended for planting after late May all have resistance to powdery mildew, drought stress, and more.

The main difference in growing field peas and garden peas is that field peas are not grown on a trellis; not that you couldn't, I suppose, in order to make space for a companion crop like carrots. You see, garden peas need to be upright and accessible for repeated picking, whereas field peas are harvested all at once when they're dry. That is probably the main

reason why commercial field pea production has largely moved to the drier valleys of the Northwest, where they can ripen the crop on demand simply by shutting off the irrigation: When vines are sprawling on the ground you don't want an unseasonable rainy spell. Because I do sometimes have problems with mould or sprouting in the pods, I've invented a compromise. I sow peas in beds, three rows each, 2 feet (0.6 m) apart. I don't mulch them until they are at least 6 to 8 inches (15.2–20.3 cm) high, at which point I first wheel-hoe between them (carefully, because legumes have shallow aerobic roots) and hill them slightly to help them remain upright longer. Then after mulching them I lay down some brush lengthwise between the rows. It matters little what kind of brush, as I am building no trellis, merely laying it there. As the vines begin to tip over I make sure the outer two rows tip inward, so they are all held above the ground by the brush. I have also done it with no brush at all; it depends on how much wet weather we have at harvesttime. Since that is usually late August I can often forgo it.

Some shorter varieties keep themselves fairly upright until harvesttime; on the other hand there are some bodacious long-vined varieties like Dun that reach 7 feet (2.1 m) and are very productive *if* you can keep them off the ground. Likewise heavily tendriled varieties are more self-supporting.

We usually think of field peas being either green or yellow, especially when they are split. In my hometown you could distinguish the Yankees from the French Canadians by the colour of their soup, which is probably why ours was usually green, but not always (Great-Grandma Poulin was Québecoise). Nowadays I grow a number of varieties of each, having hundreds in my collection to choose from. One favourite is the Russian variety Kazanskij, but others include Dashaway, Prince Albert, Lenca, Sivikka, and several other British and Baltic varieties. In general the varieties adapted to the Midwest states and prairie provinces do not do quite as well here in Maine. A few varieties in my collection have very small seeds, and while I believe they are intended to be forage peas, I find them excellent for pea sprouts, which are

Figure 10.4. Dashaway is one of my favourite dry, or soup, peas; its smallish size makes it suitable for sprouts as well.

Figure 10.5. The Raisin Capucijner pea is well named, especially as the seed coats darken with age.

better than mung beans in stir-fries. A black-eyed variety from India is exceptionally large; I like it for whole-baked peas (much like New England baked beans, but go lighter on the sweetening and heavier on the mustard powder). All of these are available from NGS, if not from my Scatterseed Project.

Speaking of large peas there's a type of field peas that are in a class by themselves. The hooded monks in the Capuchin monasteries of Flanders and Netherlands developed a food item that is much more to their credit than the cappuccino coffee invented by their Italian brethren. They bred a very large, flat soup pea we call by the order's Dutch name: Capucijners (most Dutch words sound familiar once you get around their strange spelling—who would guess that *Dertigdaagse* would sound so much like *thirty days*). These peas are either solid olive-beige or finely speckled with purple; in either case they age to a dark mahogany-purple. One variety is called Raisin, which it resembles in every visual aspect. Long cooking brings out the flavour, rich and robust, and forms a thick dark gravy. I love Capucijner peas, but to my taste they do not blend all that well with most foods, other than garlic. I might eat them with mashed parsnips or rutabagas on the side . . . .

Varieties of the Capucijner type are mostly long, heavy, rampant vines, and I have had the best success growing them on trellises. It may be worth it, especially since it frees up the adjacent space for some shorter row crops. What a surprise! Witloof chicory and brussels sprouts (iconic Flemish veggies) are especially well-suited companions.

Another class of field peas is the so-called blue peas of Northern and Eastern Europe. The seed coat colours are much like Capucijners, often to solid purple, but the shape is standard round, smooth field pea. I believe they are usually sown for fodder, and indeed their purplish pigment, anthocyanin, makes them more frost-hardy for early emergence in cold soils. Indeed I have used them plenty for green manure, but I also like them very much as a baked or soup pea.

For all field peas the timing of harvest is much more critical than for beans, as overripe peas will shatter off the vines. Squirrels and chipmunks are a major pest problem, especially if I grow the peas near a hedgerow with stone walls where they live. They can be largely foiled if I pull the vines when they are just short of ripe and spread them on an elevated drying frame to cure for a few days.

Chickpeas

Chickpeas are not a kind of pea any more than garbanzo beans are a kind of bean; they are rather a species all to themselves, *Cicer arietinum*. Unlike peas and beans, they are used only as a dry pulse; there is to my knowledge no tradition of eating shelled green chickpeas. No wonder, since chickpeas, like lentils, form one or two per pod, hardly worth the labour of shelling them out. In my experience they don't yield as well as peas or beans, and were it not for their distinct flavour and my love of diversity, I might not bother with them; how glad I am not to be growing things for the marketplace.

There are two types of chickpeas. One is the large roundish, usually cream-coloured kind (kabuli) you find in most US markets, either as dried beans or as hummus-bi-tahini (*hummus* is Arabic for "chickpeas"). Those are largely Mediterranean, and I have had generally poor luck with them.

The second type, called desi (meaning "local"), is more typical of South Asia and East Africa; those seem to be better adapted here. They are somewhat smaller and angular—the British term *ram's-head peas* is best depicted by these. They range from yellow to brown, green, and charcoal, but mostly dull red; they remind me of little chips of brick. The desi types are supposedly more nutritious; they are decidedly better in taste and texture. The only desi type I've seen in the US seed trade is Sonoran Red. Sonoran Red is actually a rusty brown, rather small but quite prolific and well adapted considering its provenance in the semi-deserts of northern Mexico. The variety Black Kabuli is a typically large, roundish kabuli type, though it has a charcoal seed coat. Black Kabuli is quite productive, with seeds fully as large as the cream-coloured types. It isn't pretty, but it is tasty.

Another of my favourite varieties is a beautiful large-seeded landrace from India; although it probably has a proper name where it comes from, I call it Clifford because . . . well it's big and it's red, and I hope to introduce it to US gardeners in the near future.

I sow chickpeas on or soon after May 10, and I am experimenting with intercropping them with chufa, or nutsedge. Chickpeas can have some disease issues, mainly ascochyta, but where I live those are largely avoided by early planting. I may be benefiting largely from the "pioneer effect," whereby a new introduction has temporarily left its enemies behind. As with field peas the main problem I've had with chickpeas is chipmunks and squirrels. Since my red-tailed hawks aren't quite thorough enough, I need to rely more on my Havahart traps and not planting near stone walls. It also helps to harvest when they're just short of fully ripe, even at the risk of losing a few late-ripening pods.

As much as I love hummus-bi-tahini I have had no success with growing sesame—and of course lemons are out. I have some alternatives (see chapter 17). My favourite use is mashed and dry-fried as patties, sometimes with mashed potatoes. For this the desi types, which are generally drier and nuttier, are particularly suitable.

Fava Beans

Those gardeners who are acquainted with favas at all usually know the "major" race: the large flat beans that are shelled out green. Those I discuss in the

Figure 10.6. Homegrown desi-type chickpeas are ideal for making hummus or falafel.

Vegetable Legumes section of chapter 8. But favas are also used as a dry bean in much of the world. In Latin America the majors are used for this, and they have mind-blowing arrays of colours and seed coat patterns. My Peruvian sister-in-law remembers a number of interesting uses, some of which I have made my own. The most unusual is Abas Tostadas, or toasted favas. These are dry-roasted in a covered clay pot until they puff up, somewhat like popcorn. I've tried it with some success but had difficulty getting them uniformly puffed, so some parts were still hard on the teeth and others scorched. Apparently there are fava varieties specifically for popping. The major type of favas are also soaked overnight and stewed, which I have done often, but I usually use the minors.

The minor race are small and off-round, similar to a peanut. The greatest diversity I've seen in this group are from Northern Europe, where the range in size and colour is dramatic. I find it curious because in those regions favas are largely used for fodder and green manure, where you would expect aesthetics to be ignored. Nevertheless we have in the collection favas that are bright green, maroon red, and dark purple. One Polish variety, Fialetovij Czyzich, is tiny, nearly black, and packed with anthocyanin and tannins that act like antifreeze, allowing it to germinate in soils that have barely thawed. I have eaten these and rather enjoyed them, but maybe that's just me. Since the pigments are in the seed coat, you might improve them for human consumption by cracking them (like split peas) and winnowing away the skins.

In the Near East and North Africa minor favas, or "ful," are used much as we would use kidney or chili beans, boiled and salted. One of my own favourite uses it is to grind the dried beans into flour, making a thick gruel to which I then add various veggies and seasonings (heavy on tomatoes, peppers, and garlic) and bake it in a long slow oven. I've also made it into burgers, or something like falafel. According to an old Aprovecho publication, favas can be used for anything soy can, including tempeh, which I have yet to try.

Early sowing is critical for favas, to minimize aphids, chocolate spot, and leafhoppers. Cutworms are so attracted to young favas that they can serve as a trap crop. If I do daily patrol of the newly emerged crop, I can spot and dispatch the pests before they do much damage. Even plants that have been felled will often regrow from below-surface buds. I sometimes grow dry favas amid my naked barley, planting what would be the paths between beds to the beans. It works best with the minor types, which I only need to access once, at maturity, when the barley is also ready. Unlike the other pulses I haven't noticed any depredation of favas by woodchucks or deer.

Figure 10.7. Lappland fava is a minor type that can be used like a dry kidney bean.

Figure 10.8. Windsor Jubilee is a major type of fava used as either a dry bean or a green shell.

Soybeans (*Glycine max*)

Many people are surprised to hear that dry soybeans can be grown this far north, yet I've been producing nice crops of them for decades. Indeed most varieties will not mature here, yet I have grown dozens of varieties of the right maturity groups. Some favourites are Kabott, Flambeau, Maple Presto, and Maple Arrow. Growing dry soybeans is similar to growing edamame, which I described in chapter 8. By sowing soybeans in my field corn I get somewhat less yield, but it is a bonus on top of the corn crop. I describe this in more detail in *Companion Planting* in chapter 14.

My main pest problem is woodchucks (or groundhogs), and my main remedy is the Havahart trap (for more on these traps, see chapter 15); however it's not always easy to lure the critters into the trap when they smell young soybean plants all around them.

Figure 10.9. Half hidden among the field corn, Kabott soy fills out the harvest.

CHAPTER ELEVEN

Growing Oilseeds

When I first began subsistence farming I was very determined to become self-reliant beyond mere succulent vegetables; I wished to grow my own grains and pulses (food legumes), and yes, oilseeds, too. One difficulty with the latter was the limited number of oilseed species that could mature on my western Maine farm. Many of the traditional oilseeds require much longer or hotter or drier summers than I have, so rule out peanuts, olives, oil palm, and probably safflower and sesame. I can grow rapeseed (canola), corn, and soybeans, but would the extraction technology be appropriate for my situation? Even though sunflowers, pepitas, flax, and poppies will grow well in Maine, how practical are they? How practical are *any* of these when I'm trying to reap a maximum nutritional value from every square foot of land? And what to do with the waste, the spent oil cake? It is ordinarily fed to livestock, but considering that the residue contains a host of crucial proteins, vitamins, and minerals that the human body needs even more than the oil, doing so made little sense to me. It occurred to me that most of those oilseeds are good foods in themselves, in their whole form, so why not use them in that form? After all, I prefer whole foods, but how "whole" is extracted oil anyway?

From that realization I went on to wonder how I might use more of these "oily seeds" (a somewhat different perspective from "oilseeds") as complete foods in our diet, instead of fractionating them and discarding some of the best parts. And so I experimented and learned how to grow and use these crops. In this chapter I discuss how to grow sunflowers, pepitas, flax, poppies, hazelnuts, and some experimental crops. For information on processing and utilizing oilseeds see part 5. I'll note that although rape is one of the world's leading oil crops, I do not grow it as an oilseed, because of our preference for using oilseeds whole. We have come to appreciate it very much as an early greens crop (as I describe in chapter 8) and as a sprouting seed (see chapter 17).

Sunflowers

Sunflowers were our first choice of oilseed crop, the main reason being their versatility: Their flavour blends well with just about anything, contributing without competing. They're easy to grow in quantity, doubling as support for pole beans (see *Companion Planting* in chapter 14). The main obstacle is removing the hulls. Sunflower milk is generally a fine substitute for dairy milk or cream, especially on cereals. It's easier to make than soy milk, without the chicken-y taste.

There are two types of sunflower seeds, each bred for separate purposes. The most common are the oilseed varieties used for pressing oil. Those are typically small black seeds, the kind you see in bird feeder mixes. There are many varieties, but the one most available to gardeners is Peredovik.

The other type of seeds, called confectionery, are what you find in the snack section of the health food store. They have less oil, though still plenty, but are larger and more hullable. The oilseed types don't need hulling, you just grind the whole seed before pressing. Since I intend to use them as a whole food, rather than just the oil, the confectionery types are of most interest to me. For people who mainly want oil and plan to feed the spent oil cake to livestock, then the Peredovik type makes sense. However, I have also developed an interest in the Peredovik types, which are useful not only for salad oil but also for industrial uses, such as homemade soap.

One reason we once ruled them out, aside from the whole-food thing, was that we saw no practical way of extracting the oil ourselves on-farm. I now own a simple bench-mounted oil press that cost less than $200, which works well for many kinds of seeds. For more about it, refer to chapter 17.

As for the confectionery type, de-hulling the seeds is the more complicated part, which I'll get to later, but growing them is fairly straightforward, so I'll start there.

I usually plant at least 150 hills of sunflowers (starting seeds first in 2-inch, or 5.1 cm, seedling trays, three plants per cell). A strict rotation is essential to minimize damage from sunflower moth larvae. I assume they have few alternative host plants in my area, because one year after an especially bad infestation, I held a moratorium on growing sunflowers, even forbidding them in my wife's flower beds. The following year I saw little or no sign of the moths; apparently their cycle was knocked way down and has never fully recovered. It's also wise to also take other precautions such as pulling all stalks as soon as possible after harvest, and putting them through a shredder. I don't always stay on top of this chore myself, but it has a significant control effect.

Another major pest is birds, particularly blue jays (aka gorbies) and chickadees. Gorbies are so beautiful and personable (meaning "like people"—not always a good thing) that I find it hard to hate them, though they certainly are exasperating. Of course chickadees are so cute and trusting, who can hold anything against them?

My main strategy for avoiding crop loss in sunflowers is timely harvesting: As soon as the heads are mature, I lop them off, often in two or more passes if they ripen unevenly. I know they're ripe when the seed coats are plumped out and darkened to a charcoal stripe and the back of the bract is yellowish. Also, if you begin to see bird-pecked heads, that's a helpful hint.

At first maturity the seeds are too tightly packed in their heads to be easily removed. To deal with this I partially dry the heads and then rub them against a screen to loosen the seeds. For a drying surface I secure a 3 × 20-foot (0.9–6.1 m) hog-wire screen to a frame of 2 × 4s. I use a couple of empty trash cans to hold the framed screen off the ground.

I spread the heads facedown in a single layer, where they dry out for a couple of days, which causes the seeds to loosen in the heads. To remove them I use a sifter made of ½-inch (1.3 cm) mesh hardware cloth and stapled onto a 24 × 30-inch (60.1 × 76.2 cm) frame of 2 × 4s. Placing this over a wheelbarrow or trash can, I rub the heads over the screen in various directions so that all the seeds come loose and drop into the container. Then I winnow away most of the dust and chaff by pouring it between 5-gallon (18.9 l) buckets in a light breeze. At this point the seeds are *not* dry enough for storage; they will mould in a tight container. I spread them out indoors on screened drying racks hanging from the ceiling for several days. I typically re-winnow them before storing in clean feed bags in galvanized trash cans. I close them tightly with bungee cords to frustrate raccoons, who can be very resourceful.

Pepitas

Pumpkin seeds, or pepitas, are easy to grow and process. The tight hulls WOULD present an insurmountable obstacle, but there are varieties of pumpkins that produce hull-less seeds. Actually, the seeds have hulls, but the hulls have been genetically reduced to flimsy membranes which are easily puffed away when dry. There are several degrees of hull-less-ness, but many varieties are essentially naked.

Figure 11.1. This assortment of hull-less *Cucurbita pepo* varieties comes from several countries and includes Yellow Stripe, Pulawske, Bush Lady Godiva, Giessener Ölkurbis, Velovske, Sweetnut, and some unnamed varieties.

Figure 11.2. It's easy to scoop the plump, buttery pepita kernels out of the shells. In most cases the shells themselves are fit only for composting, not for eating.

Their versatility is somewhat limited by their flavour. They contain sulphur-based compounds that give richness to their taste, somewhat comparable to eggs or cheese. The flavour pairs splendidly with savoury vegetable dishes, yet may clash with sweet or fruity dishes. However, this does not hinder me from using all I can grow.

One possible objection to growing naked-seeded pumpkins is that their mature flesh is useless. At least by most estimates, it's like old zucchini at best. On the other hand who complains that the sunflower "head" or bract is inedible or that the corncob is not fit for food? Indeed the densely packed nutrients in pepitas easily justify the waste, but wait—I have discovered a couple of pepita varieties from Eastern Europe that also have quite palatable flesh, even by my taste (and I am strongly biased toward *C. maxima*s such as

Buttercup squash). These varieties are not in the US seed trade, but I hope to make them available. Growing pumpkins for their seeds requires the same practices as growing varieties with regular-hulled seeds. For details about that refer to *Cucurbits* in chapter 8.

Flax

Flax is an easily grown oilseed for the Far North; much of the world supply is grown in Saskatchewan and northwest Russia. Of course flax is also the source of linen fiber, though different types are bred for the two uses. Indeed most flax cultivation today is of the oilseed type, since linen has been largely supplemented by polyester and other forms of ticky-tacky. Most of that oilseed is not used for food but for industrial applications, like paint and linoleum

(*linoleum* as in "flax oil"). Flaxseed oil has properties that make it both superior and problematic as food. It supposedly contains a high level of omega-3 oils, the type of lipids so important for heart health (again, that's hearsay, I've never seen an omega-3 in my life). On the other hand linseed oil when heated is quick to oxidize and form a natural plastic, hence its use in oil-based paints. I used to use raw or boiled linseed oil to hasten or retard the setting time of a canvas. Oxidation also makes flax oil rancid and unfit to eat (you should never chew on a Gauguin), which is why I am careful to grind only what I expect to use soon; whole seeds remain wholesome much longer. Of course this is true of all the oilseeds.

The two types of flax—food and fibre—are planted in different ways. For linen the seed should be planted thickly to encourage long, unbranched stems. Moreover the fibre extends down into the taproot, so you do not mow flax; you pull it and shake off the loose dirt after it dries and before retting. For seed—food or industrial—you grow varieties bred for branching, and you encourage that branching by sowing in drills at least 1 foot (0.3 m) apart and several inches asunder, to avoid the need for later thinning. Flax should be sown early, as it is frost-tolerant and likes ample moisture in its early growth. The plants themselves will adjust to the wider spacing by simply branching more (which of course would not do if your goal were fibre production). Flax does not need or want heavy nitrogen application, which tends to make it more vulnerable to flax wilt; however it does appreciate wood ash. I use 2 to 3 gallons (7.6–11.4 l) per bed, for strong stems.

At harvesttime remove the seedpods by "rippling." A rippling board is a large wooden comb clamped onto a bench. Fistfuls of dried plants are flopped onto the comb and pulled through repeatedly to pull off the pods, which you can then crush and winnow to release the seeds.

While the taste of flaxseed is generally not unpleasant, it doesn't blend well with just anything. The mildest-flavoured flaxes are the blond-seeded varieties like Foster. One particular way we've used flaxseed is ground into a meal, along with some flour,

Figure 11.3. I clamp this flax rippling board onto a table or bench and use it to strip off the seedpods from dried plants.

soy, and water to make a thin mucilaginous batter. Slices of whole wheat bread are dipped in that and placed on a hot cast-iron skillet "pre-oiled" with sunflower meal. This makes a rough equivalent of French toast, without the eggs. The rest of my family like it better than I do, but then I never was too excited about the real thing.

Poppies

I typically grow a couple of quarts of poppy seeds each year. To most people that seems like a prodigious amount, but in fact I could use much more, and I'm working toward that goal. The key is to think of poppies as an oilseed, as opposed to a garnish sprinkled on bagels. I use them in many of the ways I use pepitas and sunflowers, including toasting and grinding them into an oily meal that I sprinkle on steamed vegetables, much as you might add butter (more on this in chapter 17).

I usually grow poppies as a tall middle row in my companion beds, with a shorter vegetable crop on either side. Bulb onions are especially suitable as a companion, since they require full light, and poppies create much less shade than denser crops, like peas. Also poppies are very cold-hardy and can, indeed should, be planted quite early. The trick is to not bury the seeds too deeply. I don't bury them at all; I make a very shallow furrow (more as a row marker, really) and sow the fine seed thinly. Then I tamp the furrow lightly with the end of a steel rake or hoe to disturb the soil slightly, enough for some seeds to sift into the cracks.

However thinly you think you've sown poppy seeds, the seedlings will inevitably come up much too thickly and need thinning. Crowded poppies do not yield well, but a careful final thinning at this point can be mind-blowingly tedious. Instead I use my furrowing hoe, or onion hoe, to reach into the centre of the bed and casually chop out a chunk of the row every 6 inches (15.2 cm) or so. That still leaves clumps of the tiny threadlike seedlings; I let those grow until they're 1 to 2 inches (2.5–5.1 cm) in height, at which time I carefully hand-thin them to the final spacing, approximately 6 inches. It's okay to leave a couple of plants in a spot, but no more. If some are so close together that thinning will do damage, don't uproot them; simply pinch off the surplus plants, but don't lick your fingers unless you love horrible tastes and want to doze instead of work.

Occasionally I will have an empty gap of 1 or 2 feet (0.3–0.6 m) where nothing comes up. Invariably I'm tempted to transplant some thinnings from elsewhere. Silly me: Poppies are so resentful of being moved that if they survive at all, they will not thrive. Better to sow slightly thicker to begin with and then thin more heavily. Some people mix the seed with coarse sand to make the planting sparser but more even; it makes good sense. Therefore I've never tried it.

Poppies in single rows yield better than those planted in blocks, but they are more prone to wind-lodging when the pods fatten. I often put a few

Figure 11.4. Poppy in bud.

Figure 11.5. Červený Žiar after several generations of selection.

light stakes in the row and run a string on either side to help them keep upright.

The main complication with poppies is their legal status, about which you'll hear confusing and conflicting information. Some claim that the cultivation

Figure 11.6. Breadseed poppies ready for harvest (for food; much too late for anything else).

Poppy Musings

It may be delightful to experiment with mind-altering substances, but frankly I've always been so enchanted by the real world I live in that I've never felt the need to seek "spiritual enlightenment," whether by smoking, shooting up, meditation, prayer, or what have you. I'm just very dull that way. Even painkillers; I'd rather experience the discomfort and get past it; it bugs the hell out of my dentist. Not to be judgmental; I understand that other people find life difficult to face on its own terms, and substances derived from poppies, barley, grapes, hemp, mushrooms, et cetera help them get through it; I'd rather just eat them.

Whatever the legal niceties about poppy seed cultivation, they are clearly ignored—witness all the seed companies selling poppy seed for food or ornament. Once a National Guard helicopter circled low over my property for half an hour, waking the napping baby and endangering the weekend warrior who was leaning dangerously far out of the chopper door with his binoculars, cordially returning my cheerful wave as he scanned my fields. What clearly caught his attention was my gone-by asparagus patch with its ferny foliage, while he completely ignored my bed of opium poppies, whose big showy blossoms screamed to be noticed. Poor guy, probably couldn't wait to get off duty and have a joint; he didn't have any grudge against a friendly hippie.

That being said, you should know that if you eat a bagel sprinkled with poppy seed, you will fail a drug test, not because you have ingested any opiates, but because the test picks up on associated substances that are not illegal but *are* found in poppies.

of poppies is categorically verboten; others claim it is illegal only to produce and collect opium gum. (I've never tried to collect opium gum, but I understand that it's extremely simple, hence the law.)

A big advantage of opium poppies (aka breadseed poppies) over other oilseeds is the ease of processing them. The pods are harvested when dry-ripe and broken up in a bowl. After you shake the seeds loose, rake off the crushed pods and sift out the seeds from the finer trash. The fine dust can be blown free with a table fan. Remember that winnowing does not depend on seed *size*, but seed *density*; though poppy seeds are tiny, they are far denser than other parts of the poppy plant and thus easy to separate.

Actually an obstacle to using poppy seeds as ground meal is the colour. Yes, it is silly, but the blue-black seeds impart an ugly grey hue to whatever dish you use them in. Never mind the lovely flavour, it looks like someone dusted soot on everything. There is an easy remedy for that: In addition to the blue-seeded varieties, I also grow a white-seeded variety which looks much nicer in things. I originally got both varieties from my friend Peter Vido, who brought them from his hometown of Ziar, Slovakia (pronounced *zhar*). He says folks there often used the white-seeded type as a cheaper substitute for walnuts in pastry recipes.

Both varieties have a decided advantage over others in that they are not dehiscent, or shattering. Most poppy pods have little windows or vents in the upper rim that open as the seed ripens, allowing the seeds to be easily knocked out. A great dispersal mechanism, but not helpful for those of us who prefer to harvest and re-sow the seeds ourselves. A windy spell can leave the pods nearly empty; therefore it's quite nice that these Slovakian varieties have been bred to keep their vents closed, giving us some flexibility about when we collect them. The white variety is called Elka and the blue-seeded one is just Ziar. The latter has a light greyish-purple blossom that is not attractive. (Hey, I'm going to spend much more time seeing the flowers than eating the seeds, so I want them to look pretty as well as taste good.) I have been able to breed a magenta-red strain I call Cerveny Ziar.

Hazelnuts and Pine Nuts

There are some "oily seeds" that are not generally thought of as "oilseeds." That is, they are usually not grown for the purpose of extracting the oil, but rather used whole, which of course is what I'm looking for anyway. Nuts are a good example—not so much chestnuts or acorns, which are more starchy, but oilier nuts like butternuts, walnuts, hickories, and hazels. Butternuts are easily grown where I live—indeed, they grow wild—while black walnuts and hickories are easily naturalized. However, growing and harvesting these on the scale to make them a significant part of the diet seems problematic to me, especially given the great difficulty in extracting the meats. Not so with hazels, however, also known as filberts or cobnuts. Our native beaked hazelnut is rather small for practical use, while the large European filbert is not hardy here. There are some interspecific hybrids such as filazels and hazelberts that are supposed to have larger size than the American hazelnut (*Corylus americana*), yet are hardier than the European types. While I have had no experience with those, I found that *C. americana* is perfectly hardy here (its natural range is the Upper Midwest), yet the kernel size is quite adequate for my needs. I've been growing these by the bushel for many years now, with little attention and not much waiting (three to four years after setting out tree seedlings), and the processing is quite feasible with a common hand-squeeze nutcracker for small quantities. However, if I wish to consume these in large quantities, I need a more efficient device, and I believe I've found that in the Davebilt bench-mounted hand-crank cracking machine (see appendix A). It supposedly works well for most kind of nuts, but I can vouch for its efficacy with hazels and dried acorns, plus it is affordable and simply constructed. That still leaves the kernels mixed with broken shell fragments, but even picking out the nutmeats by hand is still a huge improvement over cracking nuts individually. It has enabled me to use my own hazels for nut butter and as an oily sprinkle on fruit sauces or steamed vegetables. As we say in Maine, that ain't too shabby, deah. As for growing the nuts I'll discuss that in more detail in chapter 12.

My experience with pine nuts so far is inconclusive. Although the piñon pine of the American Southwest is not hardy here, and I believe the European nut pines (*Pinus cembris*) may be marginal, the Korean nut pine seems fully adaptable—I have a few trees that have survived over 30 years here in fine shape. They have not grown very tall in that time nor borne a cone crop, but that is largely due to their being overly hemmed in by surrounding forest trees. If I release them I will expect much better results. They are closely related to our native five-needle white pines (which do not yield usable seeds) and supposedly can be grafted onto them, which would be a great asset if seed is unavailable or if I discover a particularly promising individual and wish to propagate it asexually. I have tried grafting them myself and had some tentative success with it at first, but all grafts eventually failed. I believe there may be some trick I don't know to grafting conifers (which have a totally different cellular structure from deciduous trees), and I certainly would not give it up.

Curing and removing the ripe seeds with slow heat is another process, and even then the "nuts" may mould in the shell if not carefully dried. I haven't tried this myself. That is not to discourage anyone from trying it, but so far I have some uncertainty about their potential as an important food source. There's no question about the delightful flavour of pine nuts—think macadamia with the slightest nuance of pine; as a crunchy salad garnish they're hard to beat. And while it may or may not be realistic to have a whole grove of these in anticipation of nut production, if you have the space you may just as well have growing a stand of nut pine sawlogs as of white pine, with at least some prospect of getting a valuable food crop as well.

Chufa

Another high-oil food crop, but one not ordinarily used as a source of pressed oil, is chufa, aka garden almond or tiger nuts. If you've never heard of chufa, that's not surprising, although the tubers were cultivated and eaten in ancient Egypt, and it has been a minor food crop ever since, mostly in parts of Africa and Spain. It is closely related to wild nutsedge, though it is not invasive and will not survive where freezing temperatures are common. Despite its grassy look it is a sedge, having a three-edged blade instead of two-edged, like the true grasses. Chufa dissolves soil silica and stores it in the foliage as a microscopic armour against insect attack. It is the silica (read "glass") coating that makes sedges so nasty to grab and yank.

Chufa needs 120 frost-free days to mature, but our days are not quite as hot as it prefers, so I start plants indoors (in 2-inch, or 5.1 cm, deep seedling trays) a couple of weeks early, for better yield. Well, also to get any crop at all, because wild turkeys, raccoons, squirrels, and other wildlife love the tubers so much they can clean out an entire planting in one night. I've successfully protected new-laid seed by unrolling hex wire over the rows until the young shoots emerge, then removing it immediately.

Figure 11.7. Chufa plants may look like harmless grass, but watch out for the sharp sedge leaves.

Figure 11.8. This chufa has over 80 nutty tubers hiding in the dirt.

Figure 11.9. Apprentice Isaac tries not to annoy the frogs as he washes the tubers in a screen.

I must make it clear, in case it isn't already, that the dried tubers *are* the seed. Sort of like potatoes except that, being dried, they are dormant and can remain viable for a few years.

So far I've found a spacing of 2 feet (0.6 m) by 10 to 12 inches (25.4–30.5 cm) to be ideal; I am experimenting with companioning chufa with a food legume like chickpeas, but so far results are inconclusive. Chufa requires moderately rich soil with steady and adequate moisture. My soil is naturally very rocky, and I grow chufa only in older beds that have been de-stoned to a considerable depth. When the first shoot emerges it is rather inconspicuous, but soon it forms runners, or tillers, which send up yet more shoots, so that by late summer each plant has formed a soddy clump with 50 to 100 marble-sized tubers. Anytime after the first frost, they can be dug, though I prefer to leave them a couple of weeks in case it helps them cure better. I harvest them by loosening around them with a spading fork and gently pulling the whole clump by its "scalp."

There will be several tubers left in the soil, and it is probably worthwhile to paw around in the dirt to find those, at least the larger ones. I haul them down to my backyard, where I spread them out on the same hex-wire screen that once covered the new seed. The screen is held up off the ground on poles set on a pair of inverted trash cans. There the clinging soil dries

out somewhat so I can take each clump and rub it roughly over the screen, pulling off most of the tubers, which drop through to a tarp on the ground, along with the loose soil. I put the mixture in buckets and carry it to my little pond, where I dump it onto a 2 × 3-foot (0.6–0.9 m) rack of ¼-inch (0.6 cm) mesh screen stapled onto a frame of 2 × 4s. Holding this just beneath the water I slosh it about some, working with my fingers until the muddy dirt washes through, along with tiny tubers I don't care to save. This is the same 2 × 3-foot screen I use to sift compost for potting mix and to grade hazelnuts. I also have a ½-inch (1.3 cm) mesh screen, and the two can be used to sort out different grades of the chufa. When I think I have washed them clean enough, I spread the chufa on drying racks (nylon window screening) suspended from my kitchen ceiling. I give them many days to dry-cure before storing them away in tightly covered metal barrels.

To use them I may roast them lightly like peanuts for a snack, but mostly I use them as a fine-ground meal mixed in porridge, or as the sweetened milky beverage "horchata". I haven't a recipe for that, but basically you grind the dried tuber (or blend the soaked tubers), add water and boil it, then strain it through layers of cheesecloth before adding sweetener and cinnamon (I prefer coriander or anise). Serve chilled.

CHAPTER TWELVE

Growing Permacrops

The most eco-efficient way of using any crop is to eat it directly, but that often isn't possible. We can't eat grass, so we must choose between putting it through an animal or composting it—the latter is more eco-efficient, but either use is indirect. Likewise the trees: We cannot eat the leaves, bark, wood, and so on, but we can parlay them into table foods, as I've described. However there are some human foodstuffs that can be harvested *directly* from the forest. They may never replace the cultivated foods on our plate, but permacrops such as chestnuts, acorns, and hazelnuts have a very great potential as important components of our diet. Since they can all be grown with chips or leaves as their main input, what better way to "spin straw into gold" than by parlaying that "rubbish" into basic foodstuffs?

The Permaculture people have an idea that seems at first blush like an obvious, hands-down, take-all winner: If we were to rely less on cultivated annual row crops and focus instead on woody perennials—especially tree fruits and nuts—how cool would that be? I mean, forget cultivating, spraying, weeding, fertilizing—get over the whole notion of disturbing the soil and just "manage" the plant communities: a little pruning here and thinning there and some mulch once and again. It is just so patently sensible, at least as far as it goes. My only objection is that it doesn't go far enough; that is, in terms of eco-efficiency. One aspect of Permaculture is the use of trees and shrubs to produce fodder crops (acorns are a great example) to fatten livestock (such as hogs), which saves us the labour of harvesting by having the critters forage their own. To my thinking this compromises the very advantages of permacrops. First the eco-efficiency gained by using these perennial foodstuffs is squandered by putting them through another animal and recovering less than 10 percent for our own use. Second forestland may be easily harvested by humans with little impact on the system, but running livestock into such areas is disastrous; generations of farm extension advisers have been urging farmers to keep livestock—especially large hooved mammals—*out* of their farm woodlots. It may be possible to minimize damage by understocking, but that only reduces the production to a level untenable for the farmer.

I often hear it said, "But nature farms with animals." True indeed, but just *how* does nature do it? On an area comparable to my farm (85 acres, or 34.4 ha), you might expect about one or two large mammals like deer—nature's stocking density. You might also expect to find several dozen smaller pawed critters—rabbits, woodchucks, and porcupines—along with hundreds of chipmunks, mice, and voles, and uncountabillions of worms, beetles, centipedes, moths, et cetera. *That* is how nature farms with animals, *not* with dozens of large grazers per acre. So if the carrying capacity of my farm allows for one big dumb mammal—and that's me—then any livestock must displace my wife and kids. If I insist on having them all, I am not sustainable—I am not living within my resources. I can only *appear* sustainable by bringing in lots of inputs from off the farm, and leave all that out of my accounting.

Prairie scenes of vast bison herds give the appearance of a highly productive landscape, but the picture belies the fact that those herds are occupying that relatively small area only for the moment. If they didn't keep constantly moving over a much wider range, they would quickly destroy that range.

Anyway, I've seen enough woodlands gnawed, stripped, trampled, and uprooted to want no part of that secondhand food. However, that by no means dampens my enthusiasm for Permaculture—I rather prefer a Permaculture that is less regressive and more progressive. Such a Permaculture would fit very nicely into my garden-without-borders.

The logical way to divide permacrops, in my mind, is to separate those that can serve as staples—crops such as tree nuts that are rich in protein, fats, or starch—from the fruity permacrops, like apples, hardy kiwis, and more.

Protein-Rich Permacrops

Many permacrops are hard to envision as anything *but* animal fodder. In some cases that is only because they have not undergone the same centuries or millennia of breeding and selection by interested humans as have our annual row crops. To date they are largely the result of selection by birds, rodents, and the plants themselves, and their breeding objectives are often not the same as ours. What if, for example, honeylocust (*Gleditsia triacanthos*) had been selected for several centuries for pod size, sugar content, and seed retention (non-dehiscence)? We might today have something comparable to carob for the North (more on which later). Of course such a breeding programme would require much longer regeneration cycles (because many tree crops require 15 to 20 years before their first fruiting). Also, because undomesticated species tend to be highly heterozygous, they don't segregate with any

predictability, which means multiple generations of selection would be required to create a stable result. How do you fund projects that far outlast the usual research grant cycles, that span the careers and lifetimes of individual researchers/breeders? And yet this is exactly the perspective of the workers at the Land Institute in Salina, Kansas, where they're working to domesticate perennial grasses and legumes into a kind of Permaculture. A similar effort with tree crops would require ever so much more patience—or would it? Many tree crops already have a demonstrated potential, much more obvious than, say, bundleflower or perennial wheatgrass. Some of them are useful as they are now.

I'm not thinking of stuff like maple syrup, as marvellous a food as that is, nor the more conventional orchard fruits—which are, after all, not forest trees. Moreover the more conventional fruits and berries are more in the dessert/condiment category. I'm all for expanding their place on the table, but they won't replace the staple grains and legumes. I'm looking for the entrée, that big belly-filling something in the centre of the table.

From my own limited experience I believe that chestnuts and acorns may be the most promising candidates for extending our gardens-without-borders. They give formidable yields per acre (except in some years when they give none at all); they require no ploughing, fertilizing, or watering. Nor weeding, at least not in the short run, though some population management is required to get maximum long-term results. I'm tempted to say "and no pest control," but that of course is ignoring the peskiest of all pests: squirrels.

Chestnuts

My introduction to eating chestnuts came while staying on a farm in Pennsylvania Dutch country, where dried-corn-and-chestnuts is a traditional dish. Later I had them with dried green beans and leeks—superb. There are dozens of bearing chestnuts on my land, but I harvest few if any nuts from them (my squirrels are fat and happy, though). The problem is that as fast as the nuts ripen, they drop to the ground and are stolen (that's my perspective, of course; a squirrel would say they are "rescued"). I have yet to harvest a mentionable crop of my own chestnuts, but my friend and kinsman Tom Vigue has eaten (as have I) lots of *his*. Therefore his experience will be more elucidating for you than mine.

At first Tom went out early every morning and carefully picked up everything there, missing nothing that might make a squirrel consider his visit worthwhile. Again at the close of the day, because the night shift is taken over by the deer who are no less interested in the sweet starchy kernels. Doing this every day he filled a 5-gallon (18.9 l) bucket with very tasty nuts, which had fallen free from their prickly burrs, which some cheeky wits call "tree-urchins." His method required no scare-eyes, noisemakers, or traps. Eventually, the squirrels figured out that they needed to beat Cousin Tom to the nuts *before* they fell. So Tom went to shaking the limbs daily, to knock down any that were ripe. It still requires relatively little of his time; what it does require is that rare commodity: perseverance. As Woody Allen put it: "Eighty percent of success is showing up." When, as in my case, you tackle too many things you don't excel at any of them, and in this matter of chestnuts a day late doesn't mean less, it means nothing. Ignore my example, learn from Tom.

About now, you may be asking yourself: "How is it that these guys have chestnut trees at all? Isn't there such a thing as chestnut blight?" Indeed there is. Before 1900 the stately American chestnut was a pillar of climax forests throughout much of the eastern United States. A shipment of logs imported from China carried in its bark some spores of a lethal fungus to which the Asian chestnuts had some immunity. Not so the Americans; within decades the native species was hammered to the brink of extinction; to this day the American chestnut shows no sign of developing any resistance. It might have fallen over the brink were it not for its tenacious ability to send up fresh sprouts from the died-back stump. These youngsters may live for several years, but typically as they reach fruiting age they become susceptible and succumb to reinfection. Today they are partly protected by their rareness, especially at the extremes of their range where there may be no other infected chestnut trees within miles.

I have about 40 trees from parent trees in Minnesota (they in turn came from Windsor, Ontario). I planted them in the early 1980s, and they are very healthy, as of this writing. Some are several inches in diameter and over 50 feet (15.2 m) tall. To my knowledge they are pure American, so why haven't they succumbed? I can only assume that my isolation has preserved them thus far and that a stray breeze might any day carry in the lethal spore that could doom the whole grove.

Meanwhile, I'm not waiting for doomsday. The late Elwyn Meader, a plant breeder at UNH, collected some Korean chestnuts and crossed them with the Americans, creating a fertile hybrid. He back-crossed that to pure American, and got an F2 hybrid (called NH4) that is three-quarters American, one-quarter Korean. I saw the parent tree on his lawn, and it definitely was not blight-*free*, but it certainly was blight-tolerant: There were numerous lesions that healed as quickly as they formed. He gave me seed of this NH4, from which I started a second grove, a few hundred yards from my pure Americans. They are my backup; to date, both groves seem perfectly healthy, but I know that could change abruptly, in which case I would still have the hybrids, infected but unfazed. They themselves will not infect the Americans unless the blight blows in, which would expose them all. I'm merely hedging my bets.

The American Chestnut Foundation has been dealing with this blight issue for several decades. Their main approach has been to introduce resistance through creating hybrids with Chinese and American chestnuts and then repeatedly back-crossing to the American. You see, the Asian species is fully blight-tolerant, but it tends to be an inferior tree form with multiple and stunted trunks compared with the towering columns of the noble American. That may be of little importance for nut production, but you see, the American chestnut is also a formidable lumber tree; its long, straight logs are incredibly rot-resistant. The fine-grained light-brown wood is harder than dry bones and takes on a lovely finish in furniture.

I find its toughness unsurpassed for tool handles and hay rake teeth. To combine these traits has been TACF's objective, and they are hot on the heels of success; after several tree generations of back-crossing they now have trees that are $^{31}/_{32}$ American, having all the desirable traits of the full American but also complete resistance conferred by the $^{1}/_{32}$ Asian parentage. In time they hope to offer tree seed to the general public so this noble tree can reclaim its rightful place in the American landscape. The significance of this work is hard to over-appreciate and deserves our full support.

Others are working on blight-resistant crosses such is the Dunstan hybrids and the Douglas hybrids (which Cousin Tom has). These are all, to my knowledge, quite acceptable for food production.

Just by the way, most of the world's chestnuts are grouped into four species: American (*Castanea dentata*), European (*C. sativa*), Chinese (*C. mollissima*), and Japanese (*C. crenata*). They are *all* interfertile, which is to say they can readily cross-pollinate. The Asian species (if indeed they are several) have larger nuts, whereas the Americans are said to have better flavour and higher fat. I consider all these species and hybrids to be promising tools for putting chestnuts back in our forests and on our plates as a dietary staple.

Once past the squirrels, there is still the problem of drying. The fat kernels contain lots of moisture, and left in their thick papery shells they may mould before they're dry enough to store. It is very helpful to at least tear into those shells to let moisture escape as they cure. Of course, they are quite easy to shell out by hand if you only want a handful for fresh use, but I'm assuming you're using these as a staple food and have bushels to be processed. There is specialized machinery for doing this on a commercial scale, but I don't know what options the backyard grower has; I've yet to harvest a crop with which to experiment.

Acorns

Acorns have every bit as much potential for staple food production, but they have their own hurdles to surmount. There are no blights or other disease problems to contend with, as far as I know, but as with chestnuts squirrels are to be reckoned with. Fortunately for us the oak tree itself has evolved a strategy for coping with this. In natural oak groves, if the trees all bore heavy crops of acorns consistently, the squirrel population would simply expand to equal the supply, and the oaks would be hard put to fulfil their own imperative: propagating themselves. Instead they restrict their nut production to a certain modest level, which supports a squirrel population of some size but in effect holds down the rodents' numbers. Then every so often the oaks have what is called a mast year, when they produce so many acorns that the squirrel population can't begin to consume them all. In fact the squirrels will hide more acorns than they can possibly remember to find, and many of those will sprout to become more oak trees (which is the main idea, though don't try to tell the squirrels that). What an elegant strategy, devised not by any one oak tree, nor even the oaks as a species, but rather by a whole forest community, including even the squirrels who play an unwitting role in their own self-discipline. Isn't life marvellous? I do have to wonder: If humans got involved in this, say by breeding oaks with consistently abundant yearly crops, would we thereby sacrifice this natural strategy for outwitting the acorn predators? I'm not suggesting that such breeding would not be a good idea, just that we would then have to deal with the squirrel problem ourselves.

In my state we have three major species of oak: red, white, and burr. Only the red oak is endemic to my particular area, but the other two are perfectly hardy and easily introduced. I have a grove of burr oaks and have harvested a few nuts from them. The burr oak is the lowest in tannin, the red oak is the highest. However, my friends Chris and Ashira Knapp (who consume literally bushels of acorns every year) say the reds are bigger and easiest to separate (tight shells), and Chris considers the reds to be less bland-tasting once you remove the tannin. He has much more experience with acorns than I have, so I defer to his opinion. On the other hand, if the burr oaks and whites require much less leaching, might their nutritional value be higher? I don't know.

Acorns for Civilization's Sake

There is something very important we can do to improve the usefulness of acorns, indeed to make them a domesticated crop. We can improve them by selection. Not "we" as in you and I, but "we" as in civilization as a whole. None of us will live to accomplish much, but we can make a beginning. There are at least two things we can be selecting for right off. Low tannin seems like the obvious one, yet it may be the least important one, since whatever we do, we will still have to leach the acorns anyway. Perhaps more important is to select for large and consistent yield. When collecting wild acorns from the same trees year after year, we will notice certain individuals with consistently better yields than their neighbours. We should select seed from those acorns to plant in our own groves, and in turn select and reselect from those trees as they come into bearing (precocity might be another good trait to select for—who wants to wait 20 years to see the first crop?). You may only live to select one or two generations, but that will create a base on which others can build.

Of course this assumes someone else will be interested in following up, and I absolutely do assume that. A garden-without-borders is also a garden-without-calendars; it sees not only beyond this immediate place, but also beyond this immediate time. Look beyond the current season to centuries and millennia to come. Are our gardens merely something to occupy the spaces until they get paved or are they the beginning of a sustainable future, one with room for snail darters and spotted owls?

Chris and Ashira do know a good deal about acorn nutrition, however, as behooves anyone who makes any item such an important part of his diet. At considerable expense they have had a sample of their acorn flour lab-tested, and the results seem to confirm their high opinion of acorns' adequacy to nourish humans. *Their* acorn meal processed by *their* method contained 55 percent carbohydrate (mostly starch, no sugar), 26 percent fat, and 7.5 percent protein; not too shabby, I say. It would have been interesting to know the B-vitamin content, as those are more water-soluble and could be adversely affected by the long leaching.

By all means plant oaks on your land, but not for yourself, rather for your children's children. Then raise your kids to appreciate such great natural tastes (as Chris and Ashira have), so that someday they will value this easy-gotten bounty. Meanwhile, scout around your area and find some old bearing trees that you can share with the squirrels (don't be afraid that you'll leave none for them; they'll see to that).

Keep an eye on your oak trees. If you notice when they first begin dropping acorns, you get to keep your share from the squirrels. Not that you want to start collecting the very first ones—it's probably not worth it—but as they begin to litter the ground in quantity, you should be aware of it and act accordingly. Prompt collecting gets you ahead of the squirrels and also leaves you with fewer mouldy nuts to cull out. Some years there may be wormholes (perhaps from a curculio) in some of the shells. It's good to pick those anyway and burn them; you'll knock down the pest cycle for future years, especially if these are trees you intend to return to each year.

Chris treats his acorns like I do my hazels: He spreads them out on racks away from rodents to dry for several days. Unlike hazels, there are no outer husks to thresh and winnow away, so the thoroughly dried acorns can be simply stored away in squirrel-proof barrels in a cool place until we're ready to continue processing. Chris does his in large batches throughout the year, as his family uses them.

When I helped him do a batch we ran a bucket of dried nuts through his hand-cranked Davebilt cracker (see appendix A). Very quick, very effective, and only a very few nutmeat halves stayed stuck in their shells. By dumping in a bucket of cold water we quickly floated off the shells and most of the papery inner linings. After re-drying the kernels and picking them over for any defects or debris, we ground them in his Corona mill. At this stage the meal is a beautiful tawny colour and smells like pecans, but it is still completely inedible, trust me. Next we mixed the meal with enough cold water to make a thin batter or slurry. This we poured onto a screen drying rack lined with pre-filter material used by the maple syrup industry. I assume any fine clothlike sheeting would work well. As the water drained out of the slurry, we placed a piece of the same material on top of it and set the tray under a faucet opened to a slow trickle. The material spreads the water out so it seeps through everywhere, instead of just washing out a hole under the faucet.

Left like that for 12 to 24 hours, depending on how cold the water is, the acorn cake will become thoroughly tannin-free and pleasant-tasting. Chris and I had a discussion over the relative merits of ice-cold and tepid water; it seems to me that the longer colder leach would have less risk of losing water-soluble vitamins and minerals, whereas Chris wondered if the shorter processing time might compensate for the slightly warmer temperature. The only way we can ever resolve the question is by more lab tests. It is hardly a trivial question for someone who makes acorns a very significant portion of his diet.

After leaching the drained meal is spread out on other drying racks (no more than ¼ inch, or 0.6 cm, deep), cloth-lined to keep the meal from sifting through as it dries. When the meal is half dry Chris breaks it up into a lumpy mass, which greatly accelerates further drying. Then it can be stored until use. If the original grinding was on the coarse side, the meal can now be reground for any use requiring a finer texture.

Some sources caution against storing acorn meal for too long, as its relatively high fat content

(compared with, say, wheat, but low next to most other nuts) makes it prone to rancidity. For Chris that is no problem; what he processes in a single batch will easily be consumed by his family in two to three weeks. Now, his family members are not die-hard wild food enthusiasts who will eat anything as long as it's good for them. His kids know what they like and what they don't like, and they like acorn dishes. These include acorn cakes, little unleavened burgerlike things that are two-thirds acorn meal and only one-third oat flour (no wheat), and are richly delicious, and acorn porridge that may be combined with cornmeal or other things.

Hazels

Although I see chestnuts and oaks as the power players in a staple tree crop system, lots of other woody plants have an important contribution to make. American hazelnut (or filbert), for example: While not native here in Maine (unlike the prickly beaked hazel), it is fully hardy. I have about 40 shrubs, each 12 feet (3.7 m) or so tall, and almost every year I get plenty of nuts from them. Not evenly, however: Perhaps half a dozen plants account for most of that yield. Therefore I have been propagating those and crossing them to replace the less productive ones. I have allowed too many of them to grow into thickets, which they can easily do from their many root suckers. I'm in the process of cutting those back to fewer than six stems/trunks and mowing more diligently between them. Their wont to produce those root suckers makes it easier to propagate them as clones rather than rely on seedlings, which are a less certain proposition. Like the other tree and shrub crops, they require nothing fancy by way of fertilization. In fact I have never fertilized them beyond some wood ash (when I could spare it) and mulch with leaves or chips. They are a forest shrub and would live quite happily in the shade of taller trees; indeed I have planted many as an understory in my burr oak grove, although without enough sun they won't fruit well. The ones I'm eating from are out in the field. They do lend themselves very nicely to planting on forest edges, such as on the perimeter of a newly made clearing, or the margins of a hedgerow where garden crops would never thrive due to the tree root competition.

American hazels scoff at the coldest winter, although the teeny scarlet blossoms can be wiped out by an unusually *late* freeze. They're not easily duped, but I've lost one or two crops that way. The plant itself is intrepid, but the nuts themselves can be decimated by curculio, which also pesters my apples, black walnuts, and plums, though it may be a different species. No doubt the remedies used for orchard trees (kaolin clay and the like) would work for hazels, but I find I get satisfactory control in the hazels by scrupulously picking all the nuts, including *all* the fallen nuts, and burning the ones with the little hole in the shell. I'm not sure *why* this helps, because it appears that at this stage the curculio has hatched out and left, but it seems to help keep them from building up. You'll occasionally find a little pucker-scar on the husk of an *un*holed nut. That just means that an adult oviposited there (Latin for "laid an egg") and it failed to hatch; no worries.

Other major pests include red squirrels and gorby birds (or blue jays, for folks who know better). These can wipe out an entire crop, but here vigilance pays high dividends. When the crop is just coming ripe, get right out there and strip the bushes. How do you know when? Watch the gorbies; they'll let you know. Red squirrels are devious sneaks; the crop will be gone before you realize they are around. But the gorbies are not subtle—after all, they're just crows in blue raiment. They will start a bit early, picking off the precocious nuts, and if you don't notice them you're just not paying attention—they're very raucous. More than once I've heard and thought: "Uh-oh, I'd better collect those sometime pretty soon." I get what I deserve. Other years—when I've paid attention—I've collected several baskets of nuts in their husks, so let's talk about those.

Some of mine appear to be a superior clone, because they not only bear heavily per bush, but bear in clusters of two to four nuts rather than singles. Those are the plants I propagate to sell as nursery stock. The husks that earlier were green and sticky (not prickly like the long-beaked native hazels) are

Figure 12.1. This young hazel is bearing in its third year, barely waist-high.

Figure 12.2. This superior hazel clone bears four large, full nuts per cluster.

Figure 12.3. Stomping a tubful of dried hazels is the quickest way to thresh them clean. Photograph courtesy of Yaicha Cowell-Sarofeen.

Figure 12.4. Winnowing hazels in a stiff breeze. Photograph courtesy of Yaicha Cowell-Sarofeen.

now just beginning to turn crispy brown on the edges. If I am timely at picking I'll collect lots of sticky husks full of hard brown nuts. If I am *not* timely I'll find scads of empty husks on the ground. A warm dry attic floor is the ideal place to spread hazelnuts for curing, but only if you can exclude (or trap) rodents. In several days to a week the husks will have gone all brittle brown. I put them in a washtub to thresh by treading.

At several points I stop and pour the whole business from bucket to bucket in a breeze or in front of the stream of air from a strong floor fan, repeating until most of the nuts are husk-free and the rest seem like they're not going to be. Then I pick them over to remove any with holed shells and burn those. Some years I've harvested such large amounts that I resorted to a less work-intensive method: I dumped the nuts into a tub of cold water and raked off the floaters.

It's quite effective. The sinkers can be air-dried and stored in galvanized barrels until they're wanted.

So far, so good, and if you only wish to process a handful now and then with a nutcracker, well, you're all set. But if you aspire to using those scrumptious kernels as an important item on your table, not just a snack, your work has just begun. The cracking technology used to be the mainliest thing that kept me from eating every hazelnut I could grow. There are various machines on the market for cracking nuts, including hazels, and their videos all look very impressive; however, the one I have actually tried is the Davebilt, which seems to me a great device for the price. An assortment of spacer-washers let you fine-tune it for various nut sizes.

With any device it would be very helpful to pre-grade the nuts, which vary considerably in size, but

Figure 12.5. Clean hazelnuts ready for cracking.

Figure 12.6. Using the Davebilt to crack hazels efficiently.

that's simple enough: I had pretty good luck just shaping ½-inch (1.3 cm) screening into a round trough and rolling the nuts down it. Once cracked you still must pick the nuts out from the broken shells, which is tedious enough in itself.

Anyone who is thinking of hazelnuts (or any nuts) merely as a snack food is totally missing the point. Do we not use nuts as staples because they are too expensive, or is it the other way around? Hazelnuts can be homegrown in many regions where they are not commercially viable, just as home-scale wind energy can be produced in many locations where those 300-foot (91.4 m) towers wouldn't be profitable. The marketplace is a tyrant that keeps us from living as well as we might. I assume that if the market demand for hazels were to skyrocket, so would commercial production, until the market price would

actually be lower than it is now. If not, it wouldn't matter to those of us who don't eat there.

One of the most exquisite hazelnut products—indeed foods of any kind—is hazelnut butter, with a texture like peanut butter and a flavour like nothing else. After you've eaten a slice of hot whole wheat toast with hazelnut spread, you may conclude that orgasm is greatly overrated. I've always used hazelnuts fine-chopped in salads (especially coleslaw), or ground as meal and added to steamed veggies, especially greens. Any amount of hazels—as much as you can grow or afford to buy—can be readily used in various entrées: burgers, roasts, or casseroles, adding fat, protein, and most of all rich flavour. Of course, granola without chopped hazels is like showering with a raincoat—I mean, what's the point?

Caragana/Peashrub

Nut-bearing trees and shrubs can be excellent producers of oils and starches, but what about protein? Those are also there, but consider: Among annual row crop plants an important source of protein-building amino acids is the legumes. There are many leguminous trees as well, and some of those are food producers, notably carob, but that doesn't grow here. Caragana, or peashrub, is a Central Asian woody plant that is fully hardy, although most of the ones I've planted have succumbed due to being in an excessively wet area. They certainly don't mind cold, or drought, or sterile soil, so I'm replanting them with that in mind. Supposedly they have been used as famine food, and even then mainly for keeping poultry alive. That hardly sounds like a resounding endorsement, but I've eaten them and found them underrated. I happened to come upon a landscape shrub on the Farmington campus of the University of Maine. My timing was just right: Although many had shattered onto the pavement, I was able to collect a cup or so to bring home. I boiled them up like mung beans, which they somewhat resembled in size and shape, and ate them plain. Some sources say they have a slight bitterness—I tasted none whatever, though I don't remember pouring off any cooking liquid. If anything, they were a bit bland. I could easily imagine them replacing dried peas in patties, soup, and the like.

It was a scanty trial, but enough to convince me of their potential as a staple food. The question is: How do you collect enough to be useful? They have one of the flaws that many of our cultivated plants had to overcome before becoming cultivated: dehiscence. As the seeds, or peas, mature, the drying pods spring open and scatter the seeds all about, a very effective means for a peashrub to disperse its offspring but a fatal flaw for anyone wishing to develop this into a horticultural resource.

Obviously caragana as a staple food plant is not without its shortcomings, and a lot of time and research may be required to surmount those. On the plus side, this is a legume: a pea or soybean that grows on an undemanding woody shrub with no tillage. It is only about 12 feet (3.7 m) tall, and the foliage isn't very dense, so it could easily be integrated into an alley cropping system where cultivated row crops get planted in between. The time and research needed to realize this species' potential as a cultivated crop will not come from agribiz or the USDA—the effort is far too long-term and the prospects too uncertain. Like many other permacrops, this doesn't lend itself to the industrial farm model that those entities are geared to serve. The effort will rather be done by steadfolk—gardeners and subsistence farmers who can devote a little space to half a dozen peashrubs, and who have the passion to mess around with these specialty crops, trying out random ideas and sharing those ideas with one another.

Honeylocust

Another northern tree-legume that may have some food potential is honeylocust (*Gleditsia triacanthos*). With this one I'm relying largely on hearsay, as the four young trees I once planted got choked out by weeds and brush; no reflection on them, my bad. However, before succumbing to my neglect they were surviving just fine. They're not native where I live, unlike the black locust (*Robinia pseudoacacia*), but whereas black locust is supposedly toxic, honeylocust holds its "beans" in a long pod whose pulp is sugary and edible. During the Great Depression the Tennessee Valley Authority (TVA) investigated using honeylocust as a possible alternative crop to corn, on mountain slope fields where erosion was threatening to silt up the newly constructed reservoirs. They came up with a few native varieties whose pods had exceptionally high sugar content; however, the onset of World War II pulled interest away from that research and it was never resumed. All of the superior selections were hardy in the southern Appalachians, which is of little use to me, and our northern honeylocusts do not seem to develop the high concentrations of sugar. I do not know to what extent that is genetic, or whether our low heat-units simply will not develop the sugars in any variety, no matter where it originates. That is not to say that the more northern honeylocusts aren't usable, especially if you're extracting (and thus concentrating) the sugar from the pods. The Ashworth honeylocust in Canton, New York, is supposed to have exceptional sugar content for a northerner. (When I tasted a lingering pod in winter, there was little sweetness, but that's to be expected.)

Even if we achieve hardiness and sugar content, we still have the question of *how* to use the pods. I believe the TVA focus was on grinding up the whole pods (dry, obviously, and including seeds) to mix with corn for livestock feed, not for direct human consumption. I have read (read-say is essentially the same as hearsay, although the printing process seems to confer more authority on it) that the early settlers on New York's Long Island ground up whole pods and mixed them 1:3 with corn for making porridge that required no further sweetening. That's encouraging, but I have no personal experience to back any of this up.

Fruity Permacrops

The permacrops we've discussed so far are notably sources of starch, protein, and fats, as opposed to the fruity permacrops, which are rather sources of sugar, vitamins, and minerals. We mustn't disdain those more succulent foods that, after all, assist in the digestion of other foods: They supplement grains in a different way than legumes do. Although we tend to use them as dessert or condiment foods, that is really neglecting their potential contribution to the diet. For example, rather than merely jamming or jellying them, I use most fruits as purees or dried (and reconstituted in compotes and the like). I add little if any sweetener to these purees (I find they don't need much) and I use them in much larger quantities: a slab of whole-grain bread or griddle cakes, or waffles slathered with puree as thick as I can pile it on. For purees like rose hip, of which I never have enough, or oliveberry (autumn olive), which by itself can be a bit overwhelming, I usually stretch them with a larger quantity of pear or apple puree (I say "apple puree" as distinct from "applesauce," which for many implies added sugar, cinnamon, et cetera). Because my purees are made from whole fruits (without first

coring or peeling; they get strained out after), I get much of the health benefit from pectins and other ingredients, plus flavour locked in the skins and seeds that would otherwise be discarded first; even after those coarse parts are sieved out, much of their goodness ends up in the puree. I freeze or can most of my purees.

Of course most purees can be dried into leather for snacks. I do that a little, but if I intend to dehydrate fruits (which lend themselves to dehydration better than most other foods), I prefer to dry fresh fruits directly without first cooking and pureeing. Some berries, notably elderberries and lowbush blueberries, make exceptionally fine "raisins," and are superb in baked goods, fruit stews, and more. I dry plenty of apples by the "schnitz" method (cored and cut into 12 to 16 wedges, like orange segments), but I also do a lot as chips: I put them through the coarsest cutter of my Saladmaster, cut into cored quarters with the peel against the cutter. It comes out as coarse shreds, each with a tiny bit of peel on the end. I typically use these with coriander in oatmeal, just as some might use raisins with cinnamon. Pears also make a delectable dessert when dried, but their syrupy succulence makes them trickier to dry than apples. For that we neither peel nor core them, but slice them into thin cross sections. If you dry pears when dead ripe they will surely go mouldy on you, but if slightly unripe and firm, they will ripen somewhat as they dry. We ignore the cores when eating them; most of the tiny seeds fall away and are easily discarded. Anyway I don't waste too much time or effort drying pears when the puree is so versatile. It is not so much pulpy as syrupy, nearly colourless, and very sweet in a bland way that lets it blend well with tangy fruits without clashing flavours. Pear puree on pancakes completely replaces the need for any other syrup or sweetener. Seckel pear in particular is an all-sufficient pancake syrup; add sunflower meal and skip the butter. It's easy to see why pear juice, like white grape, is used to sweeten canned fruit products. Moreover it is very easy to produce and bottle in quantity. Both hardy kiwi and oliveberry (autumn olive) are very tasty as puree, as I describe in chapter 17.

Most of these permanent crops are very minor, little-known species that have only a sort of cult following among rare-fruit enthusiasts. Some are wild or have seen very little improvement. Which raises the question: If these things are so blinking tasty, why aren't they in all the supermarket produce coolers, along with the apples, pears, and bananas? Indeed some of them may not find great acceptance among the general population, especially among the white-food types; however I think the greater problem is that, for one reason or other, they don't lend themselves to the marketplace. They may not store or ship well or cooperate with mechanical harvesting and processing technology. Okay, so they may never break into the mainstream cuisine; does that mean they don't deserve a welcome place on our own plates? Hardly, but they'll never get there if you're the sort who eats off-the-shelf—talk about restricted

diets! Even locavores will rarely get to taste these delicacies. You just have to grow your own.

I'm not going to write extensively about growing apples or berries or any of the fruit crops that are widely covered in the literature. I'm loath to cause the deaths of more spruce trees just so I can repeat what others have already said better. Instead, I'll focus on lesser-known information or an unusual perspective about popular fruit crops and on niche crops whose usefulness is generally neglected.

Cane Fruits

I grow blackberries, raspberries, and black raspberries (that's in my order of importance), and the method is essentially the same for all, so I'll use blackberries as the sample case. When I first set up my blackberry patch (some unknown old neighbourhood heirloom), I did nothing by way of site preparation. I simply

Figure 12.7. As blackberries begin to ripen and put on weight, it's important that they have strong support.

spaded a slit in the sod and inserted the plants. No compost, no ashes. I watered them once and mulched them with sawdust and chips from the wood yard, hardwood mostly. Nowhere near heavily enough; lots of grass poked through as the blackberries got established. The second year, a much heavier application of chips and sawdust nearly annihilated the grass, though never completely. I'm okay with a little grass anyway; when cut it provides a bit of nitrogen-rich mulch for the berries without being allowed to compete. Moreover, as years pass the lushness migrates into the grassy aisles, where it thrives, borrowing fertility from the adjoining mulch (remember the cover-the-earth principle from chapter 3?). That grass does get mown and some ends up on the mulch, but the big player here is clearly the chips. Every spring I have added 1 or 2 inches (2.5–5.1 cm) of woody junk. Every year has seen the canes growing taller than the

Figure 12.8. There are actually people who don't care for blackberries; please pray for them.

previous year, until their tops are mostly beyond my reach without a stepstool, and for each of the past 25-plus years they have never failed to inundate us with more finger-licking, mouth-staining berries than we could possibly consume without help from friends and neighbours.

The main adjustment I've had to make was not piling all the woody stuff in the centre of the row. It doesn't actually smother the blackberries, but it does encourage them to sucker at the outer edges of the bed, in effect making it a double row, which I don't want.

The other adjustment I had to make was in the height of the supporting trellis. I originally spaced four rows 8 feet (2.4 m) apart, which seemed very generous at the time; oh well, can't change that now, but I also thought that 6-foot (1.8 m) scorched-butt cedar posts would carry three heavy cross-wires to which I could tie each season's new canes. I soon had to add extensions to raise the top wire to 8 feet, and even now some canes overshoot that wire so much that a heavy crop makes them break over. I've also had to devise a system of braces to keep the posts from collapsing toward one another under the great weight of the loaded canes. I remember seeing blackberry canes in Washington State that would actually pull down small outbuildings; okay, I get it now, and this is not a rain forest climate.

Most of our *black* raspberries get eaten fresh, although if we paid more attention we could surely get some jam or something. The *red* raspberries are used fresh or frozen, maybe for smoothies. But the serious eating is in the blackberries. Unlike the juicy hollow-cored raspberries, the blackberries are solid meaty things, bursting with sugar and flavour that makes the others taste wimpy. We pick *many* gallons of these and freeze them whole, unless there is time to puree them. A bowl of plain or lightly sweetened thawed-out ripe berries is not too shabby, but since I always get blackberry seeds wedged in my teeth, I personally prefer the strained puree as a topping on waffles, sweetened with maple syrup and thickened with toasted sunflower or poppy seed meal. In this case I freeze the whole berries and then make puree after thawing some. It makes a grey-purple mess that's ugly as sin, but the taste is to die for. I find we utilize a lot more blackberries that way than as jam smeared on toast.

Here is a heads-up for any of you who think you don't care for blackberries: You've never had them fully ripe. They should look dull black and sun-gorged, not firm and glassy purple like you buy them. If you've tasted properly matured blackberries and still fail to appreciate them, I really can't think of much to say in your defence.

Blueberries

Blueberries are really two different crops: highbush and lowbush. They are treated somewhat differently. Highbush are the only kind actually domesticated and cultivated; lowbush are technically wild and are merely managed at best. I'm not aware of any named cultivars of lowbush blueberries. They are generally not planted, but are "grown" in certain regions where there are naturally adapted, such as the thousands-of-acre blueberry barrens of Washington County in eastern Maine. If the soil is not already suitable for wild blueberries, there is little you can practically do to make it so. They require dry acidic soil of glacial origin. In moist neutral soils they will soon be choked out by lush grasses. I have an area in my woodlot that was a beautiful fir grove before the previous owner logged it off, leaving a lot of resinous slash and coniferous needle residue. There were no blueberries growing there, but I tried the unusual strategy of setting out a lot of plants from a worn-out patch of my neighbour's hay field, where they had established themselves naturally. I slitted them into the acidic duff between fir stumps and slash, making a slight effort to remove the latter into piles. I did nothing further to tend them, relying completely on the natural suitability of the site to accept them or not. I calculated right; when I revisited the site a few years later, they had spread from a 3 × 3-foot (0.9 × 0.9 m) spacing to fill the area quite solidly. They did and do require a modicum of management to keep brushy suckers of grey birch and pin cherry and hardhack from crowding them

out. Wild blueberries are traditionally "managed" by fire, but that's probably not a realistic option for me, given the nearness of the forest. The size of the area does make mechanical brush suppression feasible, as with lopping shears. I accomplished quite a lot in a few hours with loppers, though that was long ago. Even more efficient is short rotary mowing, which not only cuts and grinds up the woody brush and any grass or weeds that volunteer, but also prunes back the old woody blueberries and encourages fresh new fruit-bearing growth. I have not yet done that with my Gravely-powered mower, because the patch is difficult of access, but I know that the rotary mower would also grind up half-rotten old fir stumps and any protruding old limbs. That practice was very effective in the patch of my neighbours, who mowed the surrounding hay field at least once a year and the blueberries themselves once every few years. If

Figure 12.9. Patriot, a highbush blueberry known for its size.

I laid low all that coniferous slash in my own patch, it would become an acidic semi-mulch, encouraging the blueberries and thwarting everything else. Again, any of those strategies would be futile if the site were not naturally very dry and acidic.

Highbush blueberries are a different species and have a somewhat different ecology. They like richer, moister soils, though still acidic; in fact whenever I have found them growing wild it is usually under pine trees near a lakeshore. They are not managed by fire or by mowing close to the ground, but rather by periodically cutting out old, diseased, or weak branches. Highbush blueberries appreciate compost. That of course promotes grass growth, and so we mow around ours at least thrice each summer. Ours are on a plot that used to be excessively wet until I dug a drainage ditch all around about, in effect creating a huge raised bed. Only slightly raised, however;

the dampness and richness of the ground would never be tolerated by lowbush blueberries.

For years we underestimated the need for acidity, thinking an occasional application of pine needles was adequate. We attributed the meagre growth and constant dieback to a lack of nitrogen and tried to remedy that with mulch and buckets of urine—the very things it did not need, which only increased the grass competition. One year we gave in and applied some aluminum sulfate. The effect was dramatic: vigourous new growth with no dieback and, for the first time, plenty of blueberries! Since then we occasionally reapply the aluminum sulfate, but the main focus is on mowing and either removing the grass or letting it rot in place. I want something more home-grown than the aluminum sulfate, and so I am shifting to a partial mulch of conifer sawdust and chips, including pine and cedar (which are highly

Figure 12.10. Long after the season is over, blueberry bushes continue to be highly ornamental.

dis-recommended for use elsewhere). I don't mulch thoroughly; I'm not trying to kill out the grass there, only to give the blueberry plants some preferential advantage over it. I continue to mow.

There are a lot of named varieties of highbush blueberry, most of them better adapted for New Jersey than for Maine. I have a number of them, but my two favourite varieties are Patriot and Northland. Northland is a prodigious yielder of small (for highbush) berries. Patriot (released for the 1976 Bicentennial) was bred by my friend Elwyn Meader and has a more moderate yield of exceptionally large and tasty berries. It is also perfectly hardy here.

Elderberry

Elderberry is getting a lot of cachet lately. Its fruit is supposed to be rich in vitamin C, all kinds of antioxidants, anticarcinogens, immunity enhancers, and I

don't doubt any of that, but what I like most about them is that they make my spit wicked gorgeous. I have a row of elder bushes along the hedgerow. I was concerned when I read that one source says the plants repel witches—I do so hate to be inhospitable— but another source claims they attract them, so until there's more clarity on the issue the elders stay. Despite their slight astringency I nibble a fair amount of the ripe raw berries—I mean, who doesn't want lovely purple saliva? Most of them I dry for use in Squirrel-Folk Biscuits, a traditional Esperian treat (that I describe in my novel, *Through the Eyes of a Stranger*). We occasionally make elderberry pie, which others in the family appreciate more than I. I like elderberry raisins better in compote, with blueberry raisins, and maybe schnitz (dried apples), all boiled in cider with yeast dumplings plopped in toward the end. Seriously, I do appreciate the

Figure 12.11. Cultivated elderberries: Eat them to stay healthy and hock pretty.

wellness-enhancing qualities of elders; I just don't get as excited about foods whose main claim to fame is that they are healthy. I will surely eat those, but I will fill up on the stuff that is, first and foremost, scrumptious. Like kiwis.

Hardy Kiwis

Visitors are often surprised to see our rampant kiwi vines—everyone knows they grow in places like New Zealand. Ours are not that sort, not *Actinidia chinensis*, the fuzzy egg-sized fruits in your chain grocery. Actually the New Zealanders don't call them kiwis—that's what they call themselves (except for the Maoris, who originally coined the word to describe a very un-British-looking bird). So anyway, the Kiwis call kiwis Chinese gooseberries (you're still with me on this, aren't you?), but there is a much smaller and hardier species, *A. arguta*, which is indeed from

Figure 12.12. Fully ripe hardy kiwis are immorally delicious.

northern China and does vaguely resemble the gooseberry (I'm truly sorry about this, but I didn't do it). *A. arguta* has come to be known as hardy kiwi, which would clarify things nicely, except that there is yet another species, *A. kolomitka*, which is even hardier and supposedly sweeter (that's hearsay; I've yet to taste kolomitka).

A problem with kiwis is that they are dioecious, which is to say male and female sex parts occur on different individuals. One male is sufficient to pollinate a large number of females. That's fortunate because these vines do require lots of space—mine are 16 feet (4.9 m) apart in the row. On the other hand, if anything happens to that single male, there goes all the fruit production. The danger of losing a whole kiwi plant is pretty minimal once they've gotten established for two or three years. Mine have repeatedly tossed off –15°F (–26.1°C) and worse and gone on to produce fine crops. The greater hazard is a late-spring freeze, once the leaves and flowers have budded out. In 2011 a late freeze wiped out all the young leaves on several of my old vines. Unfazed, they popped out a new batch of dormant buds and forged ahead; however there was a greatly reduced crop of fruit that year.

I really like those fuzzy green-brown eggs from the store, but no way do they compare to the hardy kiwis right off the vine, a complex blend of evocative fruity essences. I have suggested that members of opposite genders should avoid eating these in mixed company because of the intensity of their lush aromatic decadence. When not fully ripe the kiwi can burn the mouth, sort of like the rind of fresh pineapple, but even if you have to harvest them a little unripe they can be held in a basket in a warm place until they get soft and mellow. When I've left them on the vine until overripe, they drop to the ground, usually still good though sometimes the flavour has gone blah. I should mention that kiwis are not improved by cooking; in fact it brings out the oxalate in their skins, which is rather unpleasant. I prefer to store them as a raw puree, much like oliveberry. Squirrels and blue jays are cause for some concern; let me know if you figure anything out. The good news is that kiwis usually

Figure 12.13. Even the vines of hardy kiwis are full of character.

produce so much, you can afford to stand some pilfering. Since the pests can't store them like nuts, they seem less aggressive in raiding the vines.

I grow eight varieties of female kiwi, but my hands-down favourite is the Meader female (there is also a Meader male). The late Professor Elwyn Meader discovered those in the hills of Korea in the late 1940s. There already were some hardy kiwis growing in estate gardens along the Maine coast where they were popular as an ornamental, with their clean glossy dark foliage and reddish stems. In fact, it's uncertain whether their fruits were even used or appreciated. Some taste very good, though not up to the Meader. There has also been some kiwi breeding in more recent decades. Michigan State is a larger elongated variety that Cousin Tom considers even tastier than Meader; I demur.

I have three separate kiwi trellises using two different support systems: the overhead bower and the vertical fence. Proper pruning gives better and more consistent crops. Mine are rarely pruned enough nor promptly. That works for me because I also grow rooted cuttings to sell as nursery stock; the long unpruned vines are very valuable as cutting material. But sooner or later, pruning must be done.

As for propagating new kiwi plants see chapter 7.

Oliveberry

Autumn olive (also called oliveberry or autumn berry) is one of those fruity permacrops that people either swear by or swear at. Everything I find in the literature agrees with my own experience in both ways: It is an incredibly productive food source and it is potentially an invasive nuisance. I find overstatement in both camps: Those who say the berries are consistently delicious have very different taste from mine. Most plants bear small, slender berries with lots of bronze specks and a single large seed. Only the most uncritical enthusiast could enjoy eating those. Tear them out, say I. However, every so often I come across a bush with plump all-red berries, few or no freckles, and a relatively small, easily chewed pit (25 percent of the total). When fully ripe these are truly delightful to snarf right off the bush. They're worth keeping and even propagating. Their flavour has been compared, quite aptly I think, to pomegranate. The freckles are filled with something disgusting, tannin I assume, so fat and red is what to look for.

It's true that oliveberry can become an invasive nuisance. Given the fruit's popularity with birds and animals, and its ability to fix atmospheric nitrogen (though it is not a legume) in sterile soils, it's not surprising that it would spread. Some states considerate it an aggressive invader and have eradication programs involving the herbicide Roundup (which must be just wonderful for the native species they're trying to preserve). I have mixed feelings about it: Apple is also an imported invader that goes feral in many neglected hay fields, but who suggests we shouldn't have planted apple orchards? I think the key word is "neglected." I find it easy to reclaim places in my fields where autumn olive is encroaching simply by regular short mowing, at least twice per year, preferably more. Suckers will resprout, but as long as I'm diligent with mowing, they'll starve out, even sooner if I improve that grassland so it competes better. Include the edges, especially the edges; oliveberry is the quintessential "edge species"—don't let it hide out there. I mow right to the borders of the woods and even into them a bit, letting no feral plants escape and reinvade the field. The same applies to wild apple and chokecherry. Young oliveberry plants are particularly thorny (actually I believe they are leaf spurs, but they hurt just as bad), but less so as they age or as they are pruned—after all, that's part of domestication.

Speaking of creatures being attracted to the berries, the flowers are no less of a draw. For many days the vicinity will be redolent with a hard-to-describe perfume. Every bee, moth, and butterfly in the neighbourhood will be attracted to the nectar-heavy blooms and walking past them will stir up quite a buzz.

Using oliveberry is easy enough. Aside from downing them by the fistful, seeds and all, they make an elegant raw puree when mashed and put through a food mill. The pits are left behind, while the skins seem to largely disappear. If the berries are all red and plump and you *still* find them too tart, let them ripen on the bush for some more days until they are at least as sweet as pomegranate. I do not cook them at any time, as it seems to bring out the acidity in them. When I thaw out the frozen puree, a thin, watery yellow liquid separates from the solid gel. One source suggests discarding that liquid to remove the acidity, but I assume they are using generic unselected berries. I prefer to remove the acidity by removing those plants that make puckery fruit, rather selecting for palatability. Whenever I discard any part of a food, I have to ask myself what nutrients I may also be discarding. We certainly select preferred portions of all food plants, but I must at least give it thought; my garden-without-borders extends into my kitchen and my doctor's office.

Many recipes I've seen involve added sugar, pectin, and lemon juice, the jelly maker's staples, but I prefer to use my own foods with less outside input, and in this case it hardly seems necessary anyway. Jellies are nice little condiments to smear on a slice of toast, but I'm interested in giving fruit, particularly these minor fruits, a more prominent role on my dinner table. If I find the taste a tad too robust for eating in large quantity, I can mix it with my apple or pear puree and heap it on the bread or spoon it out of the bowl. My preference for keeping the fruit, or its product, unheated commits me to dependence on the freezer, which is a mixed blessing: Canning isn't the most nutritious way to process food, yet the freezer links me to the grid with all its vulnerabilities, a risk my own photovoltaic power supply only partly mitigates.

Oh yeah, they say there's an awful lot of something called lycopenes in oliveberries (as in tomatoes, of the genus *Lyco*persicon) and that it is wicked good for you, preventing dyspepsia, heartburn, tennis elbow, melancholia, and other bad stuff. That's really neat, but I kinda dig that they taste great, too.

Dogwood Cherry

I try to avoid writing about stuff I don't actually do myself, and when I do go astray I mean to at least warn you that I'm giving you indirect information, or hearsay. Well, I have some hearsay that's too good not to share, and my source is the most absolutely reliable: Cousin Tom. Now, Tom and I have been diligently researching the dogwood cherry (*Cornus mas*), aka cornelian cherry, which is to say that Tom has been growing these things and processing them, while I've been tasting his results and generously sharing my opinions. My share of the work has been quite demanding (as when I demand more of them), but, hey, someone's got to do it. Tom has two named cultivars: Elegant and Pioneer, both of which are excellent. Elegant is a long slender fruit, while Pioneer is plumper, but still not very cherry-shaped. Either is delicious when fully ripe, but therein lies the main difficulty with this fruit: Its window of maturity is small, so you must be diligent. A day early and its tartness can be off-putting; then for a week it is sublime, after which it soon becomes blah and loses texture and character.

Fruit Leather

Cousin Tom has two preferred ways of eating dogwood cherries: fresh out of hand, which is great while they last, or pureed and dried into a fruit leather. He further snips or chops the leather into raisin-sized pieces for use in corn bread and other baked goods. This is an excellent way of using many of these minor fruits whose size or shape or texture or pits make it challenging to use them in the fresh state.

Although these varieties are commercially available they are rare. The breeders have thoughtfully patented them in the public domain, so that someone like Monsanto can't steal them, but they're willing for others to propagate and sell them freely. (That's only helpful if you know how; see chapter 7.) When I recommend various crops and varieties of crops, the fact that they may be available in the marketplace at the time of publication says nothing about the following year or decade. It's all up to you; you must learn to propagate your own.

Although some sources refer to difficulties in getting *C. mas* established, Tom found them seemingly indestructible from the outset, even under neglect. His responded well to a permacrop regimen of woody mulch. My own experience fully confirms his observation, although I did find that one of my nine plants was susceptible to mechanical damage: It died when I mowed it down.

Rose

For many permacrops to attain a more accepted position in our food system, it would help if they were subject to the same rigorous breeding and selection as our more established crops. In various cases the removal of dehiscence or increased sugar content or reduced tannins or annual bearing would all help us to popularize these wonderful food-givers. One

species is already there, but it isn't utilized as much for food as it might be if it had fleshier hips. Now, "fleshy hips" usually evokes images of Renaissance goddess art, but I'm talking about the thin dry wall of the rose fruit, in particular *Rosa rugosa*. I can think of nothing that would contribute more to the edible usefulness of this species than a thicker pulp in proportion to the seeds. No improvement in flavour is needed, yields of fruit are already acceptable, hardiness is a given, eco-efficiency is remarkable. I would just like to get more puree per plant and per hour. And puree is what we're talking here. Rose hips are nothing much to nibble on: the seeds and fuzz feel like you've been gnawing on a rabbit's foot, and a scrawny one at that. Sure, you can dry the hips for vitamin-rich tea, but if I'm healthy and hungry I want food, not medicine. Puree is the way to go.

Like so many shrubby permacrops, roses are highly eco-efficient: They require little and give much. I don't recall ever fertilizing my hip rosebushes, unless you count a heavy mulch of woody debris—sawdust, bark, chips, and old boards—as fertilizer. Of course the rotting sod releases a onetime boon of nutrients, especially phosphorus, which roses love, but perhaps the woody mulch is best thought of as an herbicide. In any case, that alone causes the roses to proliferate; any grass that pokes through can be eradicated by yet more chips. *R. rugosa* does have a potentially annoying habit of sending out shallow runner roots (à la quack grass) that sucker and spread. However, the woody mulch causes those runners to grow near the surface—in the mulch rather than in the soil—where they are easily yanked out and destroyed or used to propagate more.

A rose without thorns would be something other than a rose, but *R. rugosa*'s thorns are more like prickers, thickly spaced but easily avoided if you don't glom right onto them—more of a warning. This as opposed to some rose species whose long spines are designed to rip and tear; like Aesop's fox, I assume those probably aren't very good to eat anyway. The flavour of rose hip puree always seemed to me like a blend of apricot with a hint of tomato, which may not sound so appetizing to some, but I find it delightful. That is to say, delightful if I could just get enough to use by itself. Failing that, I stretch it with an equal or greater amount of apple or pear puree, and the blend is mutually complementary. Rose hip puree seems unharmed by heating, so canning and freezing are options.

And yeah, rose hips have astronomical levels of vitamin C, but don't hold that against them; they really taste great.

Ribes

Whenever the conversation comes around to currants and gooseberries, the question always pops up: "Isn't there something about those being illegal?" Yes, indeed, so let's go right there first.

Long ago it was discovered that there was a connexion between certain species of the genus *Ribes* and the parasitic fungus that causes blister rust in white pines. It turns out the fungus has two stages in its life cycle, either of which is crucial for its spread. The *Ribes* is described as an "intermediate host," although I suppose the white pines might just as well be considered thus if their commercial importance were less (my native Maine has never been known as "the Gooseberry State"). Given that if you remove either host, you'll stop the disease in its tracks, it seems only reasonable that the *Ribes* must go. With that laudable intention, generations of forestry students and WPA crews have worked summer jobs eradicating all currants and gooseberries. The only problem with this scenario is that decades of plant workers have reached the conclusion that gooseberries and native American red currants (which I believe includes the "whites") are not only guiltless in spreading the fungus, they are themselves immune to it. The real culprits are the European black currant and some wild American species, possibly not limited to *Ribes*. At least one black currant variety, Consort, is claimed to have immunity, although I am not certain whether it is still capable of being an asymptomatic intermediate host. Loving black currants as much as I *don't*, that's rather a moot point. I have seen many healthy plantings and wild thickets of red currant thriving in the immediate vicinity, even in the shade

of old white pines, all parties seemingly content, though well within the range of fungal infection. Where I *have* seen white pines afflicted with rust, I have invariably found a wild relative known as skunkberry growing nearby. Nevertheless, no one in any legislative capacity seems to have gotten the memo, though efforts have been made, in Maine at least, to clarify the ban, which prevents a potentially useful crop from entering our commerce.

The good news here, especially for those of us who do not rely on commerce for all our nutrients, is that no one seems obsessed with enforcing the law about *Ribes*. About *Ribes*, I find no record of anyone actually being hauled in for making their own gooseberry jam, although it may be an issue Homeland Security is just waiting to pounce on.

Given that commercial vacuum where do you obtain red currant or gooseberry plants? While there are some limits to my indiscretion I believe anyone with a modicum of resourcefulness will manage that for themselves. Be advised that any nursery will not knowingly ship *Ribes* to "white pine states" (including Maine and New Hampshire). If you live within one of those states perhaps you should be lobbying your state legislator.

I have no unique insights into growing currants and gooseberries beyond my usual mantra for shrub crops: *ramial chips*. American agriculture is desperately trying to reinvent itself, and some of the most encouraging successes have been in niche crops. At least three such crops—*Ribes*, breadseed poppies, and hemp (as opposed to marijuana)—are being denied their contribution because of legislative intransigence.

In terms of nutritional impact I don't see red currants or gooseberries having a very great role, beyond vitamin C, though what do I know? Gooseberries are primarily for jam, although there are several little-known varieties that are quite tasty out of hand. Fresh red currants are a tart, zesty pick-me-up on hot muggy days in late July, as they segue into the raspberry season. Many folks have made wine from red currants, which seems to me a shameful waste. However, for folks like myself who fail to appreciate the medicinal taste of alcohol, bottled red currant *juice*

Figure 12.14. Red currants help us endure the humid heat with their sprightly tartness. Whoever grew these knows a good thing.

Figure 12.15. After being stored all winter in layers of leaves, these Followwaters will be even nicer than when they went in.

has a sprightly friskiness that makes it an elegant beverage for special occasions, sort of a wine analogue I suppose.

Apples

I have no particular insights into apple growing, beyond the recommending of certain varieties. In the last several decades a number of new varieties have been introduced with the welcome feature of scab resistance. That has seemed curious to me, because among the older varieties scab resistance is all over the place. I've come to realize that what they really mean is scab-resistant versions of Macintosh, which is a hopelessly scab-prone variety. When many of the older orchards were killed out in one exceptionally severe winter, those apple growers who replanted all turned to the newly popular Macintosh, since by then everyone was resigned to using fungicides anyway. In my opinion Macs are edible for the first few weeks off the tree, but far too bland for any of my purposes. I have many apple varieties on my farm, but can't find space for a Mac. Most of my varieties, however, which taste little like Macs, are wholly or largely resistant to scab, and so I ignore that problem. I use no fungicides, not even organic, and I see little if any scab. Borers are a much greater headache, especially in newly grafted trees.

Here then is a summation of some of my favourite apples in order of preference, and some of their special traits:

Baldwin. Once the all-around favourite market variety it is superb for all uses: fresh dessert, sauce, cider, drying, baking. Hard and good keeping, it is delightful off the tree, but improves until well into the new year. At some point it develops brown internal spots and from that begins to deteriorate. Bruises are slow to spread.

Black Oxford. A smallish tree and smallish fruit, but magnificent quality, especially after a couple of months in storage. The very dark purplish-red skin is underlain by green chlorophyll; as that fades out in storage a yellowish-orange background makes the fruit even more beautiful. Very hard and bruise-resistant, the skin and flesh are tough at first, but by the holiday season are fine eating and continue to improve through midwinter, by which time a noticeable banana flavour develops. Best for fresh eating, but excellent in sauce blends and cider.

Golden Russet. The finest of the russets, a heavy-yielding but relatively small-sized fruit. Very hard and excellent keeping; superb in cider and sauce. Great fresh eating after mid-December and far into the winter. Reliable cropper for me.

Followwater. Aka Fallow Water, which I suspect is the correct name. A tall, spreading tree which thinks nothing of 12 to 15 bushels, most of them blemish-free. Apparently long-lived tree. The rather large hard fruits are dull green at picking, much like a black walnut fruit; even bruises are slow to decay. Best eaten after New Year, especially as the green fades to yellow, sometimes with a faint blush. We also put a lot into cider, as they are such a generous yielder and make very tasty juice even when still green. Good quality into April. I use a lot for winter sauce, although the colour is very poor by itself. Very reliable yielder.

Stark. A long-lived tree with very good yields in some years. Good for many uses, including fresh eating, it is exceptional for sauce and cider. The skin is underlain by green and the flesh is yellowish, which makes for a rather ugly-*looking* juice that's still very aromatic and full-flavoured. Also high in pectin; cider syrup will begin to gel with little boiling.

Ben Davis. An old and much-touted variety; an incredible keeper and reliably heavy cropper to fill out any cider blend; a slight hint of blackberry in the fresh fruit, though that is certainly not its highest use; after long storage it sometimes has a hint of coconut. There's a tale that deals with an old codger who had a reputation for knowing all the old varieties; once at a Grange meeting next to the cider mill, someone challenged him to identify varieties by their pomace (fresh-ground flesh awaiting pressing). He tasted one variety after another and named each with unerring accuracy. Finally some skeptic grabbed a fistful of sawdust and splashed some vinegar on it. The taster paused a moment after swallowing, seeming a bit confused. "Well," he offered, "for a minute there I thought it was Ben Davis, only it's too juicy."

Yellow Bellflower. Great taste for fresh eating, quite a good keeper.

Winter Banana. A particularly beautiful apple with it bright-yellow waxy skin and a slight blush. Very good for fresh eating and most other uses. Keeps fairly well.

Fireside. A very attractive apple with a strong blush toward the sun; fair keeper; excellent fresh eating.

Northern Spy. Slow to come into bearing but slower to quit. Some susceptibility to scab, good keeper, excellent sauce, fine cider, good drying.

Blue Pearmain. Excellent fall apple, for fresh eating and sauce especially, but most uses.

Snow/Fameuse. Very old French variety; fall apple with very red skin, very white flesh; fresh eating and sauce.

Orange Sweet. This is a bland soft sweet apple; for a short while it is a delicious aromatic dessert apple, but it goes blah very quickly; its main redeeming feature is that it is extremely early, when the only other apples around are tart Transparents and Lodis.

Tolman Sweet. Excellent early-fall apple for fresh eating, mediocre for other uses except early sauce blends, not by itself.

Wolf River. Very large beautiful fall apple, mediocre for fresh eating or cider; excellent for dried snack chips, stays white; superb for stuffing and baking.

Dolgo Crab. Elongated dark-red crab apple with red-kissed flesh; tart but tasty as a fresh snack, but

Figure 12.16. Ben Davis is unfazed by an early-November snowfall.

mainly valuable as a highly aromatic and colour-enhancing addition to sauce and cider blends. Sap-suckers apparently find the sap very tasty, too.

Plums

I love plums, but they are very trouble-prone. We gave up long ago on the European, or prune, plums, as they are constantly decimated by the black-knot fungus that is endemic on our native wild cherries, but far more severe on the plums. A perfect solution to that is the Japanese-American hybrid plums developed by Niels Hansen in the late 1800s. He was concerned about the lack of fruit for homesteaders on the midwestern prairies. European plums were no way hardy enough to stand prairie winters, and the native American plums were totally hardy and totally useless for eating. However, he was able to create an interspecific hybrid between delicious but cold-tender Japanese and the hardy native wild plums. Some of the offspring had the best of both parents. Many superior plums have been bred from those and none of them has the slightest susceptibility to black-knot. I have some problems with pollination of these, even though they bloom heavily and are well worked by several pollinators. I have wondered whether that reflects some lingering incompatibility in their inter-specific crossing, although Cousin Tom has observed that close planting (12 to 15 feet, or 3.7–4.6 m), so that their limbs actually intermingle, seems to help, which would cast doubt on my theory.

Figure 12.17. This plum tree is loaded with newly set fruit, but thanks to curculio we will see very few ripen.

In any case some years they set fruit like crazy, and we proceed to lose most or all of them to the plum curculio. The most effective remedy seems to be Surround spray, an extremely fine clay that coats the leaves and tiny fruit, making it difficult for the curculio moth to oviposit (lay eggs). You need to have an extension wand for the sprayer and not mind the ghastly-ghostly look of the milky-coated foliage, and of course you must accept dependence on the marketplace. It may be worth it since that marketplace input is merely a form of clay and perhaps can be replaced by a local substitute, such as a decantation of local clay. I plan to try that. Other remedies include spreading tarps on the ground and knocking the trunk with a rubber-wrapped baseball bat, then gathering up the fallen worms and destroying them. That requires more attention to timing than I'm known for.

Pears

Pears seem to be one of my least difficult fruit crops, although some of my varieties are reluctant to bear. The old variety Clapp's Favourite is quite reliable; I had an ancient tree that Lucian's grandfather grafted. Even as an old hollow rotting hulk, it had a couple of high branches that persisted in setting a dozen or so excellent fruits every year, even though their weight threatened to tear down the frail remnant. My favourite pear by far is the small Seckel, or sugar pear. It yields heavily and consistently, and if I can keep the raccoons at bay I can count on a few bushels of dense, sugary fruit that keeps longer than any other pear. It dries better than any other pear, too, and the puree is a rich syrupy sauce that can be used to sweeten tart fruits. A trick with most pear varieties is that they should not be allowed to ripen on the tree. Not only will they turn mushy in the interior, but they will

Figure 12.18. It's important to pick pears before they turn yellow; otherwise they will quickly turn mushy inside.

have hard little grit cells scattered through the flesh. Just before they start to turn yellow, we pick them very carefully and place them in shallow trays on a dark shelf, checking them as they turn buttery soft and yellow.

Medlars

No, a medlar is not someone who interferes with someone else's business, it's a weird fruit from the Middle Ages, when out-of-season fresh fruit was not the norm. I mean, they've been around since then, but folks don't talk about them much. Nowadays they are wicked trendy among the rare-fruit cult, but do they deserve it? After growing them for several years, I finally had enough of a crop to do something with, and that "something" does *not* include picking them off the bush and popping them in your mouth. They are kind of like persimmons that way: however

Figure 12.19. This medlar is ready for picking, but it will be weeks before it's fit to eat.

Figure 12.20. This is the yield from two young medlar plants.

ripe they get on the bush it's not enough; you need to further ripen them indoors (it's called bletting) for several weeks until the hard greenish fruits get soft and turn a yellowish brown. In fact they have to get almost rotten before they're fit to eat. And then? Well, they are nice, though I'm not sure they're quite as exquisite as some aficionados describe them. All I did was mash the bletted fruit through a sieve and freeze the rusty-brown puree. I added no sweetening or spice to them and they were enjoyable. Now, I've heard talk about making gourmet jellies and jams, and I don't doubt they might be more sensational than the humble no-frills prep I gave them. However I suspect their main attraction in days of yore was that very fact that you must blet them; they not only can store for months, they really *have* to store for months to be at their best. Back in the day when you couldn't buy grapes and oranges off-the-shelf in February, they probably filled an important role. It certainly wasn't their appearance; they look sort of like a giant rose hips, or small green apples with persistent sepals. And the blossom-end has a decidedly banal look, so much so that Shakespeare's readers knew them as the "cats-arse fruit."

The botanic name *Mespila germanica* makes them sound like some Hanseatic sort of thing, and indeed they were widespread in Northern Europe, but in fact their native homeland is the Black Sea coast of northeast Turkey.

I must say that they are very easy to grow, have no pest problems that I've noticed, and need little pruning. They are compact—less than 15 feet (4.6 m) tall—so they can be mostly picked from the ground.

There are few commercial sources of medlar plants, but you can graft your own. The USDA Clonal Repository in Corvallis Oregon maintains a collection of 15 or 20 varieties, including Nottingham and Breda Giant. But what to graft them onto, assuming you have no medlar rootstocks? Knowing they are a pome, like apples and pears, I grafted mine onto wild hawthorn in a corner of my hay field. Not knowing how long the two parts would remain compatible, I nurse grafted—that is, I grafted low on the haw, and when they took I dug them up the following spring and replanted them where I wanted them, only in a deeper hole so the graft union is now below grade and the medlar can strike its own roots if it feels insecure. (For more on this refer to chapter 7.)

The medlars are very cold-hardy, and should remain on the bush until at least mid-October, probably considerably later. You're supposed to spread them out on trays in a single layer and keep them in a cool, dark place until they are soft and tasty; however I just stored them loosely in a cardboard box several deep and they kept just fine. I assume that if you want them ready to eat sooner, you could try bletting them in a warmer place; still, the previous year when I had only a few fruits, I left them in a basket in the kitchen, where they became dry and withered before ripening.

One recommended way of using them is as a sort of pudding with eggs and milk, which doesn't interest me. The jelly idea involves adding sugar and pectin, which doesn't excite me too much, either. So far, I use them as a thick jamlike puree, much like apple butter, with or without some added spice like coriander. I've only eaten the raw puree and don't know what cooking will do to the flavour. I definitely like them enough to keep experimenting. Wildlife don't appear to bother either the bush or the fruit, but of course, they haven't learned the secret of bletting.

Figure 12.21. There's snow outside and these medlars are finally rotten . . . er, ripe.

Mulberry

I have three mulberry (*Morus alba tartarica*) trees, which are incredibly variable. Since they began as ungrafted seedlings, this is typical of most tree/shrub crops, which is why we must be cautious in evaluating the quality from a single specimen. Only one of those three trees bears fruit regularly and has a robust dignified stature; the others are scrubby runts with poor or no fruit. The nice one is very nice; with its wide-spreading habit we keep a bench under it for workers to take their nooning in its dappled shade. That tree is ever loaded with fruit, which everyone nibbles on in passing. The quality is not superior; the berries are smallish with a bland sweetness, off-white with a purplish blush at maturity, and a bit rubbery. For all that we enjoy them, and I've often wished I could grow one of the really superior eating mulberries, like Illinois Everbearing, but assumed they were not hardy here. Having recently learned otherwise I'll give them a try.

Part of mulberry's claim to fame is that it attracts birds away from other berries. That fits with my experience totally: the cedar waxwings show up when the first fruits just start to ripen, and they don't leave that tree until it's over. How fortunate that mulberries ripen over such a long season, else the waxwing flock would run out of food just as our other berries came into fullness, and we'd be stripped bare. As it is the lovely masked birds cannot even keep up with the mulberries, so we have as many mulberries to eat as we wish. Whether that would be the same if we had a superior fruiting variety remains to be seen.

So far the feature of mulberry that has impressed me most is its roots. My favourite tree is about 20 feet (6.1 m) tall and the crown has a 30-foot (9.1 m) spread. One year I was digging a rock out of a tilled garden when I came upon a root the size of my big pinkie and 60 feet (18.3 m) away from the trunk of that mulberry. The distinctive bright-yellow root left no doubt about whence it came, and it continued several feet beyond before petering out. A general rule of thumb used by foresters and landscapers is that a tree's roots extend about to the drip line or perhaps to the equivalent of the tree's height. This root was three times the tree's height and 45 feet (13.7 m) beyond the drip line. I still marvel at what it means.

Non-Woody Permacrops

There is a small group of perennial food crops that are valued by Permaculturists for the fact that they are perennials, although they differ from tree and shrub crops in that you have to disturb the soil, if only to harvest them. Whereas tree and shrub food crops require no tillage at all, substituting the much less invasive procedures of mulching and pruning, members of this group—which includes terrasols (aka sunchokes et al.), groundnut (*Apios americana*), Chinese yam (or cinnamon vine), crosnes (*Stachys afinis*), and skerrit—rely completely on digging the soil; they are all root crops. I don't object to that any more than I object to potatoes; it's just that they don't fulfil our criteria for low inputs quite as well as the tree and shrub crops. I *can* say that sunchokes (called terrasols around here and henceforth referred to as such) are a highly efficient cropper with modest inputs; the others less so. None of these crops requires rotation; rather they defy it by their persistence. You plant them where you want them; changing your mind is possible but difficult. They are not very common in the marketplace, which is the very reason you should have them in *your* garden: let the off-the-shelf types wonder what they're missing.

Terrasols

Terrasols (aka sunchokes, sunroots, Jerusalem artichokes) are the most promising of this group, as indicated by their rapidly growing popularity (I sell several bushels of them every year, one of my few cash crops). They're actually a tuberous sunflower; in fact during the Brezhnev era Soviet breeders worked to develop an interspecific hybrid that would yield both a root crop and oilseed. I don't know how far that research got, but the Soviets did produce a number of terrasol varieties with superior agronomic traits.

I maintain a collection of over 80 varieties, and there is enormous variability among them. Of all the food crops grown in this country, nearly all of them did not originate here; even corn, beans, and squash came here from Central America. Terrasols and their sunflower cousins are among those few species that were here before the Pilgrims arrived and even before the Native Peoples arrived. They were not brought here; they evolved here. That and the fact that they are largely clonal (not having a great need for sex) may help to explain why many of the wild patches show few signs of domestication. I say "wild" to include truly wild clones (which I suspect are rare) and feral patches, which are often found around abandoned farmsteads. These have usually been "bred" by nature and "selected" by humans, so the improvement has been relatively haphazard compared with, say, corn. Especially in the North

terrasols rarely mature viable seed, although most clones flower at some point, usually too late in the season. That's not so important for a species that has trained humans to move it around.

Recent decades have seen some systematic breeding programs with some very encouraging results. One of the major improvements has been tuber set. For the sexually challenged terrasol plant it makes sense to send out long runners with tubers at the end, extending a plant's vegetative offspring into new habitat. For human cultivators that's all quite a nuisance; we'd much rather pull up a stalk and find the tubers all right there, not too far off and not too deep. Breeding has done much to accomplish that. Another breeding objective has been number and

Figure 12.22. Terrasols exhibit a wide range of tuber variation.

Figure 12.23. Skorospelko is a highly improved Soviet variety of terrasol.

size of tubers; again solid advances have been made there. It would also be nice if the tubers were smoother instead of knobby. Knobbiness may increase yield, but tell that to the cook who tries to scrub off the mud.

One of the Soviet breeds, Skorospelko, embodies most of those improvements to a high degree: The tubers are set close to the stalk, they are numerous and large, and they are relatively knob-free (some cooks feel they have to *peel* the tubers—tee-hee-hee!). Another superior variety is called Clearwater, named by me after a nearby lake. It is one of the more distinctive varieties in my entire collection, which is all the more unusual since I discovered it in my own backyard (my neighbour's actually). That neighbour, Lucian, said his mother had gotten it very long ago from someone else in town. Curious, that, because I've investigated old homestead patches all over the area and never seen anything quite like it. It has roughish buff skin, but few or no knobs. It is not an exceptional yielder, but an excellent keeper, in the ground or in the cellar, remaining crisp and dormant when all others have gone pithy and sprouting. I have even pulled up plants (to thin them) in spring, after the sprouts were aboveground and green, and found the attached tubers to be still crisp and palatable. It is unlike anything else in the collection, and no surprise, it is by far my biggest seller.

This is as good as any time to talk about farting. A major complaint about terrasols is gas. Many people experience flatulence from eating raw terrasol tubers, but none when they're cooked. Others have just the reverse experience: okay raw, but gassy when cooked. I'm not aware of having any "issues" (*sic*) either way, but perhaps my friends are just being polite. It has to do with the fact that terrasols store much of their energy in the form of inulin, a carbohydrate that digests differently from starch, making it harmless to diabetics and causing others to make chamber music. The good news (or bad, depending on which category you fit) is that the inulin begins to convert into other simple sugars in storage. After a few months the inulin content has plummeted to levels that give little benefit to diabetics, but make

the rest of us more comfortable. Come to think of it I guess that's when I eat them mostly: late winter and spring.

Although it is gratifying to see the growing interest in terrasols, it's unfortunate that everyone focuses on them as a fresh vegetable, which ignores their larger potential. My main food use for them is as a grain extender, a chestnut-flavoured meal that gives porridge a whole new perspective. (See chapter 16.)

Aside from food the crop is being investigated for alcohol production. As in methanol for gasohol? Perhaps, but what I heard about was high-end alcohol for the pharmaceuticals/cosmetics industry. Yummy! I've always heard that the stalk fibre may have some textile potential, but that's purely hearsay. I can vouch for the food potential.

Terrasols are not difficult to grow; indeed some say it's difficult *not* to grow them. Some friends once bought some tubers at a farmer's market in Toronto and asked the vendor if it was possible to plant and grow their own. Oh yes, it's possible, they were assured, "but you have to watch them." Thinking that meant a certain amount of coddling was required to keep them alive, they put them in a particularly choice spot where, 40 years later, they're still "watching them." That said, terrasols get a bum rap for being invasive when in fact they're just very good at holding their own. Untended they might "invade" at the rate of 10 inches (25.4 cm) per year. Of course if you plough or rototill a patch and spread the broken tubers all over, all bets are off.

The application of high-nitrogen fertilizer, even organic, will produce a prodigious volume of top growth (up to 14 feet, or 4.3 m), but it will *not* improve the quality of the tubers. What is far more essential is a balance of minerals, particularly phosphorus, potassium, and boron. A surplus of nitrogen at the expense of these "root elements" will produce big tubers with hollow or brown hearts, poor flavour, and poor storage quality. I've had my best crops when I had lots of leaves to spread on the plot after harvest, tilling just enough so they didn't blow away. Wood ash is always a profitable additive applied at the rate of 1 gallon (3.8 l) per 50 square feet (4.6 sq m), though probably not every year. And water; even though terrasols can survive in some pretty arid places, the tuber yield is greatly increased by ample and steady moisture.

Since the soils in my region are predictably deficient in boron, I have occasionally added a light dusting of 20 Mule Team Borax (sodium tetraborate, not the detergent). This is mined from a desert lakebed in California (I've seen the place, and it looks no worse for the countless tons of borax that have been hauled away), and if it seems incongruous with a self-contained, closed-cycle food system, it is. Of course so is selling off half a ton of high-value biomass: the tubers. I'm sending those minerals off to the marketplace, and they're never coming back. My regrets are somewhat assuaged by the fact that they're being sold for seed, not food. Its goodness in the form of genes is the main product being exported, enabling others to grow their own food, hopefully with their own boron and stuff.

Since my terrasols grow in the same place year after year, they are always coming up thickly from last year's missed tubers (we're always *so* sure we cleaned them all out). I always hold back a few culls to replant any gaps, which is purely for my emotional security; if there *are* occasional spots where nothing comes up, there *certainly* will be several replacements nearby that need thinning. Many people don't bother to keep their patches thinned, counting on harvesting to do that. It probably works okay, but folks who use terrasols haphazardly usually leave much of the patch unharvested. Those areas will come in much too thickly, and the tubers will become run out and stunted. When starting a new bed I plant the seed pieces in rows 2 feet (0.6 m) apart with at least 12 inches (30.5 cm) within the row. In subsequent years I try to thin them to that density—not easy when new ones keep popping up in between.

Apios

Another tuber-forming perennial is groundnut, or *Apios americana*. Several other species have been called groundnut, including the peanut; moreover *Apios* is also called potato-bean, which confusion leads me to call them simply *Apios* (which is sort of what Linnaeus had in mind?). Native Peoples ate them, but the crop has never been systematically improved until very modern times. The first real attention given to *Apios* as a food crop was by William Blackmon at LSU, whose work has been continued by Bert Reynolds. Their object has been to develop *Apios* breeds that would be compatible with conventional agricultural methods. There is one obstacle to this, which may actually be an advantage in *unconventional* agriculture. In the process of producing a new crop (unlike most biennials), *Apios* tubers just keep getting bigger from year to year. They form strings of tubers, with the newest one at the far end, which sends out new growth while the older ones just continue to fatten up. They don't necessarily become tough and pithy, just bigger. Therefore, with conventional annual row crop systems, especially in the North, they might never realize their full potential. In a Permaculture system of some sort they might be just left to grow for two or more years for a proportionately greater yield. This might be especially productive if they could be combined with a companion crop species with a similar growth habit.

I once established a patch of *Apios* on the inner part of a terraced plot where the soil was poor in organic matter yet comparatively wet. Since wild *Apios* thrives by riverbanks and ditches, I reasoned that they would be happy there. Indeed they grew and spread well, so well that when I finally elected to move them out of there, it took two full seasons to get thoroughly rid of them. Unlike terrasols, *Apios* really could become a nuisance unless they are planted in a place where they are easily confined. Another reason at the time was that while they were growing tenaciously, they didn't seem to be sizing up a good crop of tubers. In hindsight I suspect it was because *Apios* is a legume, and the terrace soil was a bit too wet and compacted for their nodules to fix nitrogen. I did move them to a different place, where they continue today with little attention. I occasionally dig and eat a few tubers; however they are not close to my heart like terrasols. They have a rough-textured skin that holds dirt unless you scrape it off; you can peel them, but only if they size up enough to be worthwhile; and the tubers are not very exciting to eat, having a dry, pithy texture and a bland, starchy flavour. They are definitely improved by including them in stir-fries. I certainly do not rule out their further potential as a good permacrop, but as yet I am not overly impressed with them.

Odds and Ends

Cattails. As a kid, while gallivanting around the countryside in hot summer weather, I learned to appreciate the cool succulence of inner cattail shoots from roadside ditches that were, like me, of dubious hygiene. When I finally got to dig my own first farm pond, I was delighted that cattails wasted no time at all in moving in. Now that I have them at my convenience whenever I wish, I only occasionally think to eat the young shoots, or the golden pollen, or the green seed heads, or the starchy rhizomes. The cultivated crops demand more attention, and that which is freely given tends to be ignored, though it is not necessarily inferior.

Chinese yams. What little experience I've had with Chinese yam, or cinnamon vine, has been encouraging. I especially like the fact that it is opportunistic: If you leave the root in the ground a second year, or even a third, instead of harvesting it, it will just keep growing larger and larger (like *Apios*). That's a real advantage where I live, as many of the young plants do not size up well at first, but more than make up for it in their second season. The tubers can in fact grow huge and are a fine food, the only objection being that they are a bit bland and starchy and are best accompanied by other stronger-flavoured foods.

Crosnes (*Stachys afinis*). This interesting novelty crop is good as a salad garnish, but little more. It's actually in the mint family (succulent square stem), with crispy, crunchy little poppet-shaped tubers that apparently reminded someone of Jerusalem artichokes, because they're also called Chinese artichokes. How embarrassing: a crop plant misnamed for another crop plant that's also misnamed. They're not very productive, but they're not too demanding, either; somewhat bland yet always refreshing.

Daylily. Another splendid addition to veggie mixes is the common daylily. I especially enjoy the unopened flower buds tossed in at the last minute. Although I have a patch of wild daylilies planted for that specific purpose, I've discovered that most of the cultivated varieties in my wife's perennial garden have much larger and more succulent buds (we won't discuss how I discovered that; she still hasn't gotten over it).

Figure 12.24. Whether wild or cultivated, daylily buds are a splendid addition to any stir-fry.

Figure 12.25. These nannyberries, or Indian raisins, were meant for the birds, but when the berries are ripe I sometimes snarf a handful on the way to the pond.

Nannyberries. Nannyberries are an undomesticated viburnum that I planted near the lower pond as a privacy screen for bathers. I intended their fruit as a wildlife enhancer, but I am not wholly disinterested myself, especially since I tried one of Cousin Tom's innovations: canned nannyberry puree. If you could get past the appearance, a shiny purplish grey easily mistaken for axle grease, you would be in for a pleasant surprise; though a little bland the dry, sweetish pulp is a bit reminiscent of dates, though not as intensely sweet. I don't grow them for my own use, but neither do I refrain from helping myself on occasion.

PART IV
The Garden in Context

Up to this point in the book I've described my methods for building and maintaining soil fertility, my techniques for propagating seeds and plants, and my experiences with growing specific vegetables, grains, pulses, oilseed crops, and permacrops. All of this is important for creating a garden, but now I want to step back and put all this information into its context of a larger farm/garden system. That includes consideration of the land itself: the rocks, water, and topography. Within that my gardens are not simply collections of individual beds of plants. As you've gathered by now I set up most of my beds as interplanted mixtures of crops, and I want to explain the principles and strategies I follow to create those mixtures. Also, the context of my gardens—and yours, too—inevitably includes pests and diseases. So I want to tell you about how I manage my gardens to prevent some problems and cope with those I can't prevent.

CHAPTER THIRTEEN

Rocks, Water, and Land

Rocks and water: They're either a liability or an asset, depending upon where they are and what you do with them. I choose to view them as a resource. Water can be drained and stored; rocks can be "cobbled" into a number of useful infrastructure features including culverts, wells, and walls; land can be sculpted to better serve our purpose. These are all highly labour-intensive and time-consuming projects, but I am not building for today. Gardens-without-borders must also be gardens without calendars.

Rocks and Water

My neighbour Lucian, whose family had worked this land since 1803, had this insight about rocks: They are an expensive proposition, whatever you do with them. Removing them requires a huge amount of backbreaking labour, yet leaving them in place leads to worn and broken equipment and reduced crop yields. I agree with Lucian's observation and would add a couple of my own: Removing them is a *one-time* expenditure—those particular rocks will never trouble you again; and rocks are also useful building material. The two main objectionable characteristics of rocks—they're durable and they're everywhere—can be turned into their main assets, especially for water management projects.

Stone Walls

The traditional use for rocks in my region was for building stone walls, and I use the word "walls" advisedly. The iconic New England laid-stone walls were built partly to delineate boundaries ("good fences make good neighbours") and partly to confine livestock. No, let's just say cattle—a stone wall that would safely restrict a cow (not the most agile creature) would serve as a highway for a goat or sheep or even a pig, even with a rail fence on top of it. And those lovely neat-laid structures that adorn postcards of more prosperous areas are rather scarce around here, where hard-strapped hill folk had more pressing demands on their time. On my own land the only rock wall that could remotely be called "laid" is a section along the town road; elsewhere the "walls" are more like long rock piles, vaguely defined in places, but presumably adequate to restrain a curious bovine. That they were not wholly adequate for that is witnessed by the existence of at least two "sheriff's pounds" in our town—small corrals intended to impound runaway cattle until the owner paid a fine.

Those haphazard walls were obviously more of a place to get rid of rocks, which supposedly had no higher use. That is generally how I have treated them; when I had a particularly in-the-way rock and no immediate call for it, I have parked it on the nearest wall partly in an effort to raise the wall higher. I

Figure 13.1. This large rock is interfering with good drainage here, but it will be quite useful elsewhere.
Photograph courtesy of Yaicha Cowell-Sarofeen

keep no livestock, but I would like to think they discourage deer. Now Mama didn't raise no fools, and I know perfectly well that any self-respecting adult deer can leap over a 9-foot (2.7 m) fence in a single bound. But adult deer are very often accompanied by younger deer, and the group will not venture where the young cannot follow. Moreover, while deer *can* easily clear high fences, they will not try to if they cannot see where they will land; therefore a much lower solid obstacle may suffice. Anyway I will not live long enough to bring my perimeter walls up to 5 or 6 feet (1.5–1.8 m), and meanwhile I have other uses for many of those rocks. There is no competition for them; I have an ample supply, and every season produces an abundant new crop. I get endless amusement from viewing those suburban stone walls laid by some landscaper using slate slabs trucked in from Pennsylvania on a plastic-wrapped pallet.

Since I don't keep livestock most of the stone walls I have built myself are not freestanding, but retaining walls, either interior (cellar walls or wells) or exterior (terraces). The rocks out of my ground are tailor-made for that. Although the bedrock 80 feet (24.4 m) below my feet is Silurian-Devonian slate (or phyllite), the rocks in the soil immediately beneath me are very different, a hodgepodge of igneous granitic material carried here from far to the north and dumped by the last ice sheet. Supposedly when the glacier edge reached Maine, it just got fed up and dropped them all right here. Where is that ice sheet today? Some say it went back for more; others say it's gone for good, along with the polar bears.

Those rocks in my field are not flat slabs like the bedrock slate; they are an assortment of shapes and sizes, although I see a tendency toward tetrahedrons, which work very well for building. Some friends in the neighbouring town of New Sharon have only slate (from the local bedrock), and after rebuilding an old cellar there I came to appreciate my quirky igneous rocks.

Robert Frost observed that "something there is that doesn't love a wall." He was of course referring to his namesake frost, the inexorable power of expanding ice. Here is how it works: When two touching stones freeze they do not expand, but the wet surface between them does and it pushes them apart imperceptibly. Stack up 3 or 4 rocks or 10 or 20, and all those wet surfaces will freeze and expand, though not so's you'd notice. But in a column of dirt you have hundreds or thousands of particles (and thus wet surfaces) per inch, and that exerts a tremendous push outward, collapsing the most carefully laid wall if it is backfilled with dirt. Using larger facing rocks will not avail; the greatest boulder is a toy in the hands of frost. The trick is to backfill with more stones, which will budge and shift before the thrusting frost, but not collapse. Carefully placed chock stones are ideal for locking the face stones together, but often you can simply dump in buckets of assorted rubble, tamping to settle it in more securely. I figure the walls should be at least 2 feet (0.6 m) deep (or thick), counting face stones and backfill; if the face

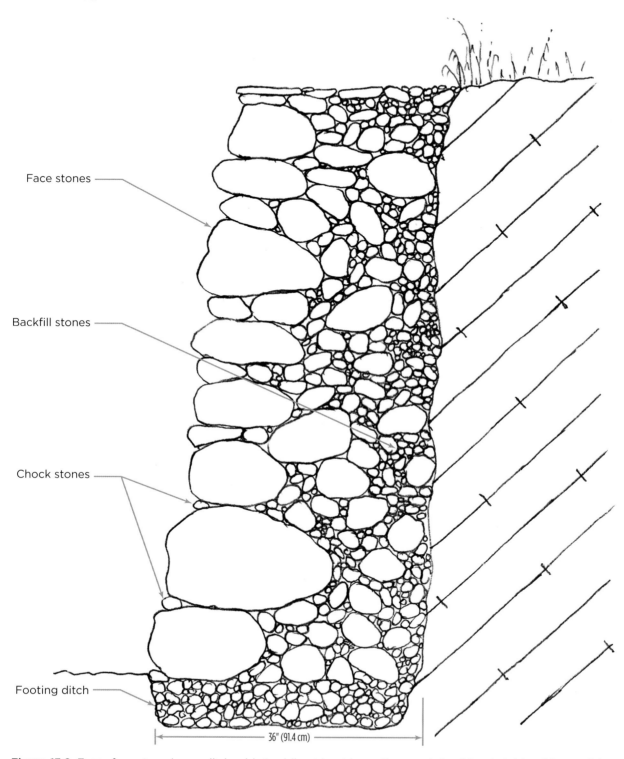

Face stones

Backfill stones

Chock stones

Footing ditch

36" (91.4 cm)

Figure 13.2. Every face stone in a wall should straddle at least two others and should rest stably without relying on shim stones, which may wiggle out of place over time. Add chock stones for appearance and to deter rodents, but don't rely on them to supply support.

Figure 13.3. Stone retaining wall for the perennial flower garden.

stones are particularly large then 3 feet (0.9 m) may be necessary to ensure that several fill stones separate the face stones from the dirt.

Although I rarely do it, creating a shallow footing ditch filled with rubble and tamped down by foot may be helpful. It must drain properly, however, or standing water will make things much worse. I generally tend to place larger stones on the bottom layer, partly for stability but mostly because it's harder to lift them up as the walls grow higher. That being said, I also save large rocks for farther up, provided I can lift them and if the shape is such that I can jut them forward a bit. You see, the old-timers usually made their dry-stone cellar walls slope backwards slightly for stability, but I prefer to go as vertical as I can. Sometimes a recess is formed by the shape of the lower rocks, so that I cannot go straight up unless I have a large and well-shaped rock to bridge that void. I watch out for such rocks and set them aside.

For interior walls, such as the cellar of my house, I have the built-in forgiveness factor that my cellar rarely goes below freezing and never freezes deeply. That wall is over 9 feet (2.7 m) tall and carefully laid. After the bottom 4 feet (1.2 m), I increasingly lowered rocks onto it from above rather than lifting them up. Large rocks required special care not to jar the wall as I rolled them down. I kept an exceptionally deep backfill, so that I had a broad shelf on which to lower them, then rolled them around gently until they were in place. If I were to do it again I would use some cement, or at least clay, to grout the top 6 inches (15.2 cm) or so, not for strength but just to seal the interstices from cold outside air penetrating through the backfill to the wall.

Exterior (terrace) walls are more challenging due to full exposure to frost heave and rainwater. On the other hand errors in workmanship are more easily repaired without a building sitting on it. Over the years I've had two or three small sections collapse, and I've simply rebuilt them; with no cement involved, it is easier to incorporate the new work into the old adjacent sections.

I have found wells to be the easiest of all for dry-stone projects. Everything I've said about retaining

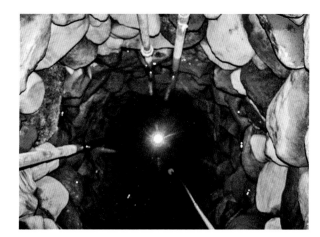

Figure 13.4. This 25-foot (7.6 m) deep well was built only from nearby stones.

walls applies here, with the huge advantage that the well wall is a cylinder: All the forces pushing inward only tend to lock it more tightly together. On the other hand it is impossible (and undesirable) to use such large rocks as you might in a less restricted space. I never used anything larger than what someone could lower down to me with a rope and bucket.

Ditches and Drains

Just as stones in the garden can impede proper drainage, they can also be made to improve drainage. Although the average slope of my cropland is between 6 and 8 percent, which allows excess water to drain off readily, there are some areas where it stands too long and interferes with working the soil, especially by excluding air. I want the excess water to leave, but I do *not* want it to move across the surface, carrying with it the soil particles as well as soluble nutrients. Ideally I want it to sink slowly into the soil until it joins the water table, leaving behind its burden of soil and soluble matter as it filters down through the dirt. However, when that process is too slow to remove the excess, or else the water table is too shallow, some additional drainage may be required.

I have been able to improve some of my wetter areas by simple ditching. For example, one area was so waterlogged it only grew polypody fern. Neighbour Lucian assured me that polypody was an indicator of

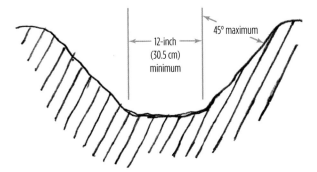

Figure 13.5. When a ditch is necessary, it should be wide and gentle-sided enough that it can be maintained with a scythe, mower, or weed whacker. Weed-choked ditches are unacceptable, as they impede the flow and harbour ticks and other pests.

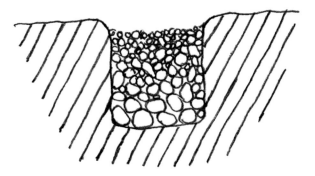

Figure 13.6. This cross-section view of a stone drain shows the gradation in the size of stones.

Figure 13.7. When completed this drain will carry water from where it is a problem to where it is needed. Photograph courtesy of Yaicha Cowell-Sarofeen

everything bad and that I should just abandon that piece. But at the time I had little enough cleared land and could not afford to leave any part of it unused. So I did several things, including digging a trench (up to 12 inches, or 30.5 cm, deep) all around it and leading to a nearby farm pond where the water was more appreciated.

You see, as eager as I am to remove the water from my cropland, I don't wish to unduly hurry its journey to the sea. Rather, I'm glad to impound it (which by the way is the origin of the word *pond*) for a while so that some of it can sink down to the water table while yielding me a welcome crop of cattails. That ditch is still there and functioning well. The area now produces an abundance of highbush blueberries (which

do not tolerate poor drainage) and several cuttings of grass every year. Indeed I wish the ditch were deeper yet, but that brings up a problem with ditches. If they are steep-sided canyons they do not last long; the sides slough off from frost and erosion and they tend to fill in. Plus they are a hazard to foot traffic, an obstacle to wheelbarrows and equipment, and difficult to mow. Making them wide with gentle sides helps remedy these problems, but also requires much more space. That's why a stone drain can work as well or better in cases where the volume of water and the rate of flow are small. And once built a drain is vastly more convenient than a ditch.

To install a stone drain I begin by digging a ditch, vertical-sided and as deep as or deeper than

Figure 13.8. Seepage accumulation at the bottom of a new drain in August shows why it is needed.

Figure 13.9. This half-filled seepage drain will need no culvert, just the many passages among the stones.

necessary to allow the seepage to sink down and be carried away. If it is merely a seepage drain I can simply fill it up with stones, starting with the larger stones on the bottom to leave a maximum of passageways, and smaller stones on top to fill in spaces where dirt might work down in and clog the drain. I stop filling it slightly below the grade, to leave a slight depression where a mower can pass right over without kicking out loose stones (see figure 13.6). This type of drain works well to intercept water coming from uphill or to allow water in habitually wet areas to seep down and away.

In situations where the volume of flow may be excessive for a regular stone drain, a stone drain with an open channel running through the centre can work well. To make a drain like this I dig a ditch wide enough to allow for two rows of parallel stones with an opening at least 4 to 6 inches (10.2–15.2 cm) wide between the rows. As I work I tuck in smaller stones between the larger, fitting them somewhat to lock everything into place. I don't want to have to redo this drain in the future. As with retaining walls I never use dirt to backfill the space between the face stones and the wall of the drain, because frost can work on the dirt. Instead I use more small stones: the more the better. Then I select some cap rocks large enough to span the open central channel of the drain. I rest these securely atop the two parallel rows of stones. As I fill in the trench above the cap rocks with stones, I'm careful to use large enough stones so

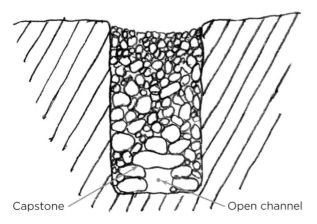

Figure 13.10. This stone culvert drain has a large capstone shielding the open channel.

they won't drop through the gaps between capstones and block the open channel below. I usually install a channel with capstones in a drain that may see occasional heavy flow, as in springtime.

Over time, even if the channel itself should collapse, water may still be able to pass through gaps between larger stones. However, if I use a lot of small stones and they end up collapsing into the channel, that could choke off the flow altogether. This is an important reason to never fill a stone drain with dirt, which would certainly filter down and clog the channel over time.

To install a larger culvert under a road, I create the same type of drain, except that instead of merely

Figure 13.11. This roadway culvert has survived the passage of numerous logging trucks without collapsing. Note how massive the capstone over the channel is.

laying two rows of stones, I actually build two low walls, two or more stones high and carefully laid, chinking in between them with smaller stones. A stone culvert works best if laid deeper than necessary. One of mine has frequent car and truck traffic over it, and the heavy vibration could knock loose any carelessly laid stones. It's 2 feet (0.6 m) deep, so any jarring from above is likely to be absorbed by the deep layer of rubble above it.

Ponds and Wells

My goal is not merely to remove water from where it is unwanted, but also to collect it where it is more useful. I have two small ponds, both built in areas that were formerly useless—not properly wetlands, yet certainly not fit for crops. The process of digging out the ponds and using that fill to raise the banks and form the dam concentrated the water in one body and left the surrounding land far more enjoyable. The smaller one is about 40 feet (12.2 m) long with a 35-foot (10.7 m) dam. I don't think the water is ever deeper than my height (6 foot 2, or 1.9 m), and the sides are as steep as I could make them without erosion. They are anchored by mats of cattail roots, which would spread out into the pond and choke it out if I let them (fortunately they are also a source of several foodstuffs; as I write this I am about to harvest a crop of cattail pollen to enrich some egg-free waffles).

In early spring a raucous chorus of several frog species makes sleeping in the overlooking bedrooms a challenge, until you decide that it's not a racket but a lullaby. We are visited by mallards, Canada geese, and great blue herons, plus a host of songbird species that are not considered common in our area. Damselflies, tree swallows, and bats swoop over the ponds, thinning out the mosquitoes and blackflies. They also swoop around our heads, which is a delightful if unnerving distraction. Much of the enchanting diversity that surrounds us is a result of our conscious and unconscious effort to create a great variety of habitats; the ponds are an important part of that.

The ponds offer more specific benefits, too, including irrigation and fire protection. The larger lower pond is spring-fed and maintains a more stable level, deep enough and cool and clean enough for swimming.

It is ringed by comfrey, lupines, and Siberian iris, and—except at the shallow end where you enter—the banks are too steep for cattails to take hold. I keep willows at bay since they transpire so fast, they tend to dry up a pond. I do have a row of Japanese fantail willows on the lower bank of the dam for stabilization and beauty. Elsewhere I have a plantation of basket willows, near a swaley place where the east rill tarries.

I have two wells on the land, one a 23-foot (7.0 m) deep well that I hand-dug in the cellar to supply the

Figure 13.12. A dense covering of comfrey, lupines, and Siberian iris prevents any collapse of the steep banks of my larger pond.

house. Up at the top of the hill I have a 160-foot (48.8 m) drilled well (80 feet, or 24.4 m, through glacial drift, and 80 feet through slate bedrock), which serves mainly for irrigation. A pair of solar panels power a 24-volt submersible pump at the bottom of the well. As it leaves the well the water is further pumped up and into a 2,000-gallon (7,570.8 l) storage tank or down to another 500-gallon (1,892.7 l) tank, from which it can gravity-feed to any of the croplands. I have a network of 1¼-inch (3.2 cm) pipes running to various faucets, where conventional garden hoses can be connected.

I have dug or built a number of wells and ponds, for myself and others, and have never bothered to dowse. Anyway, I was assured that dowsing doesn't work unless you believe in it (which I don't), so I just dug a well wherever it was needed and—what do you know?—there was water. In the case of ponds I'm less random: I look for a low place where wetness is a problem and dig it out, spreading the fill around the edges. Instead of an ill-defined slough, I now have an actual all-season body of water with high, dry banks.

Standing in the bottom of a well is itself an interesting experience. For one thing you can often see numerous stars even in broad daylight. I once heard the roar of a jetliner that sounded like it was coming right in to join me. I yelled out in alarm to my helpers at the top. They looked up and barely spotted a plane

Figure 13.13. Solar panels power the irrigation pump.

Figure 13.14. This 500-gallon (1,892.7 l) tank provides auxiliary water storage for when the sun doesn't shine.

straight overhead, easily 10,000 feet (3,048.0 m) up; they could barely see it, let alone hear it. Regarding anything that is *not* straight overhead, it is as quiet as a tomb. Actually I've never hung out in a tomb, but have been 4,000 feet (1,219.2 m) inside a mountain, which can be very tomblike until an ore train comes rushing by; then it's kinda like that jetliner down in the well with me.

The Land

Land is the basic assumption for gardening. Creating the proper infrastructure is one of our most basic tasks. And one of the prime questions that gardeners

must decide is whether to till and when. So that's what I start with here, and then I continue with a description of another land challenge that I face—gardening on slopes—and how I've dealt with it by creating terraces.

To Till or Not to Till?

Among all the things that may be said for and against tillage, the weakest claim is that it's "unnatural." Keep in mind that most of our food crops, especially the vegetables, were domesticated from the Band-Aid species—those pioneer species that leap to life wherever a fire, washout, uprooted tree, or other localized catastrophe causes an "owie," an exposed

Figure 13.15. New-laid stone steps make accessing the upper pond a lot easier.

wound that needs immediate protection until more stable long-term plant communities (ultimately a grassland or forest) can reestablish. So those various no-till systems (Ruth Stout, Fukuoka, Lasagna Method, and so forth), whatsoever may be claimed for them, cannot rightly be called natural. In fact those wild plants from which our cultivated crops are derived are truly creatures of disturbed soil. In their natural grassland context they would just lurk there in the shade of tall grass, chronically repressed by the aggressive grass and its dense sod. They only attain their real fulfilment when that eco-niche gets overturned, perhaps literally as by a ploughshare. Fortunately most of those vegetable crop species are fairly adaptable to no-till/permamulch systems, but let's just not kid ourselves; it's not what they are evolved for.

Given all that, there are indeed some serious problems with tilling the soil. It reverses the normal soil layers of organic matter near the surface and minerals deeper. The deeply buried organic matter will not decay properly, while the minerals may be more vulnerable to erosion. Rototilling also destroys the natural texture of the soil, breaking apart the ionic charges that bind the soil into "crumbs." Lacking that physical structure there is no longer the capillary action that allows soil moisture to wick upwards to within reach of the shallower-rooted plants. Young transplants can suffer drought shock unless you water heavily, yet heavy watering will cause the fluffy soil to slump down into a dense muddy mass, paradoxically devoid of air.

I should point out that *shallow* tilling, as with a hoe or wheel hoe, can have an opposite, positive effect. It is much like the method F. H. King describes in *Farmers of Forty Centuries* as dust-mulching: By breaking that soil crust, you disrupt the capillary action in the top 1 to 2 inches (2.5–5.1 cm), allowing the soil moisture to rise to the plant roots without being further lost to evaporation. This will be highly beneficial whether or not you add an organic mulch on the surface.

Tilling, especially rototilling, is a kind of soil genocide. Huge numbers of earthworms and other soil creatures, most of them beneficial, are wiped out by this "osterizing" of their ecosystem, not just once but again and again. They will replenish themselves, but it may take a while to reestablish the more balanced soil community with which you started. The more you can avoid it, the better.

Another drawback of rototilling is often seen by some as a plus: The incorporation of so much air accelerates the rate of oxidation, or decay, of organic matter in the soil. This releases more nutrients for the crop, which is good, right? Except this creates a glut of available nutrients, faster than the crop can appropriate them. In essence it burns out the soil in a flurry of bacterial activity, here today, gone tomorrow. In a more natural soil environment both air and organic matter work slowly into the deeper soil layers by leaching, worm burrows, et cetera. Decay is accomplished by soil fungi as much as by bacteria. As a result you find organic matter at all stages of decomposition, from crude vegetable matter (like newly shredded leaves) to stable humus.

Although I occasionally resort to rototilling, especially for incorporating green manures, I have other tools that accomplish those ends without the drawbacks just described. Those favourite tools are the wheel hoe, the broadfork, and the broadhoe or mattock. Let's discuss those now.

Cultivation Tools

The wheel hoe, or push cultivator, is well named, as basically it does what a hoe does, though a completely different shape, while the wheel gives it much more speed and control. Mine is an ancient artifact, having been given to me four decades ago, at which time it was already an heirloom in someone else's family. I have replaced several broken or rotten handles, but the metal parts were surely crafted by some contemporary of Methuselah. It only has moving parts when I move them, such as to flip the tool head over to use the tines instead of the crossbar. Actually, I haven't done that in many years, since I discovered that the crossbar, or stirrup, is the only attachment worth using; indeed I think most of the tines are missing by now. I don't know where you could buy a similar model today, but I would not be without mine. I can imagine no improvement, although a larger wheel might have some advantage. For someone shorter than me an adjustment for handle height would make a more efficient leverage. At times I have slung a little basket from the lower handlebar to toss in pebbles as I turn them up.

The crossbar is only 10 inches (25.4 cm), which is good for getting into tight spaces. It is lightweight and very maneuverable—I can twist the crossbar sideways to cut out some in-the-row weeds, while twisting back out in time to avoid hurting a crop plant. I can till as shallowly as I wish, skimming over the tops of bean roots, but digging deeper up close to a carrot row. I can incorporate fine compost and all shredded leaf mulch, not so much by turning it under as by crumbling and mixing it in with the upper soil, which is where it should be. I can loosen the soil to a depth of at least 3 inches (7.6 cm) with comparatively little time and effort. The wheel hoe won't break up stringy debris like grass or twigs; the debris just tangles up in the crossbar.

When I need to chop in stubble of grain or green manure, I turn to my broad*hoe* (aka Italian hoe). This is a relatively lightweight mattock, forged not cast, designed for use in European vineyards. It should always be kept fairly sharp, which isn't easy in my stony ground. I must file the far side only to maintain the ideal bevel. The handle is a trifle short for maximum comfort and efficiency, but since I make my own tool handles that's easily corrected (finding an ash blank with the proper S-curve isn't always so easy). It can even do a respectable job of breaking old sod for new gardens; indeed I think of it as the poor man's rototiller. You wield an Italian hoe much as you wield a conventional garden hoe, but since I find that very many gardeners do not use any hoe properly, let me discuss that for a moment.

I constantly see people make little chops with a hoe, and then step back a tiny bit and chop again, and so on. The problem with this technique is that you are constantly cuffing dirt back onto the very spot you're going to hoe next. You should rather chop a little divot, then step or reach forward a tiny bit and

hoe the next bit into that divot, thus creating a new divot, and so on. In the case of the broadhoe you're basically mimicking the motion of a rototiller, but much more coarsely and doing less harm. Moreover, as the hoe, or broadhoe, arcs downward with each chop, it clips off the tops of further weeds and buries them. A taller weed may get chopped at several heights as you advance, so that it is not only uprooted but shredded as well. A broadhoe accomplishes this at a much greater depth (perhaps 4 or 5 inches, or 10.2–12.7 cm) than a conventional hoe and with much denser weed growth; in fact I often use it for chopping under green-manure stubble, especially clover. It is much less destructive of the crumb structure. I have even used it on the densest old quack grass sod; it's just a matter of keeping it sharp and biting off corners instead of straight cuts. A rototiller would just buck and bounce over that.

Whereas the wheel hoe will do a very superficial tilling quickly and easily, if you need somewhat more depth, the broadhoe will answer nicely. But sometimes it is helpful to loosen the soil to a much greater depth than either of these will reach; loosen it, that is, but not churn it all up. The idea is to break up any hardpan left by excessive ploughing or rototilling, not to bring up subsoil or bury organic matter. You wish to leave the layers about where they are, only open them up so that taproot crops can penetrate more easily. That is where the broadfork comes in.

The broadfork has longer, narrower, rounder tines (less easily bent by rocks) than a spading fork. The broadfork tines do not move the soil much, but rather break it up, at least 12 inches (30.5 cm) deep, for easier root penetration. Few if any earthworms or other creatures are destroyed; indeed you have enhanced their work by loosening the soil and letting oxygen enter (but not too much). Any vole tunnels are thoroughly disrupted. A side effect of the broadfork is that most of the organic fertility stays near the surface and more of the mineral fertility remains at depth, while more accessible. Nor is the tool difficult to use; despite its daunting size, a person of moderate strength can readily lift it and plunge it in anew, then tread on the crossbar down to the soil level and pull

Figure 13.16. A broadfork lets Mike Bouchard loosen the soil deeply without inverting the natural layers.

back on the long, stout handles. No need to push them all the way to the ground; the goal is to loosen the soil, not turn it over. This works well only on soils that have already been deeply dug and rocked. If I feel the tines hitting against something that might conceivably bend them, I stop right there and go fetch a spade and/or pry bar. That is not the broad-fork's job, although like the spading fork its tines are easily hammered straight. It is a really efficient tool, forking a 20-inch (50.8 cm) swath. I step back every 6 to 10 inches (15.2–25.4 cm), so it moves right along. At two or three passes I have finished the bed in 15 or 20 minutes. Even if some compaction were to result from occasionally stepping in the beds, the broadfork would more than compensate for it.

Terraces

Where I live in the western foothills of Maine, flat farmland is at a premium. My cropland has an aver-age slope of 6 percent, in some places as steep as 16 percent. In my first years farming I recognized a serious threat from erosion. The glacial soil was none too deep to begin with, and five generations of clear cultivation hadn't improved things.

This led me to think about terracing. A big advan-tage of terracing, in addition to preventing erosion, is that it manages water in both directions. Terraces hold water longer so that more of it can soak down into the thirsty ground. On the other hand any excess can eas-ily be drained off by ditches into a tank, holding pond, or something where it can be stored until needed.

I had seen charming pictures of Asian mountainsides sculpted into rice-growing terraces, but I didn't have a team of water buffalo or a large family to put to work on it, and anyway I preferred not to spend too many centuries accomplishing the feat. What I did have was my neighbour's old 1946 John Deere tractor with a single-bottom mouldboard plough and trailer-hitch double-gang disc harrow. My neighbour Lucian, who himself had never dreamt of terracing, now came up with some useful ideas. He was well aware that he and his ancestors had been exacerbating the problem by ploughing downhill year after year, constantly exposing the land to the washing away by heavy rains. He pointed out how the best soil was on the uphill side of the hedgerows, in some places nearly burying the old stone walls that were the occasion for the hedgerows. He noted that at least the soil did stop there, unable to be carried any farther. In those areas there was already a tendency for terraces to form.

How could the same destructive downhill ploughing be used to *create* the terraces that would prevent further movement of the soil? Lucian reminded me how ploughing a larger piece is often done in sections called "lands." In turning the soil a mouldboard plough scoops out a shallow trench and lays the sod (or soil, if it has been ploughed recently) to one side of the resulting furrow. Thus you immediately form a high ridge (called the "live furrow") and a shallow trench called a "dead furrow," as shown in figure 13.17. As you continue ploughing you continually create a new dead furrow by filling in the previous one. When you finish a land there is a little ridge on one side of the field and a shallow trench, or dead furrow, on the other. These irregularities are not dramatic; indeed, when you proceed to harrow (or disc) the piece to smooth out the plough ridges, the dead and live furrows pretty much go away. In fact to avoid building up an unwanted ridge, you usually commence and finish ploughing at a slightly different place each time.

But, suggested Lucian, what if you deliberately started and stopped at just the same spot every time? *And* what if you in effect made several narrow lands,

each time setting the tractor wheel up on the high live furrow and quitting before filling in the next dead furrow? If the narrow lands were laid out on premarked contour lines, you would form terraces, or at least the beginnings of terraces. (See figure 13.17.) Repeating this operation several times would create a series of nearly level terraces.

Of course, before I could construct these terraces, I had to decide where they were going to be and what final shape they would take; I had to survey them. I am not a trained surveyor, nor could I afford to hire one, yet I needed to somehow determine the contour and slope of the land with enough accuracy that my finished terraces would be largely level and with a minimum of labourious earthmoving.

To begin with I improvised a giant "spirit level" by taking a 75-foot (22.9 m) garden hose and tightly inserting a 2-foot (0.6 m) length of clear glass tubing in either end. I filled the hose with water coloured by beet juice and fastened each end to a 4-foot (1.2 m) hardwood stake; I corked the glass tubes so I could move the hose about without sloshing. With a friend holding the other stake upright and me at my end, we each moved to positions on the slope that appeared to be roughly level and about 70 feet (21.3 m) apart. Uncorking the tubes we slowly moved the stakes up and down the slope until the coloured-water line was the same height above grade at both ends (if we moved too quickly, the water sloshed out one end and had to be replenished). By moving around on the hillside we were able to find a number of points that were all on the same contour and place marker stakes there. I didn't much care *what* contour; any contour would serve as a reference line from which I might measure uphill or downhill to establish other contours. That is, assuming the slope was equally steep at all points, which it was not. On the steeper part the contours would of course be closer together than on gentler parts of the slope, and thus the terraces narrower.

That raised a number of questions: Should I shape each terrace to follow the level of its contour, regardless of how irregular, or was I determined to have all terraces with straight, parallel sides, even if that would require a lot of earthmoving?

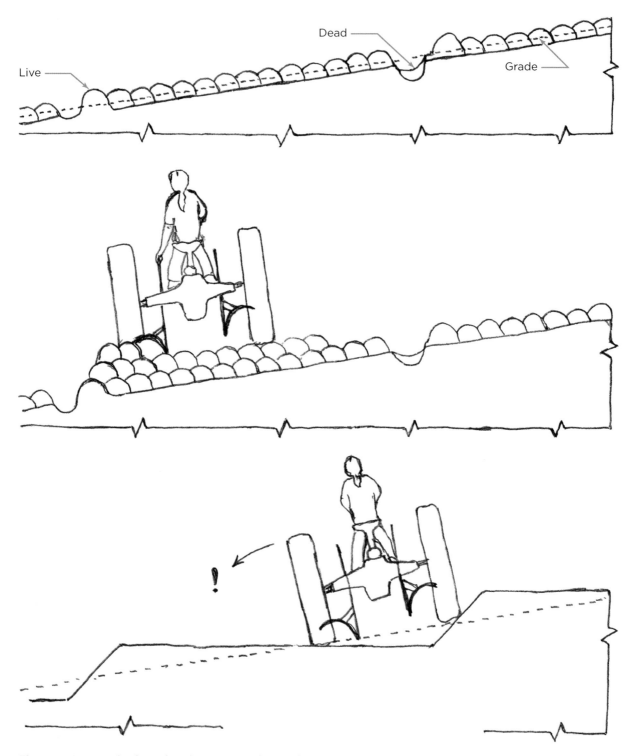

Figure 13.17. *Top,* the first ploughing creates live and dead furrows. *Middle,* repeated ploughing shapes the terraces. *Bottom,* there are two reasons to stop ploughing: if there is danger of tipping, and if the plough no longer contacts the soil.

For my first set of five terraces (the Lower Garden) I chose a compromise, insisting on parallel but allowing curves. Since all of my beds are 6 × 40 feet (1.8 × 12.2 m), each terrace must be a multiple of that, plus an additional access path on the outer rim. I also desired straight beds—helpful when stretching a string to mark terraces—but that was impossible, given the way the terraces wrapped around a curved slope. To follow that curve I had to place a number of stakes, every 10 feet (3.0 m) or so, and wrap the marker string around those. It was a minor nuisance, but I must admit that the curving rows, especially with crops like grain, are visually quite appealing, emphasizing the sensuous form of the land.

My obsession with parallel terraces cost me in this way: If the top terrace is on a level contour, those below it must slope somewhat downward as I approach the steeper part of the slope (since I'm not allowing them to narrow up). I adjusted this largely by moving many wheelbarrow loads of dirt from the upper end to the lower. However, the bottommost terrace was so far off level that I opted to break it lengthwise into two shorter terraces with a smaller lynchet extending crosswise. "Lynchet," by the way, is a little-used term for the edge of a terrace, whether a retaining wall or, as in this case, a steep sod-covered slope. Keep that in mind for your next Scrabble game.

Figure 13.18. The terraces in my Lower Garden are parallel, with each wide growing area separated from the one above it by a sod lynchet.

I love my Lower Garden terraces, but when I terraced elsewhere I took care to avoid such back-wearying effort as much as possible. For example, in the Back Garden I had a slope that was fairly uniform for 240 feet (73.2 m; six beds long), so the terraces could each be parallel two beds wide. Moreover 160 feet (48.8 m) of that terrace was naturally straight enough that I could remove any slight curve with a minimum of wheelbarrow work. The last section (80 feet, 24.4 m, or two beds long) had a pronounced curve, or dogleg, from the rest of the terrace. I chose to break it at that point into two separate sets of terraces, both straight but at a slight angle to each other. That gave each of the longer terraces (three of them) 4 double-wide beds, or 8 beds per terrace; the shorter terraces are each double-wide, 2 beds long, or 4 beds per terrace, giving me 36 beds on that slope.

At the top of the slope a flatter grade allowed me to have similar a terrace, also 6 beds long but 6 beds wide for another 36 beds (also divided into two sets because of the dogleg).

Since then I have laid out more terraces by other equally crude methods, and the results were no less satisfactory. For example, I used a 36-inch (91.4 cm) carpenter's spirit level on a 16-foot (4.9 m) 2 × 4, laid on the ground and held on edge lengthwise, on the contour. Of course I had to reposition the 2 × 4 five times in order to establish an 80-foot (24.4 m) terrace, so I was very careful lest any error be fivefold, but of course it was not: Since the error was random rather than cumulative, if I erred in one direction for one segment, I was just as likely to be off in the other way for the next segment, so it averaged out to be well within my needs for precision.

I've already explained how I used the tractor and plough as a bulldozer, but that had its limits. Each time I ploughed and harrowed the outer ridge grew higher and the inner depression became deeper—just what you want, as shown in figure 13.17. However, in order to keep doing this repeatedly, as you plough into the inner part of the terrace, you have to place the uphill tractor wheel on the next terrace above. That's no big deal at first, but as the shaping progresses your downhill wheel sits ever lower and the uphill wheel sits ever higher, until you are perched at an angle much steeper than the natural grade ever was. Not only is the tractor in very real danger of tipping over, but the plough can no longer reach down to full depth; in fact at some point you are ploughing sky (see figure 13.17).

So, well before I reach this perilous and fruitless situation, I abandon the tractor and resort to shovel, pry bar, and wheelbarrow. I should emphasize the pry bar; all of those iconic Asian terraced hillsides appear to be either some fine windblown silt or loess or some long-decayed volcanic lava soil; not a glacial erratic in sight, mind you, not so much as a doorstop. By contrast, I and my fellow upland Yankees have an abundance of stones ranging from pebbles to Volkswagen-sized boulders. To build terraces we have to work a lot harder (and there isn't a century to spare). We must not only remove all the shallow stones to get enough dirt to grow stuff, but now we must make that rock-free dirt deep enough so we can shift around and still have dirt. However shallow my topsoil was to begin with, I've moved it sideways (can't call it "downhill" anymore because it isn't) so that the outer half of each terrace is now deep, fertile, rock-free soil (assuming the top few inches have been de-rocked all along). Such deep friable soil feels very un-Maine-like. I might almost feel like I'm back in California (and I mean that in all the best ways), but the inner half of each terrace dashes any such delusions. It is largely subsoil, rich in minerals, but quite devoid of organic matter. I can immediately grow fine crops on the outer half, which is at least as good as I started with, plus much deeper, no rocks, no hardpan. Lowlanders farming flat bottomland won't have those rocks to contend with, but then neither will they need to terrace in the first place, which is a mixed blessing, as I'll explain later. I must remove rocks, many tons of them, to as great a depth as I wish to farm.

I also need to do some further earth shaping, since the plough did not finish its work because of the tipping danger I described earlier. The subsoil, picked clean of rocks big and small, must be spread around and smoothed to match with the outer half of the terrace. That done, it is ready to be converted to good topsoil, by huge additions of humus-building organic matter. However what the new "soil" does *not* need, or even want, is lots of high-nitrogen fertilizer like animal manure. At best it will be wasted, since that excessive concentration of nitrogen without humus to absorb it will only form anaerobic compounds that will toxify the soil or pollute the atmosphere. Far more useful, indeed essential, is the addition of oodles of high-carbon material like tree leaves and chipped ramial wood. This is the stuff the forests used to rebuild the soils of Maine after the last ice sheet retreated; the occasional deer or moose poop was purely incidental. These materials do not lack nitrogen, but they will build the soil structure and develop humus with the nitrogen in a sustainable balance.

It is not a simple matter of tilling those masses of residue deeply into the soil; indeed that would be counterproductive. Rather they should be shallowly incorporated into the top few inches where conditions are more favourable for aerobic decay. Paradoxically, as they break down they will make the soil much more aerobic—by creating interstices in the soil particles and nurturing microbes—so that yet more organic matter can be added, life begetting life and so on ad infinitum. (This is discussed in detail in the *Ramial Chips* section of chapter 3.)

Sod Lynchets

When I look online to learn more about how other people terrace, I am impressed by how different my method is. Most of the buzz is over terraces as a modern landscaping feature; for example retaining walls of cut stone or pressure-treated wood are fully as important as the perennial flowers and shrubs that are within them. I am not building terraces for ambience (although mine also look wicked cunnin'). First and foremost I am after food. Also, I need to terrace larger areas. I'm already investing a lifetime of labour; the materials at least must be cheap and at hand. In fact, the material I use is the earth itself; only one of my many terraces has a retaining wall. That is the one containing my wife's perennial flower

garden, and it is made wholly from the igneous and metamorphic rocks that came out of there. I have no shortage and the price is right, but such walls can have a high price in maintenance, as the rubble backfill is an ill-defined weed haven. Mice and snakes also love it there, but as they pretty much take care of each other, they're not much bother. The maintenance is not unreasonable for a small kempt flower garden, but for cropland I needed something quicker to build and easier to maintain, and that is sod. My terraces are 2 to 3 feet (0.6–0.9 m) high, and the lynchets are formed to a 45-degree slope. Some I carefully covered with sod and others just sodded themselves. The sod must be mown. If I'm the one doing it I use a scythe; everyone else uses a string trimmer, which is probably less awkward on the steep slope. I usually haul the first heavy cut of grass to compost; the skimpier later mowing usually just lies at the foot of the lynchet where it is a welcome fertility boost for the relatively sterile inner terrace below.

I don't much begrudge the space my sod lynchets occupy, especially since they make the rest of the land surface so much more productive, but I do have a particular gripe with them. As any gardener knows the grassy or weedy margin of a cultivated area is an invasion zone, an edge-effect place where only diligence prevents the surrounding area, full of aggressive highly adapted life-forms, from overwhelming our wimpy attempts at cultivation. It is challenging enough when the plot is a square or a circle with a minimum proportion of edge effect, but when a series of lynchets surround a series of terraces, the situation becomes more precarious, particularly when the sod contains something called quack grass. Its long white runners will be constantly probing your defences. In fairness, with a bit of vigilance it's not such a big deal to till or chop them back, but who wants to increase that chore 10-fold? Fortunately the incursion is mainly into the terrace *above* the lynchet; the sod is quite slow to aggress downwards. Although I insist that the benefits of terracing are well worth the nuisance of edges, there is no point in increasing the latter unnecessarily. For that reason I have always preferred to build a smaller number of wider terraces, although many narrower terraces would require a lesser and shorter commitment.

Of my nearly 30 terraces, most are two beds long (80 feet, or 24.4 m), and two beds wide, which including three access paths is about 14 or 15 feet (4.3–4.6 m). For excess water to drain harmlessly away, most of them are transected by a drainage ditch. Where the amount of water justifies it, they are open ditches, up to 12 inches (30.5 cm) deep with sloping sod-covered sides. They are designed to lead the runoff into one of my ponds. Like the lynchets, these ditches must be mown occasionally, lest they become choked with old grass and so people don't stumble into them. Where the seepage is small and infrequent all I need is a stone drain, as described earlier. Any significant rush of surface water can simply sink into these stones and be carried away by the culvert, but most normal seepage will simply settle into the ditch and further sink into the ground where I want it.

The stone drains, with or without built-in culverts, have a big advantage over open ditches. The latter are obstacles to moving wheelbarrows, rototillers, and the like. I am always having to fetch some short planks to span them, and they are rather awkward to mow, yet wherever there's a volume of flowing water to move quickly, there's really no alternative. They are also one of many ways of using my overabundance of rocks to do some good.

When building another set of nine terraces still farther up the hill, I opted to just work with the natural slope and let the terraces take whatever shape they would. I made them all an arbitrary 26 inches (66.0 cm) high and had help from a neighbour with a transit. They are all crooked and irregular, and I love them all, as I love many people who are crooked and irregular; they all have their purpose. With those terraces my original expectation was to use them

mainly for improved grassland or perhaps in rotation with certain field crops like corn or potatoes where I might fret less about nicely spaced straight rows. So far I have not integrated them into my standard bed configuration.

Thus far I have created enough terraces to hold about 140 beds (about ⅞ acre, or 0.4 ha). About the only part of my cultivated land that isn't terraced is my original (Upper) garden, which is naturally less than 2 percent slope; so far I have not seen fit to terrace it.

There is another way of building terraces that avoids the need to rebuild the stripped inner halves. There is a mountainside in Austria where Sepp Holzer has built terraces by first scraping off the topsoil and pushing it aside (not sure where) while he then levels the underlying dirt before spreading the topsoil back over that. Although he does not get the extra-deep rich soil on the outer half of each terrace, neither does he have to spend years restoring the inner strip to useful fertility. I am far from insisting that my way is superior; there is much to be said for and against either method. Mine was the only apparent option given the resources available to me (Holzer had earthmoving equipment; I didn't).

I mentioned earlier that low-lying flat or alluvial soils are a mixed blessing; that's because it doesn't *need* to be terraced, but on the other hand it *cannot* be terraced. Floodwaters may sit on it for days or weeks, having no place to go, while crops are submerged or washed out. Terraced hillsides, where the slope is compressed into either walls or sodded lynchets, with strips of well-drained flatland in between, give us the best of both worlds. I am 65 and have spent countless hours of hard labour to create them, but if you ask me how I'd feel if I had to start all over again, I would say, "Hand me my level, please, no time to waste."

CHAPTER FOURTEEN

Smaller Footprints

There are various strategies we can apply to get more food out of a given area of land. But why bother? Especially if you have plenty of land to spare. "Intensive" is not just about space. Land is not the only resource needed to produce food; fertility, mulch, water, your time, attention, labour, are all required, and those tend to be roughly proportionate to the area involved, *not* the quantity of food produced. Therefore anything you can do to reduce the footprint of your garden—and all those other inputs—without reducing the yield will result in a much more efficient use of all resources.

Old World Versus New World Crops

A basic principle that's critical to my thinking about the context of my gardens is applying what I know about the differences between the ancient agriculture of the Old World (Europe, Asia, and Africa) and the New World (North and South America). The differences between these two agricultures were so profound that they might enlighten us with ideas for our own methods, especially since we have borrowed crops from both of them. Although agriculture in the Old World was developed independently in several locations by different cultures, those peoples eventually had contact with one another and gradually

shared their inventories of crop species and technologies. They also developed an assortment of domesticated animals, for food and labour, which, when combined with the invention of metal tools, had a profound impact on what they grew and how they grew it.

By contrast, New World agriculture arose in only two centres—the Mexican lowlands and the Andean highlands of South America—and spread from there, with peoples sharing a few crops but essentially remaining isolated from one another until the arrival of Europeans. With no metal implements and no draft animals, ploughing was never practised, for better or worse. Therefore New World agriculture—or at least cultivation, which is not quite the same—was confined to river valleys and lake plains and coastal plains, where alluvial soils could be worked with less sophisticated tools. Stony soils were relegated to hunting-gathering, with the exception of the Andean region, where terracing enabled a whole civilization to produce food on steep, seemingly unfarmable mountainsides.

Old World horticulture evolved in conjunction with the herding of large ungulates (cattle, sheep, goats), which produced manure and required large areas of permanent grassland pasture, areas that were maintained as such over many centuries. Moreover its crop inventory included many cold-hardy species, which enabled it to expand into more temperate northern regions. Many of these crops tended to be small-seeded compact plants, including many biennials. New World crops, on the other hand, were largely frost-tender tropical annuals with large seeds, with the major exception of the potato. Most cultivation was slash and burn, with periods of abandonment to restore the soil fertility. Lacking metal implements the main technology for reclearing of the jungle involved the use of fire. Those soils were fertilized largely by the buildup of decayed wood and leaves and the ash residue from burning slash. Even though the forest residues were more eco-efficient than Old World straw and manure, in the hot, humid tropical lowlands organic matter was very quickly mineralized, and so it was harder for humus to accrue over time, thus necessitating the periodic abandonment of croplands to jungle, in effect a long-term crop rotation.

In hindsight it appears that the Old World food-growing technology triumphed over the Americas. Although New World *crops* are now very widely grown (especially corn!) and eaten (imagine pizza without tomatoes or peppers), the *way* they are grown is decidedly Old World. Permanent croplands are ploughed and sown to closely spaced row crops that are intensively cultivated; fertilizer (whether manure or chemicals) is imported onto the piece from elsewhere. That's very Old World; on the other hand having fallow years and the use of green manure is basically a short-term version of New World swidden agriculture (the slash-and-burn jungle rotation). I'm not suggesting that they were inspired by the New World; Eurasian farmers were practising fallow long before 1492, as described in Leviticus 25. Both systems, but especially the Old World, tended over time to desiccate the landscape and require irrigation, which eventually destroyed those croplands, and the empires they supported, by accumulating salts. What implications do these historical practices and their consequences have, and how might we incorporate some of these observations into our own gardens and farms?

When it comes to intensive planting we're mainly looking at Old World crops with their compact plant forms and root systems. For example, peas, carrots, and onions make a perfect companion grouping, but I can think of no New World species whose growth habit lends itself to such crowding. On the other hand the very sprawling nature of so many New World crop species is conducive to three-dimensional arrangements. Pole beans and sunflowers can support each other with greater combined yield, as described elsewhere. Squash/pumpkins planted in hills with 8 feet (2.4 m) plus in every direction leave lots of space for a living mulch. Indeed the fact that they are frost-tender is almost an asset, since the cold-hardy green manures (oats and peas) can get a head start and be out of the way by the time the squash does indeed need the entire space to itself.

The various no-till permanent mulch systems (Ruth Stout, Lasagna Method, et cetera) are much more appropriate for New World paradigm species, which do fine in thick mulch with their wider spacings. Even if the patch were to become weed-infested (which mine generally don't), a crop like pumpkins would just clamber over the weeds with fairly little crop loss, whereas carrots or onions would be totally overwhelmed. Tomatoes with their 2 × 3 or 3 × 3 feet spacing are happily ensconced in deep mulch of any kind. Determinate types can just sprawl on the mulch (with less blossom-end rot than on bare soil), and indeterminates can be staked up for higher production. The only Old World crop that I find works well in this context is cabbage (or brussels sprouts), which needs at least 2 feet (0.6 m) in every direction.

Without particularly meaning to I have tended to duplicate the New World soil types by my use of shredded leaves and ramial chips to build and maintain fertility. Those burned-over jungle clearings contained huge amounts of rotten wood and leaves, half-burned stumps, and undecayed twig litter. As these broke down amid the crops they produced a largely fungal (as opposed to bacterial) soil community, which might have been too acidic were it not for the ashes. My own situation is similar in that it includes lots of coarse rubbishy material that is slower to degrade than materials derived from straw and annual residues.

I find that people are often confused about the word *hilling*. I once watched a fellow patting up a nice mound of soil before sticking in a few squash seeds, because he had heard somewhere of planting squash in hills. *Hills* and *hilling* are derived from a practice of Native Americans of using a digging stick to make a cluster of holes into which they dropped the seeds of corn, beans, squash, sunflowers, and probably other New World crops like amaranth and tobacco. The seeds were simply covered, but later when they cultivated—with bone or clamshell hoes—they merely scraped up enough soil to bury the weeds and, in the case of corn, to give support and encourage the formation of prop roots. In the case of squash mounding would encourage more rooting of the vine nodes, in case borers damaged the spreen. Burying the lower stems of beans gave weed control without damaging the shallow legume roots by chopping near the plants. The combination of scraping up the soil between and dumping it around the plants would form a slight mound or hill, which a second hoeing would further accentuate. Likewise the Quechua method of drawing earth around the stems of potatoes (to encourage the inception of more tubers and prevent their greening) formed a long hill or ridge. In all these cases the seed was not planted into a preshaped mound, but rather the mound or hill was formed by the subsequent cultivation of the crop, in a more casual manner than would have been feasible with most Old World crops.

As I described in chapter 1 I occasionally have some half-cured compost that I could apply, but it's still rather coarse-textured to use on most crops. Well, it can be used on several of the New World crop species, especially tomatoes and squash, either chopped in or as a topdressed mulch. Most Old World species would take offence at such rude treatment, but tomatoes are insensitive to insults.

New World crops (except potatoes, sunflowers, and terrasols) are frost-tender, but that's no reason to leave the land idle until the weather suits them. In fact we can take advantage of their cold-wimpiness by integrating them with Old World crops like early lettuce or green onions, or with green-manure crops

Table 14.1. Crop Origins

Old World Grains	New World Grains
Barley	Amaranth
Buckwheat	Corn
Oats	Quinoa
Rice	
Rye	
Wheat, spelt, emmer	

Old World Pulses	New World Pulses
Chickpeas	Beans, kidney
Favas	Beans, lima
Lentils	Beans, runner
Peas	Peanuts
Soybeans	

Old World Veggies	New World Veggies
Cabbage (including kale, etc.)	Ground cherries
Carrots, celery, parsnip, parsley	Peppers
Cucumbers	Potatoes
Melons	Squash, pumpkins, gourds
Turnips, rutabagas, mustards	Terrasols
	Tomatillos
	Tomatoes

they can either accompany them for a while or perhaps precede or succeed them. The first frost that ends their career is just in time for the planting of winter wheat or rye, although if the mulch is old hay the grain crop may suffer badly from weeds. Using my green-manure-*cum*-mulch method, this is not a problem.

Incidentally another way in which my system mimics New World agriculture, albeit coincidentally, is the absence of livestock manure. The aboriginal American farmers had only a few domesticated animals, including turkeys, dogs, and llamas, but none would have been a significant source of fertility. In New England at least they used rotten fish, but even that would not have been an overall, long-term soil builder. That was supplied by the forest residue (I've never heard that any New World farmers used their night soil for fertilizer, but anyway it would have been insufficient).

Here's a story that illustrates how an understanding of Old World/New World agriculture and crops can help us to maximize our use of our land. My friends Denis and Betty Culley moved to Maine from New York City and bought a piece of uncleared land, mostly fir forest, in a nearby town. Denis had no homesteading experience, but was eager to learn. By dint of great labour he managed to clear a fraction of an acre. He removed what stumps he could from a small plot in the middle of the clearing, but most of the stumps remained, to be pulled out or rot out over time. Among them were a few large piles of brush (mostly fir) waiting to be burned. Some already were burned, and their ashes lay in a heap. He wisely focused his efforts on making that small stump-free plot into a civilized garden. He removed rocks and applied some of the ash and whatever else he could scrounge in the area, including a tiny amount of compost from his food wastes, not enough of anything to accomplish much, but a brave beginning.

Wanting to make the most of the tiny plot (while he continued to expand it), he asked me my opinion on what he should plant. I told him about the Old World crops—peas, carrots, onions, turnips, lettuce, beets—and how by intensive planting methods he could meet a respectable portion of his need for those foods. The ashes he added helped balance the acidity from all the decayed fir spills, which was the main organic material in that soil. After much effort he had a bit of half-rich soil that was still not Old World soil in any sense, but was well on its way; time and the addition of more organic matter would eventually produce a soil where almost any crop would thrive. He did all that, and I believe he had quite decent results. But what about the rest of the new clearing? I suggested he plant that area in appropriate New World crops. Their new privy had a small accumulation of night soil, a few bucketfuls. Denis buried a half bucketful of raw humanure next to each brush pile, added a few handfuls of wood ash, and stuck in

a few squash seeds. Now the ground was full of dead fir roots, but all he had to do was dig up a 2- or 3-foot (0.6–0.9 m) circle to plant; the squash vines could spread out from there. He trained the vines up on the brush piles where they could proliferate without competing for space with anything else. They were exceptionally happy and healthy being held up off the ground.

Elsewhere he planted hills of pole beans in spaces between the stumps. He planted a few tomatoes among the stumps, either letting them sprawl over the stumps or staking them up. They do require plenty of fertility, and he met that need by dumping half-composted kitchen garbage around the base of the plant, covering with enough duff to keep the flies off. Many crops would do poorly in that situation, but the tomatoes were in hog heaven. All during that summer, the stumps were decaying or being pulled out, the soil chemistry was changing under the influence of sunlight and developing humus. The brush piles were either decaying or being burned for ash.

By taking advantage of the peculiar features of Old World and New World crop species, Denis was able to harvest a fair amount of food from what otherwise would have been just another clear-cut. This even though he was at a disadvantage: New World crops, which were evolved in tropical soils built up by broadleaf jungle, could not be expected to feel quite at home amid the acidic resinous detritus of a northern coniferous forest. Considering that, the plants did quite well, and did much better yet in succeeding years as cultivation altered the chemistry and structure of the soil. In two or three years his original plots were equally suitable for either Old World or New World crop species.

About that time I happened to read a book from my father's library called *Saints and Strangers*, about the early Pilgrim settlement in Massachusetts. In it the author comments how poorly the colonists did with their first plantings of field peas, turnips, and other crops of which they had brought seed from England, and which they were thoroughly used to cultivating. The author found it paradoxical that they were saved from starvation by the corn, beans, and squash that the native Narragansetts taught them to grow. By now I trust that the reader will see that it was no paradox at all; rather the Old World crop species, though well accustomed to a similar climate, were ill adapted for the type of soil they encountered and the method of cultivation which that soil required.

Natural Supports

One way to avoid expanding your crop footprint is to use the third dimension. Remember, your garden is more than length times width; it also has height and depth. We can take a cue from gardeners (and architects) in congested urban areas who deal with spatial limits by building upward. For example, given some support, pole beans will reach up to the life-giving sun and produce much more food per square foot than their dwarf forms. In fact the main reason bush beans are so dominant in the marketplace is that they lend themselves to onetime machine harvesting. Pole beans yield four times as much, but they require handpicking. Pole beans also bear over a much longer period of time, but that is merely a nuisance for commercial production. The marketplace assumes cheap land and expensive labour; if you're not working for the marketplace your perspective may be somewhat different.

Of course, growing pole beans requires a source of poles. In a true garden-without-borders you have a source, and much of my approach to gardening is predicated on possession of (or at least access to) woodland. Obviously woodlands can also provide structural materials for the garden: stakes and poles. I use young fir thinnings from my woodlot—balsam fir often grows in thickets, which must be thinned for maximum growth and quality, and those thinnings are generally the straightest and slenderest young trees, which make the best trellis material. Balsam fir is *not* the most rot-resistant wood, particularly when in contact with the soil, so every two to four years the bottom foot, the part that's in the dirt, becomes dozy and weak and gets busted off, leaving a shorter pole that is usually still serviceable for holding

pea fence or staking tomatoes. When it eventually becomes useless for any "supporting role," I feed it through the chipper/shredder and it becomes mulch for raspberries.

All well and good, but what if the support, instead of being a pole, were itself another crop? For example, we grow lots of sunflowers, but we used to plant them in blocks with 3 feet (0.9 m) between rows and 12 inches (30.5 cm) between plants. They presented a real barrier to winds, which sometimes knocked quite a few over. It occurred to me: What if I planted them in hills instead, three seeds per hill, 3 feet between hills? That left gaps or canyons through which the winds could blow more harmlessly. That part works swell, but it created a new problem. The sunflowers reacted to the crowding by leaning outward, away from one another, and that off-balance pose caused some of them to fall over with the weight of their ripening heads. Says I to self: What if I put a stake or pole in each hill to . . . but wait! These things *are* poles; I just need to loop a string loosely around them so they can support one another! But then, what if that "string" were also a crop, like a pole bean vine? It worked beautifully.

Here's my method: Because the sunflowers can (and should) be planted in early spring, weeks before the last frost, that's when I plant them, whether in the ground or in 3-inch (7.6 cm) peat pots in a cold frame. An advantage of the latter is that I needn't wonder how many the birds will leave me. I plant several seeds and thin to three; by the time I set them out they're too big to interest birds or mice. I plant hundreds of hills, so I need an entire cold frame for that crop alone (it is a major item in our diet).

Pole beans cannot be planted as early as sunflowers, as they will not abide any frost or even cold soil. I direct-seed the pole beans around Memorial Day, when the sunflowers are already at least 8 inches (20.3 cm) high. I plant four to six bean seeds per hill, poking them in around the sunflowers. Of course each bean does not climb up its designated sunflower stalk; they all clamber from one to the other in a useful sturdy tangle that binds the whole thing together. In fact it takes some attention to prevent them from leaping from one hill to the next. This would not be a problem if I didn't need to walk through there, cultivating or watching for Mexican beetles. Since dried beans need little attention until harvest, I can largely ignore them. By the time their pods are mature and dry, the sunflower leaves will be frost-killed, allowing the bean pods to be easily seen and picked (my favourite dry pole bean variety, a local heirloom called Orlando's, has red-streaked pods when mature, making them especially obvious).

Incidentally everything I've said regarding pole beans on sunflowers could be equally applied to pole beans on grain amaranth. For this I prefer a tall variety like Opopeo (from southern Mexico yet).

Trellises

I use either sunflowers or poles to support pole beans, according to the type of pole bean. If the beans are for fresh green use and require frequent picking, I grow them on an all-pole trellis where the green pods will be less hidden. Here is how I build it.

At each end of a bed I centre a heavy post about 7 feet (2.1 m) tall. Between the endposts, on the bed's centreline, I set other posts for a total of five (10 feet, or 3.1 m, apart, since my beds are all 40 feet, or 12.2 m, long). To these I tie a ridgepole about 6½ feet (2.0 m) above the ground. Then I make a shallow furrow 22 inches (55.9 cm) to either side of the centre (44

Figure 14.1. An Orlando's pole bean climbs on an Opopeo amaranth.

Figure 14.2. With the ridgepole in place the actual beanpoles can be leaned against it from opposite sides.

inches, or 1.1 m, apart). The furrows serve as the anchoring spots for 9- to 10-foot (2.7–3.0 m) long poles that I lean inward to the ridgepole. I set these poles in position 24 inches (60.1 cm) apart. I make sure both rows of poles are opposite each other, so that their tops cross at the ridgepole, where I tie them together so they won't slide sideways in the wind.

Then I plant six beans in the furrow at each side of every pole, three beans on either side. I watch the young plants closely in the beginning, making sure each vine climbs up its own pole instead of reaching for the neighbouring pole or across to the opposite side. This eventually becomes impossible, but I try to avoid an impenetrable tangle that is difficult to

pick from. Since the 9- to 10-foot (2.7–3.0 m) poles reach well above the cross rail, many of the vines will climb up and across or back down again, leaving a lot of beans out of reach. I could use a stepladder, but I rarely bother with them; since the main variety I grow is a three-purpose bean, I just let the unreachable pods go by, to become shell beans or dry beans, which can be left until the first heavy frost shuts down the whole mess and I dismantle the trellis for the season. I say "the first *heavy* frost" because the early light frosts usually spare most of the crop, protected by its own dense foliage. The lower leaves may be nipped, but they are usually senescent anyway. Bush beans are more vulnerable, growing close to the ground.

Figure 14.3. Each beanpole is tied to the opposite one and to the top rail to prevent sliding sideways.

Figure 14.4. The pole beans are ready to be mulched; they will soon cover the entire structure.

And what is the variety I use for this purpose (or rather purposes)? Well, of course it is Jeminez, the only green pole bean variety that clueful gardeners will bother with. I notice one seed company describes its flavour as "a bit too beanie for some tastes"—absolutely correct! Just as some tomatoes will be too "tomato-y" for those who are ambivalent about tomato taste. If you love rich green bean flavour Jeminez is for you. It is a heavy-podded Romano type with "carmine" (that's red) streaks. Its pods can grow very long and fairly overmature and still be quite excellent, as the thick flesh is slow to become fibrous (the red disappears upon cooking). However, the yield is so overwhelming that plenty will get by you anyway; no matter, since they make an exceptionally fine creamy shell bean, sensational with fresh or frozen sweet corn as succotash. Again, they are so overbearing that you will be hard put to keep up with them at this stage, too. How fortunate that the dried beans are superb baked or stewed, and of course you'll want plenty for seed to grow even more next year.

I'm a fan of trellising peas, too. In recent years I've seen an increase in dwarf, or bush, pea varieties on the market. More than ever, it seems, gardeners do not want to bother to set up support fences, even though that would significantly boost yields. The assumption is that the shorter plants can hold themselves erect, although in my experience (with over 1,100 varieties of peas) all but the most dwarf types are prone to sprawl without some kind of support. Leafless pea varieties are self-supporting, but they have a serious drawback—see *Peas* in chapter 8 for more about this. I prefer normal or tall peas with plenty of leafy solar panels, and if that requires external support I'm happy to provide it.

The quickest, simplest type of support is hex wire (aka chicken wire), available in 3- and 4-foot (0.9 and 1.2 m) widths from the marketplace. The vines will cling reasonably well and you can use it from year to year. Rolled up and stored off the ground when not in use, it will last for at least six years, often much longer. When it does begin to rust out and pieces break off, get rid of it before you have a nightmare mess. I've always used hex wire for my Scatterseed pea growouts—with 300 varieties per year, I need quick and cheap, and in recent years the same as applied to my own garden peas. However there is a way that is

Figure 14.5. Jeminez beans have almost completely covered their trellis.

much more organic, more aesthetically pleasing, and cheaper (*if* you don't value your time too much), and that is a woven brush fence. I hope to return to the system someday (it's also more fun), but meanwhile let me explain it. I used speckled alder because it's readily available to me, and it is flexible to work. Willow is also available and even easier to work, although the osiers will root and grow, competing with the peas for light and more. Alders will die very conveniently, although I've had the occasional branch tried to take root and leaf out. In fact alders are very quick to decay, one reason why they are so good at building the soil where they grow. They will last that season, no longer; don't even think about reusing an alder trellis for a second year, and anyway peas should be rotated.

Building an Alder Trellis

I have known people to "brush" their peas by sticking individual branches in the ground among the plants. But such pea brush arrangements flop around, the pea vines don't cling well, and they just look messy and stupid. Building alder pea trellis that is aesthetically appealing and effective requires a modicum of skill, but it is well worth it. It is a folk craft at which you will soon develop your own style.

Since my peas are always companioned (usually with carrots) in a 4½-foot (1.4 m) wide intensive bed, I first broadfork the entire bed to loosen and aerate the soil. I'll have to walk all over that bed while building the trellis, so to avoid recompacting I first lay down a row of old boards on either side of the centre row. Then I hoe a flat-bottomed furrow (about 6 inches, or 15.2 cm, wide) down the centre. Next I cut several armfuls of alders, preferably with numerous side branches, but not too old and brittle. I always try to cut the alders shortly before using them; they are just ready to bud out and are at maximum flexibility, if not allowed to dry out. If I have extras to continue on the morrow, I'll throw them in my pond.

The length of the alder cuttings depends on the expected height of the pea variety. I usually allow an extra 8 inches (20.3 cm) to insert into the ground plus about 2 feet (0.6 m) at the top for weaving a rim. I

Figure 14.6. Here I am hauling the alders to the site from the roadside where they were cut. *Photograph courtesy of Scott Perry*

stick one branch firmly into the ground every foot—if the lopping shears left them too blunt to drive them, I may sharpen them slightly with a hatchet. There are always some sections of the row that are too open near the ground because the alders were too unbranched, and for those gaps I cut a lot of small alders—usually culls from the main trellis—and stick them in between. It is not critical that they reach the top; they are not builders, only fillers. Now comes the fun part: Starting at one end I bend the alders to the desired height of the finished trellis, but not so sharply as to snap them.

What I call weaving of the tops is really more of a braiding: I intertwine a number of the supple twigs, adding on a few new ones as I move down the row, constantly using up the old ones. Frequently I have

Figure 14.7. I insert the withes firmly in the flat-bottomed furrow. Photograph courtesy of Scott Perry

a rebellious "cowlick" that is either too long or too stiff to work in easily. These I weave back down into the fence, thickening and strengthening it. If the alders have any tendency to be two-dimensional (as opposed to the twigs sticking out in every direction), then I insert them so they are in line with the row, because I want the trellis to end up as flat as possible. Sometimes if a particularly heavy branch insists on sticking out sideways, I lop it off and use it as a filler. The goal is to keep the top, or rim, of the trellis as uniform as possible, so I have to pay attention to where I'm bending the branches. If I need to insert a longer whippy piece to strengthen a particular spot, I do so. The rim should consist of a number of hoops that all run together. This is the main strengthening element, but I add to it by interweaving the low branches and twigs. If one is insisting on leaning one way, I'll make certain that its neighbour is leaning the other, so that when their side shoots are trained in and out they clasp each other in a springy mutual embrace whose tension adds strength.

When I think I need more rigidity, as with taller trellises, I take a number of weavers—long slender osiers with few or no branches—and weave them in and out horizontally about halfway high. These aren't stuck in the ground; they merely add to the length-wise stiffness and fill in lower gaps. The fence is now much like a very crude section of basketry. When finished I like to give a good sidewise whack at one end and watch the shock go down the whole length and back as a sine wave. If it does that I know I've made a good rugged fence that will last the season.

Figure 14.8. All the willowy tops will be braided into a single rim. Photograph courtesy of Scott Perry

I sprinkle pea seeds in the furrow, in and among the alders, and cover them by cuffing in soil with a steel rake. Lastly I remove my stepping boards and make two furrows on either side, where I sow carrot seed. A day of sun will cause the alders to "harden up" and make the trellis stiffer still.

Despite my best efforts to make the structure two-dimensional, it will nevertheless have some thickness, and this is one of its big advantages over hex wire: The vines have plenty to cling to. Indeed some of them are growing within the fence as much as on it.

Is it worth it? Aesthetically it is much more attractive, but it assumes you enjoy futzing with all that twigginess (which I do) and that you can spare the time (which I usually cannot). A 40-foot (12.2 m)

fence can take from two to three hours from beginning to end, and this is in late April or early May, when so much else is clamouring for my attention. It *is* quintessentially organic; I keep a corner of my field as an alder coppice. Many alders get by me and become firewood or chipper fodder. Meanwhile they are building up the soil there. As for the trellis, at season's end I pull the whole thing—brush, pea vines, and all—and put it through my chipper/shredder for compost. Don't try that with your hex wire.

Since I do not keep livestock I do not need fence posts. I do need posts, however, for numerous other uses: kiwi arbour and grape trellis, outdoor shower stall, tower platforms for my irrigation tanks, and carrying electric wire to wherever my coon fence is in any given year. Those posts are set in deep holes

Figure 14.9. A stiff rim will stand plenty of strain.
Photograph courtesy of Scott Perry

(some 3 feet, or 0.9 m, deep) which in my very rocky soil are not easy to dig; moreover I usually fill the surrounding hole with small stones to avoid frost heave and prevent decay. It's a lot of work; therefore I do not care to have to replace them soon. One option is to use wood that is naturally rot-resistant, like white cedar or black locust. The latter is best for rot—I've had black locust posts in contact with soil for 20 years and still be completely solid. Therefore I use it for my irrigation towers. A problem is finding enough of it, and it is often very crooked. Also, driving nails or staples into it is like pushing thumbtacks into old bones.

I have cedar in abundance and size and straightness. It is very rot-resistant, though less so than locust, but I have a method of extending its life

greatly. When doing an archaeology project in college I learned that charcoal is biologically indestructible, and thus useful for tree-ring dating of ancient sites. I also discovered that when a piece of wood is charred on the outside, the uncharred interior will resist rot almost as well. I think this is because of two factors: First any decay bacteria or fungus must pass through the charred layer, which absorbs and neutralizes it; second in the process of scorching it, some resins are heated and driven into the interior, having much the effect of creosote pressure treatment. I do my scorching in the same outdoor furnace I used to boil sap, but you could simply use a small bonfire. The trick is to keep turning the post, moving it in and out, round and round, so it is scorched evenly. A depth of 1/8 to 1/4 inch (0.3–0.6 cm) is ample; you don't want to weaken the wood.

Crowding

In addition to utilizing the garden's third dimension, another way to intensify production is by simply crowding the plants. Clearly this has limits, but what are those limits and on what are they based?

One of the limits is the actual spatial requirements of the plants themselves—at some point competition for space, light, water, and nutrients offsets any advantage from dense plantings. However, a further limit is imposed by *our* need for accessibility: We need to cultivate and harvest. If that cultivation is by tractor or rototiller, we may need 3 feet (0.9 m) or more to get between rows. With a few exceptions, like corn and sunflowers, the plants themselves do not require that much room. Conventionally tilled gardens are more like deserts, with most of the area consisting of dry naked soil, not for the plants' sake, but for the sake of our machines.

I deal with the access issue by not rototilling, at least while the crop is in place, if at all. Instead I cultivate by hoe or wheel hoe. That is somewhat more labour-intensive, but it is largely offset by the efficiency gained in cultivating much less area for the same amount of food. Plus, I cultivate only until the crop is mulched. My feet require about 18 inches

(45.7 cm) for walking comfortably, and even that needn't be between every row. To minimize access space (and the soil compaction caused by treading around), I arrange my annual row crops, terraced or not, in beds, each 4½ feet (1.4 m) wide by 40 feet long (12.2 m), not including the 18-inch-wide path. These are not *raised* beds—we tried that once and were thoroughly disappointed—but rather level beds. My 4½-foot-wide beds result in 25 percent "wasted" space. Some find this width too broad for comfortable reaching, but I insist on lowering the bed-to-path ratio, though it means that I always get stuck planting, thinning, and harvesting that middle row, because nobody else wants to do it. Moreover, a total of 6 feet (1.8 m) (4½-foot-wide bed + 1½-foot-wide path) between bed centres makes layout simple.

When I rotate an area into green manure, grains, or various field crops, I don't confine the crops to the beds. For example, I sow grain over the whole area, including the paths. Winter squash usually sprawls over two whole beds. Thus the bed configuration is not always obvious to the eye; however it is on the garden maps and in my mind, and so the centrelines and paths are always in the same place.

I may occasionally step into a bed, but generally I avoid putting my feet in there at all. Of course when I'm preparing a bed by wheel-hoeing or broadforking, I must stand there, but I count on that tilling to offset my footprints, and even then I take care to step lightly and wide to reduce packing.

The paths are generally tilled, though not as often or as deeply as the beds themselves. While the paths are relatively compacted and I don't deliberately fertilize them, fertility is there. That's partly because I mulch the paths as well as the beds, so my footsteps are cushioned and the soil inevitably gets some fertility. The crop plants undeniably send their roots out under my feet, which is fine. Some intensive gardeners are obsessive about never stepping on their beds, and I would point out that it is possible for soils to be *too* fluffy; after all, that's one of the disadvantages of using a rototiller to cultivate.

When I calculate crop yields, planting density, fertilizer use, and so forth, I must include the path space in my figures. Everything may be concentrated in the beds, but the paths are still there nonetheless; they are part of the planet, I pay taxes on them, and the crops occupy them, whether that's obvious or not. To ignore that area is to exaggerate the efficiency of my system by 25 percent. It is already plenty efficient; I don't need to distort things. The configuration of plant rows within these beds varies according to the crop, the extremes being half a row and nine.

Half a row crops. A single half row per bed? As mentioned before my winter squash and seed pumpkins sprawl over two beds—it's hard enough to confine them to that—and so their row ends up between two beds: in the path. That works out fine because I plant in hills 6 feet (1.8 m) apart and I can easily broadfork those areas and fertilize them heavily. (I explain in chapter 3's *Living Mulch* section how I grow oats and peas between the rows and the hills.)

Single-row crops. I plant melons and cukes and zucchini in a single row down the middle of their bed. The first two, melons and cukes, are never satisfied with their allotted 6 feet (1.8 m), so I am constantly turning the straggling vines back into the bed, or occasionally snipping them off. I envision those beds as single rows, although I usually plant some early companion crops alongside them until they start to sprawl (for more about melon companions, see chapter 8).

Two-row crops. A few crops, like corn, sunflowers, and potatoes, are planted 3 feet (0.9 m) apart, or two rows per bed, with each row 18 inches (45.7 cm) away from the bed centreline.

Three-row crops. Green beans are in a three-row configuration, with a row on the centre and two others at 24 inches (60.9 cm) to either side, near the edge of the paths, or else a single centre row of beans with the other rows occupied by a more compact crop like chard or rutabagas.

Five-row configurations. Most of the compact veggies go in a five-row configuration, one on centre and every 12 inches (30.5 cm) outward. Often #1, #3, and #5 are the same crop species or the same group, like celery/celeriac on the outside and a taller umbellifera such as coriander in the centre. In cases where the centre-row crop is especially tall and shading (as in peas/carrots), I plant rows #2 and #4 slightly closer to the outer rows.

A seven-row scheme. The only crop I always plant in seven rows is onions; the middle and outer rows are located the same as #1, #3, and #5 in a five-row bed, but instead of splitting the intervening space in half for rows #2 and #4, I divide it in thirds for what become #2, #3, #5, and #6 (or 8 inches apart).

Nine rows for grains. Grains I plant nine rows per bed, but the spacing stays the same as with onions; the two additional rows are planted in what ordinarily would be the path.

Companion Planting

The use of beds with loose soil, where the fertility is concentrated in the growing area instead of the paths and near the surface instead of deeply buried, makes possible the close spacing of plants I've just described. But there is another practice that permits me to crowd plants without their suffering: companion cropping.

There are books and charts on the very subject of companions, and I shan't endeavour to make this one of them, especially since my experience is more limited. However, I will enumerate some of my tried-and-true favourites, many of which have been discussed already in the other sections.

Companion planting is usually described as the practice of growing two or more symbiotic crop species in the same space, crop species that are convivial because one or both have some beneficial influence on the other. The usual implication is that it works because one repels the other's pests, or supplies some nutrient needed by the other, or provides physical support, or perhaps some biodynamic factor like a hormone. Well, those all may be true—some of them certainly are—but there is a more general principle at work here, one that applies to a much wider variety of plants, and it is this: Every plant has some needs and some non-needs. By "non-needs" I mean that if it doesn't actually supply the need to its neighbour, at least it meets its own requirements, and therefore doesn't need to compete with the neighbours for the limited supply. Here are some examples.

New World crops (except potatoes, sunflowers, and terrasols) are frost-tender, but that's no reason to leave the land idle until the weather suits them. In fact we can take advantage of their cold-wimpiness by integrating them with Old World crops like early lettuce or green onions, or with green-manure crops they can either accompany for a while, or perhaps precede or succeed. The first frost that ends their career is just in time for the planting of winter wheat or rye, although if the mulch is old hay, the grain crop may suffer badly from weeds. Using my green-manure-*cum*-mulch method, this is not a problem.

Favourite Companions

Peas and carrots. It is often said that peas and carrots are convivial because peas are a legume and fix atmospheric nitrogen, which the carrots need. Carrots in turn are taproots that pump up alkaline minerals (calcium, potassium, magnesium, and more) for which peas have a particular need. Sounds great, except for this: The pea rhizobacteria fix nitrogen *for the peas*; only when the peas die and the nodules decay is that nitrogen fully available to other plants. Likewise, those minerals are incorporated into carrot flesh and only become available to the soil community if we eat that carrot and return our composted poop to that particular soil. No, the reason that peas and carrots get along is more subtle. As long as the soil nitrogen is adequate, the carrots can consume it all, with no need to share it with the peas, which are self-catering in terms of nitrogen. And the peas must already have adequate calcium, et cetera, within reach or else they will suffer; however what is there is theirs alone, since the carrots will dig down to "mine" their own. As a result of these meshing needs and talents, peas and carrots do not compete with each other as much as they would among themselves, and therein lies their eco-efficiency, regardless of whether they actually *help* each other.

Of course, if I sow half of my space to each crop, then I only have half as many plants of either kind, so obviously I must plant twice as much space. Well, that's the same overall amount of space as I have before—with each species planted separately—so if

Figure 14.10. Peas combine well with carrots or with chard and lettuce; it's not so much that they like each other as that they just have little to quarrel over.

the net space per crop is unchanged, what have I gained? The *overall* yield will be significantly greater because of that noncompetition: Although I planted 50 percent of each crop in one bed, the yield from each component species was somewhat greater than 50 percent—let's call it the conviviality bonus—and thus the overall crop yield is greater than 100 percent of what it would have been with each species growing by itself. This conviviality bonus can easily be 50 percent for some crop combinations. It is never 100 percent; you can never simply double up without making some adjustment or incurring some loss. So what? Just think of it: You can increase your crop yield by half without adding any more fertility than you would anyway. Of course, you can further improve yield by adding more fertility, more water, more care, but this particular gain is got merely by mixing things up a bit.

Sweet corn and edamame. Here is a more extreme example: I grow sweet corn and edamame (green soybeans) in the same plots, even in the same rows. I plant the corn at its usual spacing, as if it were alone, which is 36 inches (91.4 cm) between the rows and 12 inches (30.5 cm) between corn plants in the row, so let's call that 100 percent planting density. Then before covering the seed I drop the soybeans in the same furrow, also about 12 inches apart. I pay no attention to position whether the beans lie between the corn or on top of them; it seems to make not a whit of difference. The thing is that the beans are about 30 percent of what they would have been alone, plus the fact that they are in 36-inch rows, as opposed to 24-inch (61.0 cm) rows if they had been planted by themselves. So the corn is what I'm considering the main crop; the edamame is more of an extra, an

Figure 14.11. Egyptian onions to be chopped and frozen; I can always use more of these, so I include them in many of my companion plantings.

Figure 14.12. Bulb onions won't stand for much shade, but they're happy sharing a bed with poppies.

add-on. Well, the corn seems to not even notice the soy—as explained before, the nitrogen thing is mostly bunk—but neither is it in any way repressed. We find we get fully 100 percent of the corn yield, probably no better, but that means that *anything* we get from soy is extra. As for the soy it is definitely repressed by the shade of the tall corn, so we only get maybe 50 to 60 percent of the usual soybean yield per row. That's 60 to 70 percent cream, skimmed off the top, even though the spacing was greatly reduced. Again, it is the result not of mutual aid so much as mutual tolerance.

By the way I use the exact same method with flint corn, except that I plant my dry soybeans among that instead of the edamame type. That's mainly because I'm not going in among either crop for frequent harvests, as with the vegetable forms, and so there's less unnecessary trampling.

Other examples of this conviviality bonus abound; indeed you can invent many of your own. You just need to mesh things in as many ways as possible. Do their fertility requirements dovetail nicely? What about the shape of their root zones? A long, skinny taproot avoids the territory of a shallow-rooted plant such as pea or lettuce or cabbage. Aboveground, tall, skinny leeks will manage well between bushy kale plants. Fortunately both crop species have hyperabundant chlorophyll, so they shouldn't need to quarrel over access to light.

Onions. What about onions? I mean the big bulb ones that you store over the winter and slice into rings. Those tend to suffer badly from reduced light, and so are difficult to companion. The problem isn't really limited photosynthesis—after all, scallions do fine among many other crops. It's that the onions supposedly need light on their shoulders to make them bulb up. Shade them or bury them too deeply and you'll get something more like a scallion. Yet everything you read about companions will recommend onions with nearly everything. Rightly enough; they are a helpful repellent to many insect pests, or perhaps their odour confuses them. I am all for growing the various green onions all over the place, but the bulbs I have usually planted alone. In recent years, however, I find they do well with breadseed poppies, a row of poppies down the middle of the bed and three rows of onions on either side. Poppies are scrawny and cast a rather scant shade.

Another combo I've come to like is table beets with various umbellifera. Not carrots or parsnips, though; their taproots are seeking the same things in the same place. In the centre row I plant a tall, lacy annual umbellifera, like dill or coriander (both for seed); in the outer rows (#1 and #5, in this configuration) I plant shorter umbellifera, like celery and celeriac, with beets in rows #2 and #4. It is very attractive and everyone seems to get along very well; and yes, I do pay attention to how things look—it has been many years since I've had much time for drawing or painting, but now I have acres of canvas to work with and I intend to make the most of it.

Timing of Planting

An important thing to account for when planning companions is timing. Generally it's ideal if both crops are planted at the same time, so that planting one doesn't mess up the other. For example, corn and soybeans go in the same furrow, so it is essential that they go in at the same time. When I do have gaps in one crop or the other, I have to drill in replacement seed, which is slow and tedious. Sunflowers and pole beans are separated by about three weeks, but since the sunflowers are several inches tall and in wide-spaced hills, it's easy to poke in the beans around or among them.

Some companion crops make good use of the different timing, but require special care. Early lettuce goes well between garlic rows, but it is best transplanted rather than direct-seeded in order to get harvested before the garlic is in the way. Since garlic is fall-planted and comes up in the very early spring, I must take care not to damage the garlic while I reach in to plant the lettuce.

There are several crops that work well as a fill-in companion for other crops. The main thing we require of fillers is that they reach a usable size very quickly, mainly in the early spring or late fall, when there is neither time nor heat enough to mature anything else. Some excellent candidates for this role are lettuce, spinach, radish, scallions, dill weed, rape greens, and other brassicas. A second thing I look for in such filler crops is that they have more than incidental usefulness, especially when planted all at once. For example, there would be a limit to how many radishes I could use on short notice, even if I

Figure 14.13. An old favourite combo: coriander, beets, and celery.

really liked them. I do really like lettuce, yet likewise I can use only so much within its narrow window of readiness. The brassica greens (rape, Chinese cabbage, and so on) can at least be frozen or salted to make some form of sauerkraut or kimchi. Scallions, perhaps more than anything else, can be used in considerable quantity. Frozen, dehydrated, even powdered, they can supplement bulb onions and leeks for some uses.

I have used scallions, or green onions, as a fill-in companion with various crops. They are either followed by another crop or else harvested so that a neighbouring crop (such as cucumbers) can spread into that space. Since I'm usually counting on the scallions being out of the way early in the season, I often use Egyptian, or top-set, onions for that purpose. They can be sown late the previous fall, and

Figure 14.14. These early green beans will be done and gone before the late rutabagas planted along the edges of the bed need all the space.

Figure 14.15. These late cabbages and sugar beets won't need all that space until much later; by then the ground surface will be covered with Dutch clover, which was sown shortly before this photo was taken.

Adjusting for My Own Needs

Unlike a market gardener, all my computations of planting area are based solely on what I and my family like to eat. Tastes and habits change somewhat over time, and if I do not adjust for that in my plantings, I am wasting time and space growing food for the compost pile.

Recently I've begun to plant late snow peas (for fall harvest) with late broccoli and Vienna kohlrabi (#1 and #3 of a three-row configuration). That works well as the sequel to an early-season green-manure rotation of oats, By the way, that bed contains 40 feet (12.2 m) of snow peas and 80 feet (24.4 m) of broccoli *in addition to* the same amount planted elsewhere as an early crop. That reflects my increased consumption of those foods with a commensurate reduction of shell peas and cabbage, respectively.

The options of canning, freezing, and dehydrating help a lot to even out the vicissitudes of poor planning or unexpected crop failures. A few years ago an overplanting of tomatoes combined with a fair yield left us scrambling to get them all processed before they spoiled. Understandably there was some grumbling about my misjudgment, but the following year a devastating blight left me feeling quite vindicated, as we continued to enjoy tomatoes through the second winter. It's harder to do that with lettuce.

make rapid spring growth before the space has other demands on it.

One particular cluster of crops all seem to be quite compatible: peas, carrots, salsify, lettuce, radishes, and scallions. I use this congenial lot in various combinations. Most of them are best planted as fillers between something else, not between one another.

Peas, carrots, and salsify are a perfect full-season companion group; any gaps can be filled with the other species.

When I step back and look at that cluster from a broader view, I notice what might be parts of a pattern, a pattern that might suggest still other candidates for the cluster and other arrangements of them. For example, peas are a tall cool-weather legume; how might favas do in that situation, a single tall row down the centre of a bed? Carrots are a short first-year umbellifera; could parsnips be substituted? Indeed they can; I always plant my carrots/salsify with main-crop shell peas, and my parsnips with early snow peas. I know of no reason why celery/celeriac might not work as well, except that, as already mentioned, I have another companion I prefer for them. Likewise, dill weed and parsley are happy among those, but since I grow them in much smaller quantities, I usually just stick them in odd empty spaces. Lettuce and salsify are short composites, so why couldn't endive and chicory substitute? Indeed they could. For example, I grow Capucijner peas on a centre-row trellis (even though they are a dry soup pea); witloof chicory (aka Belgian endive) does well on either outer side of it, with early scallions or early "forcing" onions in the inner rows (#2 and #4 of a five-row configuration).

Kale and leeks have come to be an old standard for me, with #1, #3, and #5 to kale, and #2 and #4 to leeks. That's in the approximate proportion to how I use them. In fact, a large amount of them get eaten together, often in a stew with potatoes and a bit of dulse.

That brings me to another observation: It is uncanny how often plants that go well together in the garden also go well together on our plates. So much so that I often look to culinary compatibility for hints as to companion planting. It seems to work both ways: Despite the title of one companion-planting book, many people have concluded that carrots do *not* love tomatoes; indeed, they are rather a poor combination for various reasons. Moreover, at least to my own tastes, their flavours are generally conflicting to the palate.

I have found tomatoes to be one of the more difficult crops to find companions for. They are quite a rowdy bunch, and their "other-plant skills" are not well developed, although a few herbs like basil and summer savoury seem to abide their aggressive roots better than their domineering tops. Even those need a bit of help in holding their own space and, more important, light. Recently I have had good results with growing bush beans, especially green beans, on the outer row of a bed with tall tomatoes on trellis down the centre. However, that's not so helpful with shorter determinate tomatoes.

Peppers actually yield better when somewhat crowded, at least within their row, so I plant them only 12 inches (30.5 cm) apart in rows #1, #3, and #5. I do like to alternate the rows with another crop, but there's always some controversy over it. I haven't yet figured out just what works best for a companion, although planting spring-seeded multiplier onions (as opposed to fall-planted Egyptians) is one good idea, if we can use that many. Basil is a good size and shape to fit in there, and they (peppers and basil) seem to enjoy each other's company. However, even with my wife's pesto fetish, it's hard to use two 40-foot (12.2 m) rows' worth of basil. Of course there's no problem with mixing things up within the row; for example, I've alternated scallions and lettuce, so why couldn't basil be spaced out among those? This kind of discussion is a constant feature of our companion gardening; some combos are tried-and-true, while others come and go as we play around with new ideas.

There are a few, I think only a few, combinations that are actually bad—that is to say, harmful. Fennel is said to be incompatible with almost everything, although I've grown the Florence type in several situations—anywhere I would plant celery, parsley, carrots, or dill weed—and seen no harmful effects in any direction.

I did notice that sugar beets seemed not to thrive well next to a row of runner beans; later I read that beets generally (few gardeners grow sugar beets) do not thrive around runner beans (they even specified runners); would it have been different with kidney beans or limas?). I have read no explanation for this hostility, nor have I any of my own; in fact it seems counterintuitive, but there appears to be something to it.

While many plant combinations may be an improvement on mono-cropping (having an area occupied by one crop species only), not all of them are equally convivial. Those that are have certain qualities that mesh. As suggested earlier, it may simply be a matter of plant shapes or root zones that dovetail or noncompeting nutrient requirements. Just as a pebble may actually serve as a water source by its thermal mass condensing soil moisture on its surface, plus trapping capillary moisture wicking up from the depths, plants may perform similar services. For example, succulent plants like purslane, and plants with dense systems of fine roots like chufa, may in

Figure 14.16. Late brussels sprouts replace an early planting of peas, combined with leeks and carrots; all of these are very compatible on our plates as well.

the process of conserving water for themselves make it more available to certain less resourceful neighbours, which may reciprocate in some other manner, such as emitting some aromatic oil that repels their mutual enemies. I've heard that when certain taproot species grow among corn (like red clover or alfalfa), forcing their way down through stony hardpan in search of water and minerals, the corn roots may reach down alongside them to seek their own moisture in the same tunnels they have excavated.

Some of the effects of companion cropping are hard to define and harder yet to explain, but do we need to know? It might be helpful to understand the cause behind the effect, but with all due respect you can be either a left-brained lab-coated empiricist or a New Age flake and still get the same results.

CHAPTER FIFTEEN

Pests and Diseases

From a plant's point of view there is little difference between a cutworm, a woodchuck, a blight spore, and, for that matter, us. These are all things that in one way or another prey upon it. It is an inevitable constraint of all living things: We escape one peril only to ultimately succumb to another. Of course there is senescence, the ultimate vulnerability of our own body mechanism. Even if no other organism were to attack us, sooner or later we fall prey to our own frailty: Our bodies quit on us. Not the end of the world, just the end of us as individuals. As long as there are replacements the game goes on. I feel that a desire for immortality is a symptom of pettiness and ingratitude, and I know for a fact that my cabbages feel the same way. They do not long for eternal life—at least they do not expect it—nor do they aspire to perfect enlightenment. They're just grateful to be part of it for as long as possible and as fully engaged as possible. Me, too.

There is a theory that if plants grow in a perfectly balanced soil with all their nutrient needs met and with no stress from heat or cold, too much or too little water, then they will be immune to attack from diseases and pests, since pests prefer weak and vulnerable plants. But if so, then wouldn't they also be immune to attack from *us* (we prefer to call it harvest)? Anyway the theory can never be more than a theory, because there is no perfect balance in the universe. Rather chaos and change and instability are more the norm; the best we can hope to do is exploit them to our temporary advantage. Should we not strive toward an ideal of perfect harmony? Absolutely! Should we expect to achieve it? Get real. Therefore we must deal with pests.

When we speak of "natural pest control," or for that matter of "natural farming," we must bear in mind that there is something inherently *un*natural in any kind of farming. Whenever we make a major modification in the landscape, such as converting forests or prairies to croplands, we must constantly struggle to *keep* them as croplands, because their natural tendency is to revert to whatever they were before people came along. The well-kept gardens we see as so attractive are really gross enormities, straight rows, sometimes whole fields of a single species, often of tropical origin, separated by strips of bare soil such as one would only find in the severest desert. I know of no agricultural system, including mine, that even approaches the nuanced complexity of a truly natural ecosystem, and therefore it will be chronically unbalanced, simplistic, and vulnerable to a host of pests and diseases. Given our clumsy hubristic attempts at dominating nature, it is remarkable that anything grows there at all. But apparently nature has considerable patience with us, so that even our clumsy efforts to cooperate with it are rewarded far beyond our desserts.

Plant Disease

We can do a lot to avoid problems with plant diseases simply by tweaking a crop's environment: good drainage, irrigation when necessary, high levels of humus and bioactive minerals, and so on. These alone will thin out your adversaries to something you can hope to contend with. This is especially true of diseases: Lifeless, burned-out soil is an ideal medium for promoting infections, which will have little to combat them. Monoculture is also conducive to the spread of diseases; therefore diversity is the best remedy for it. Genetic diversity, habitat diversity, crop rotation, whatever; mix it up a bit for best results.

Those are general principles for minimizing diseases; they are better for prevention than remedy, and even then they are not panaceas. The majority of pests have evolved a rough equilibrium with their prey/host. Like tapeworms in humans, they are not ordinarily fatal but rather a nuisance and an obstacle to full achievement. We can and should deal with them, but our crop as a whole is not usually endangered. Some other disease/predators have not yet acquired that stability; they are explosive, annihilating. Ignore them and there will be no crop at all. The disease has not yet "learned" that such behaviour is generally not in its own self-interest. Like the bubonic plague, which nearly doomed itself by so completely destroying its hosts, such infections tend to occur in intense waves, dying out only after wreaking nearly universal havoc.

My main approach to disease in general is contextual—maintaining a soil environment where diseases, whether bacterial, fungal, or caused by other microbes, find life uncongenial or hostile. Because I move crops around from year to year, the diseases have difficulty building up to levels that harm the crop. In humusy soil beneficial or neutral bacteria or fungi already occupy the niches in the soil ecosystem; newcomers are hard-put to find a toehold, if they aren't actually attacked. For example, soil with high levels of long-term humus from woody residues (ramial chips) is rife with nematodes that not only are harmless to the crop, but attack and consume invading pest nematodes such as the potato cyst nematode. Cabbage yellows and late blight spores cannot survive for long in a soil where their host has moved out and other microbes are moving in on *them*. The spore bodies of sclerotinia—which might lie dormant for many years in chemically fed, low-humus soils—will be destroyed in a much shorter time by competing organisms in a truly living soil. I could list one soil disease pathogen after another and the remedy would be basically the same: Create a vibrant and diverse living soil community and they will defend their own interests far better than you can. Without them your poisons are a feeble defence that must be used over and over to correct the very problems they've created.

I would reemphasize that when I speak of high humus levels I'm describing a condition that cannot be created by adding any amount of short-lived organic matter. Animal manure, and even straw-based compost, cannot by itself supply that stable soil community that is nourished by forest-based materials, especially chipped ramial wood.

Okay, but all that being said, there are other things you can do to alleviate damage; perhaps the most common is sanitation. I'm careful to remove *all* vegetable crop residues (not grain stubble) from the garden as soon as possible after harvest, regardless of whether they appear to carry any disease. Trust me, they do, including many diseases I wouldn't even recognize if they got out of hand. The point is I don't let them get out of hand; I remove them to compost, where they are subjected to the highly thermophilic (over 160°F, or 71.1°C) conditions that most soil microbes cannot survive—and those that do are soon consumed by the gazillion compost organisms on our side. That heat of composting plus the rotation of crops pretty well guarantees that diseases will not have a chance to build up. I seem to have very little problem with plant diseases, so little that I don't even recognize most of them, although I assume that they are all around me.

By the way there are numerous microbial cultures you can purchase to bolster your garden's defences. Some of them are selected strains from the laboratory that have been bred for superior vigour. I have no particular contention with them, although I have to assume that unless there is a very high soil humus content, they will soon die out and simply be a short-lived remedy for which you must constantly return to the marketplace, whereas if the soil humus level is already high enough, I wonder why they are not already present. I'm not categorically opposed to importing exotic life-forms—after all, what are my crop plants?—but like my seeds, I prefer to buy them once and know that I can henceforth maintain them on my own.

Animal Pests

I have several animal pest problems, some of which are serious enough to require preventive measures. Generally, the larger the animal, the smaller the problem. Deer are all around my land, and they are perfectly aware of my gardens, especially in late autumn when they are in a fattening-up mode. I have several responses to them, all partially effective, none completely satisfactory. Perhaps the least effective are the various repellents: bars of scented soap, sweaty shirts, and so on. Each works for a while, until the deer get accustomed to it, then they ignore it. In fact I suspect that after a while you could hang Ivory soap in the middle of the forest and the deer would gather round, wondering where the kale was that always grows near it. I've used the various noisemakers supplied by the wildlife department; they are shot from a little pistol, and the screechy shells scare the deer even more than they scare me, at least for a while. Before long they're right back, but they are much more skittish and their feeding is significantly reduced. That's the underlying secret to most deer-repellent strategies: By messing with their heads enough, you simply give your garden enough negative associations that they prefer to avoid it, even when you're not present.

Nevertheless when they are hungry enough all bets are off. I have resorted to "lead poisoning" on a few occasions with quite good results. My live-and-let-live philosophy also applies to myself, and before I will allow myself and family to go without, I'll certainly kill other things. I kill potato bugs; why should deer be any different, just because they are cuter? In fact killing one or two deer may give me perfect protection for a year or more after, whereas I must destroy thousands of "sentient" potato bugs every season to have even a modicum of success. The game warden is quite willing to issue me a special permit, good for any number of deer, buck or doe or fawn, in or out of season, day or night. The usual rules of sportsmanship do not apply; these are crop pests. Likewise there is no legal limit to killing aphids. In my case the warden is particularly cooperative, knowing that I do not value the carcass for food. He prefers I call someone from the food bank to pick it up, but I feel no compulsion to supply them with venison; I am not a game provisioner. In general, I prefer to drag the carcass into the nearby woods to rot and hopefully attract coyotes. Dogs might help keep the deer at bay, but I would have to feed and house and castrate and restrain them for their whole lives, whereas the coyotes come and go, taking care of their own needs and some of mine. Of course, the local hunters do not understand or appreciate my policy of shooting deer and sparing coyotes; from their perspective I've got it all backward. And yet hunters are not a reliable pest control; they are after the big bucks, and anyway they cannot legally hunt right around my place even if I were foolish enough to let them. The coyotes, on the other hand, are self-trained to cull out the older, weaker, sicklier individuals, and unlike the hunters they pose no safety threat to me and mine (I have stories, believe me). Furthermore, the coyotes do not need to kill many or any deer to be effective. Nervous deer are always on the move, never daring to tarry long enough to cause any serious damage.

All that being said, I prefer not to kill deer at all; I don't wish to keep a gun on the premises regularly, and slaying Bambi is just no fun, however necessary. There is another stratagem that works quite well if you are diligent at maintaining it, and that is electric fencing. You don't need several strands to be effective; I have used a single strand of electric tape clipped onto 4-foot (1.2 m) fiberglass rods. The setup includes a line charger, which takes the 12 volts from an automotive battery and sends it out in frequent short jolts of 5,000 V. The hitch is that the deer barely feel this through their thick fur, and young ones may walk under it without even noticing it. You need to bait them to put their tongues on it, using stapled strips of aluminum foil with cheap peanut butter and grape jelly smeared on it. Once trained to fear that filament-woven plastic tape, they'll stay completely away, but not indefinitely. Apparently they recheck the wire regularly, perhaps by merely sniffing it, because if your battery loses power for a couple of days, they may come back in. Therefore the current should be monitored regularly with a voltmeter (less than $40) and care taken to remove any tall vegetation or fallen branches that may short out the line.

Figure 15.1. Electric fencing is a simple way to exclude our enemies—and maybe our friends?

I should point out that the effectiveness of the system is inversely proportionate to the size of the area enclosed. An average garden will be perfectly protected; there's just too much other food outside it. However, at one point we had electrical tape encircling over 7 acres (2.8 ha) of hay fields, orchards, and gardens. Its north–south reach was over 600 feet (182.9 m), rather a nuisance for deer to walk around, especially when it did not appear to be an enclosure, but rather a puny obstacle to wherever they were heading, which might *not* be my gardens. It was still partially effective, but the psychological deterrent was definitely weaker.

Speaking of psychological deterrents we've discovered that a small enclosure, say one or two beds, can often be protected by a single strand of cotton twine tightly stretched between the tops of 3-foot (1.2 m) tall stakes. It seems bizarre; the deer could just get right over it if they chose to, but somehow they never choose to. It has been suggested that the linear feel of the horizontal barrier against their flanks feels unusual and, thus, nerve racking. I rather suspect that they are simply claustrophobic; that while they could leap over that string in a heartbeat, they are uncomfortable in places that compromise their ability to flee, however slightly. I do know that as soon as you expand the area the effect dwindles.

Speaking of string I'll share another strategy. Last year I had a large planting of sweet corn get wiped out by crows. The interplanted edamame was largely ignored by them, but I expected problems from deer and wild turkeys, who were bothering other crops. I had heard that a string stretched over the rows would deter crows, so I replanted and tried the string, about 10 inches (25.4 cm) above the ground (for eight 80-foot, or 24.4 m, rows). I immediately saw that they were undeterred, so I tried something else: I put the string 3 feet (0.9 m) higher up. I supposed that a crow walking along the ground might feel nervous about something overheard, something that might impede a quick takeoff. The rows were 3 feet apart, and I put adjacent strings at different heights, like 3 feet and 4 feet (1.2 m). A crow's wing span is about 3 feet, and the 3 feet between strings created a dauntingly narrow window for an emergency exit. To heighten the effect I fastened several aluminum pie plates to tall poles around and in the patch. Those fluttered around like tetherballs, banging on the poles and reflecting flashes of sunlight that I and my workers found very disconcerting, and we flatter ourselves that we are smarter than crows. Anyway, it worked: The third planting of corn was unmolested, and neither deer nor wild turkeys entered the piece, though there was a lush crop of green soybeans there among it, too.

The other major challenge to my getting a corn crop is raccoons. In early years we relied on just watching the ripening ears and picking them before the raccoons did. That was pretty ineffective; we can usually count on 30 percent crop loss. We finally wised up and got a fence charger—the same one used to power the deer fence—then strung a single smooth bare wire on stakes about 8 to 10 inches (20.3–25.4 cm) above the ground. It is completely effective; now I don't lose a single ear, unless I stupidly procrastinate over getting the system set up before the first ears ripen. Like the deer fence, it does require some diligence to make sure that any grass and weeds are not allowed to touch the wire, especially when it is wet with dew or rain. If short-circuiting pulls the current down and flattens the battery, the raccoons will be rather quick to notice. I assume they don't repeat the unpleasant experience of touch-testing it. Someone has suggested that, like deer, they may be able to sense the concentration of negative ions just by sniffing. It's important to keep the wire from being touched by drooping corn leaves, so I usually set it back a few feet, or sometimes I just enclose the adjoining crop as well; after all, stakes and wire are cheap. However, depending on what the adjoining crop is, I may want very much *not* to enclose it within the wire. Frequently the adjacent crop is squash, luxuriating in a dense mulch of oats, leaves, and spoil-hay. The mulch is invariably home to mice, and to date my best remedy for mice is skunks. Foxes and weasels are also helpful, but skunks are more comfortable around humans; for some reason they feel (sniff!) safe. (We get along very well; it may help

that I once rescued one of their babies, who had fallen into a 3-foot, or 0.9 m, posthole.) Yes, it seems counterintuitive to despise Bambi and love Pepé Le Pew, but we know who our allies are, and it is *not* our fellow veggie lovers who covet our food supply. The same electric wire that keeps raccoons out of the corn can also keep the other critters from doing their blessed work, so I am careful not to use the electric wire indiscriminately. I have on occasion lost a melon or two to skunks, but I don't begrudge it.

There is yet another way of excluding raccoons from corn that doesn't involve electricity. Plain old hex-wire fencing stapled to posts is nothing for these agile climbers, but they can only scale it if it holds steady. If the top 12 inches (30.5 cm) are not stapled to the post, it will flop outward when they put their weight on it and they'll fall back to the ground, eventually giving up in frustration (and trying something else).

I should mention that hex wire laid down on the ground can be more effective than upright, especially for deer, who keep getting their hooves caught in the gaps. Most animals find it unpleasant on their feet. The problem (aside from expense, if it's a large area) is that grass or weeds grow up through it, depending on where you lay it, and since you cannot mow it, you'll have a merry time taking it up in the fall.

Woodchucks, or groundhogs, are sometimes a serious problem with tender young vegetables, and especially with green legumes like peas and soybeans. I say "sometimes" because they are only around sometimes; when they are around they are *always* a serious problem. When they first show up I find the burrow and place a Havahart trap near it (or near the crop). If they are lucky they'll get caught and I will bear them off to a new life somewhere else. Larger chucks are reluctant to enter the trap cage, and so I may have to shoot them or, worse, stop up their burrows and toss in a smoke bomb. I say "worse" because I know that I would rather die from a bullet than by suffocation. Of course, when they do cooperate and get trapped, then I'm faced with another problem: where to release them? It's the same with red squirrels and chipmunks. It may "feel good" to let them go a few miles down the road, but how

Figure 15.2. This avian pest control expert deserves a rest after a long day of helping in my gardens.

humane is that really? Wildlife scientists posit that every bit of wildlife habitat is pretty well occupied to capacity at any given time, and that importing the "surplus" from elsewhere only creates stress in the new habitat. If so the established incumbents will certainly resist any intruder, by driving it away or killing it. Especially squirrels trapped in late autumn or winter; they have no hope of finding a new burrow or gathering a new supply of food. It's more humane to just shoot them, and I often do. However, that premise is not completely valid; I know for a fact that many areas are occasionally cleaned out by predators (foxes, coyotes, what have you), and animals from elsewhere are constantly moving about, looking for such gaps to establish their new territory. Ecological

niches are not static. My garden refugees have as good a chance there as any other newcomers, if the timing is right. It's not a given that they're doomed, any more than when they blundered onto my place.

There is also the matter of the human occupants of that new habitat. When I first settled on my farm there were miles of scarcely occupied land, mostly hardwood forest, in every direction. Since then a mushrooming of new homes has left fewer places where a pest animal can be released without simply foisting my problem onto others. When I do find a sufficiently wild place, I do not kid myself that I am giving that exile a secure life, far from it. He wasn't safe before I caught him and is probably no safer after. At best I only give him a chance to try again.

Birds can be real pests, at least some birds; others are real allies. The birdhouses (nesting boxes, actually) that I erect around the farm are only partly for aesthetics. I appreciate the lovely swallows and bluebirds and phoebes, but I also expect them to work for me, and they do so gladly. Watching them swoop and flutter across the sky decimating my foes is itself gratifying. The relative lack of mosquitoes and blackflies is due in no small measure to their constant foraging. In fact I have two shifts. On a calm sultry summer evening I love to watch the tree swallows stitching the vaulted heavens together with their flight patterns (otherwise it would all fall in on us). Then rather suddenly it stops, as the twilight signals them to hunker down for the night. For a very few minutes it is quiet; the sky is empty. Then a new flutter of wings appears, feathers are replaced by fur, and a phalanx of bats assumes the night shift. I am so glad to not be a mosquito.

Chickadees and blue jays are a pest with sunflowers, but I can easily control their depredations by watching the crop and—at first sign of damage—harvesting all the mature heads to cure under hexwire racks.

By the way chickadees are officially the Maine state bird, although unofficially pink plastic flamingos are far more common in every yard. Others have suggested the blackfly, more to give it protected status, since they are our main defence against infestations of tourists.

Of course you can use poison against rodents, but I'm very loath to do it, especially outdoors. For one thing again it seems like a horrible way to die, especially compared with the sudden death by mouse- or rat trap, but I'm also concerned about other animals—owls, foxes, coyotes—who might eat those poisoned carcasses. At least tourists are in no danger from that, as they generally shun roadkill. They prefer lobster, which they seem to feel is more wholesome.

At twilight I often go down to the lower pond, the one with a privacy screen of nannyberry shrubs, to wash off the day's sweat and grime. I am not alone; I am strafed by damselflies using me as mosquito bait. I'm delighted to oblige, though I sometimes move

Pest Control Philosophy

Although generally an advocate of nonviolence, I'm not categorically opposed to killing, which I consider hypocrisy, as you literally cannot walk through life without taking some life. However I am categorically opposed to wantonness: pointless violence that could be avoided with a minimum of effort. It's not about whether a critter has cute eyes and squeaks in fear, any more than some gelatinous creature which crawls around in its own slime. I don't trust my value judgment there, nor yours, either. I have somewhat more confidence in my ability to recognize a real threat and deal with it in a way that inflicts minimal suffering, though suffering nevertheless.

If I didn't know a better way to feed myself, I would have no qualms at all about killing animals to eat them, but I do know other ways, so it just makes sense, much less work, too. However, when it comes to protecting myself or my food supply, things are often less clear.

my arms carefully so as not to risk knocking these fighter pilots out of the air. No need to worry; they are all aces.

Wee Beasties

When I say that smaller pests are a proportionately bigger problem, that's partly because the smaller pests (including insects, slugs, and mites) usually arrive in vast numbers; catching or destroying a few individuals doesn't accomplish much. The wee beasties tend to come in waves, unlike warm-blooded vertebrates, which typically bother us in flocks or herds or even individually. I suppose that makes a genocidal approach to bugs and insects (*not* the

Figure 15.3. The most valuable part of the crop: A homemade extract of rhubarb keeps the brassica flea beetles at bay.

same) seem more logical, even imperative. Perhaps, but not necessarily. Often the cheapest, simplest, and least invasive remedies are those that merely exclude or repel the invaders. I suppose when a wave of pests arrives and starves because my plants are inaccessible to them, that is a sort of indirect genocide, though it causes me no angst. Another highly pragmatic reason for avoiding pesticides, even organic natural-ingredient low-toxicity-to-humans pesticides, is that they largely come from the marketplace and involve tropical ingredients or products of the laboratory. None of those is a fatal objection, but it seems that many of the repellents can be homegrown and I always prefer that. Let me give a few examples.

Flea Beetle Remedy

One race of flea beetles used to decimate all of my cruciferas: cabbage family, turnips, mustard, radish. The damage is most severe on the new seedlings; if you can get them past that stage, the vigourous plants can pretty well fend for themselves. However, if they get badly chewed at that vulnerable stage, they will be too weak to rebound and will continue to fall prey. One solution, and a very effective one, is to cover the new seedlings with some sort of row cover, like Reemay. Narrow (24 inches, or 61.0 cm, wide) strips are ample; they can usually be removed as the plants take off, perhaps after as little as two weeks. There may still be flea beetles around, but their cycle will be broken and the population knocked down to numbers that will inflict little damage. An alternative to Reemay, which comes from oil wells, is Elmer-plantex, which still comes from the marketplace but is made from cotton. Its greatest asset is also its greatest flaw: It is highly biodegradable. You can reuse it several times *if* you can keep it from getting snagged and torn, and *if* you can avoid having it in contact with damp soil. I have prolonged its useful life greatly by sandwiching (technically "shrewsburying") it between two pieces of cheap wooden strapping: the upper piece to anchor it, the lower piece to protect it from wet dirt. I add occasional stones to pin it tight and keep wind from working it. I sort old pieces into various usable sizes and use each piece appropriately. A problem I have

with any row cover is the difficulty of seeing the crop through it (Elmer-plantex is better than Reemay in this respect). If an infestation of aphids or something does break out in there, the beneficial predators are largely excluded, and I may be unaware of a problem until it's too late. Another very effective remedy is screen-covered frames. These require a much larger investment up front, but can be reused for many years, provided they're kept under cover in winter.

Removing crop debris from the garden helps control insect pests as well as diseases. I try to be particularly diligent in pulling and shredding all my corn and sunflower stalks soon after they've been harvested. This seems to be the best control for corn borers, earworms, and sunflower moth larvae.

I have another remedy that doesn't depend on the marketplace and the sea of petroleum: rhubarb. Not the stalks, but the foliage, which ordinarily goes to compost. Of course the leaves are inedible due to toxic levels of oxalic acid, which is the very component we need to repel those flea beetles. I chop up the huge leaves, cover them with boiling water, and simmer for a few minutes before straining. I used to use them raw, for fear that the oxalate might be too heat-labile and lose its pizzazz when boiled. I simply bashed the chopped raw leaves in a bucket with the end of a wooden baseball bat and added tepid water and left it to sit in the warm sun for a day before straining off the decoction. Then I read about boiling it and tried it both ways. I don't see much difference, and the boiling is more immediate and probably more concentrated. The trick to applying it is to use a very fine mist; don't spray heavily enough for it to bead up and roll off the waxy leaves (especially cabbage). On a bright sunny day, the spray will dry on in minutes and you can spray again. Usually after two or three mistings you will notice a slight glaze on the leaf surfaces and you will know that you have excellent protection, at least until it rains. Even then the effectiveness is not wholly washed off, and that is without adding any "sticker" like fish oil. It is effective for all that family of crop plants.

Another type of flea beetle attacks tomatoes and potatoes. In addition to the damage from feeding, these flea beetles may transmit early blight and various viruses via their mouthparts. I have never tried rhubarb extract on them and don't know if it would be equally effective. In fact I see no reason why the oxalate glaze, a broad-spectrum irritant, wouldn't be effective against a whole array of soft-bellied insects, especially their larvae. How silly that I haven't tried it.

Companion Crops as Deterrents

Other plants also have toxic or irritating properties. One source suggested planting horseradish among your potato plants to repel the Colorado potato beetle. We tried it and saw no pronounced effect, although we didn't plant very many and what we did plant persisted for years after, long after the potatoes were gone. I have wondered how an extract (infusion) of horseradish root might have worked; I could have applied it more uniformly to all the potato plants, without creating a future weed problem.

Cousin Tom has gotten complete protection from CPBs by spraying tansy tea, but he emphasizes that it is a preventive, not so effective once they're settled in.

We likewise followed directions to plant mint among our cabbage plants, and spent the next two years eradicating mint from successive crops. Again, what if we applied an extract of mint: the essence without the living presence? We recently tried mint oil, mixed with vegetable oil, using a window cleaner bottle-sprayer. If there was any benefit it was offset by the leaf burn caused by the oil spray. A commercial garlic extract was also of dubious value; would a spray of fresh homegrown garlic be any better? I mean, if it takes an acre of Watsonville garlic to protect half an acre of Maine cabbage, maybe I need to rethink some things. Not saying it does, but in a garden-without-borders we need to ask some pointed questions of the marketplace. Things arising from our own labour and our own ground are more obvious; for better or worse the externalities are all there to see.

Cabbageworm Remedy

Back to the cabbage family: A major pest of the maturing plant is the white cabbage moth. Actually the moths are no problem at all; they hardly think about feeding, so intent are they on laying their eggs, and *that's* the problem. Those little green eggs turn into the rubbery, squishy worms that even devout carnivores find off-putting. The worms prefer cabbage and broccoli because those offer the most concealment. Submerging broccoli in salty water for 20 minutes before cooking will force most of them out of hiding, either floating on the water or crawling up the sides of the bowl. But wouldn't it be nice if they weren't there in the first place? Again, row covers offer effective exclusion if you're content with market solutions. We've had pretty good results using cedar clippings, though we haven't done it recently, mainly due to negligence. To my knowledge all Lepidoptera are uncomfortable around cedar, as well as southernwood, wormwood, and other artemisias. Of course hot sun will dry these things out pretty quickly, but if you can go through periodically and crumple the sprigs a bit, it will release more essential oil (which is tied up in the woody fibres). Also it is easily and cheaply replaced. We mostly placed the cedar sprigs on the ground around the plants, but in retrospect I'm sure we should have laid them right over the plants, especially the developing heads of broccoli. Of course there is always Bt, which is a wonderful organic remedy, but again I am focusing on my own solutions that aren't birthed in a laboratory.

Recruiting Beneficials

Just as I recruit skunks and weasels to control my rodent pests, there are plenty of beneficial insects, nematodes, and microbes that can control our buggy enemies. I am always somewhat loath to import critters that cannot become endemic but will die out and need to be replaced. Not only am I reluctant to forge yet another link in the chain to the marketplace, but somehow I hate to think of sending my allies to their doom. On the other hand there are plenty of beneficials that aren't in my garden simply because they haven't found their way here yet or that are already here and merely need favourable conditions to expand their presence. The wet area below my pond is always a favourite breeding place for fireflies. How

Figure 15.4. Aside from beautifying the premises these perennial flowers also attract a mixed horde of bees, wasps, moths, butterflies, and birds, for a deeper layer of protection against pests.

nice that they flit about flashing their sex-lights on July evenings; how nicer still that they have a voracious appetite for slugs!

Saving seed plants of second-year biennials like parsnip is a great idea, though I need only so much parsnip seed. The plants also serve as a nectary for syrphid wasps, which at other times are heavy predators on aphids and numerous other pests. It all works together. The kind of ladybugs that are common here now are not the same species that was here decades ago. Both species are here now, but the new is larger and I've heard is a more voracious eater of aphids. Since they can survive here without any help, I welcome them both.

Bug Juice

One radical pest remedy is highly genocidal, but also very funky and homegrown—I'm sure they do this in Esperia—although I haven't used it myself, which is absurd. The so-called bug juice method requires only a dedicated blender, perhaps one of those smoothie makers you see at yard sales. You will not want to use this one for anything else. You capture a number of pests, especially beetles (although this supposedly works for a huge assortment of pests), and put them in a cup or two of water in the blender. You liquefy them, strain, add more water, and set the "juice" in a warm place for several hours, although some sources recommend applying it immediately with a backpack sprayer. I've heard that it doesn't work, and I've also heard that it works extremely well for a wide range of pests, including Japanese beetles, Colorado potato beetles, and Mexican bean beetles. I suppose for flea beetles the difficulty would be collecting enough of the tiny specks to suffice, and are they too tiny to liquefy? I guess the concept is obvious here: We all have the seeds of our own destruction within us, especially in our lower guts. Release all that, allow it to fester, and inoculate our surroundings with it, especially our food . . . well, you get the picture.

I have also heard that the bug juice doesn't actually kill the pests (other than those you liquefy), but merely repels them. I'm sure it does that, but I can't imagine it doesn't also cause them grievous bodily harm. I have noticed that when I have gone through the potato patch, squishing CPBs here and there, successive immigrants seem to avoid those areas, although perhaps they just prefer the new healthy growth.

I believe that for every tip I've mentioned here, there are 20 more that would be just as effective or better. What other aromatic plants, herbs, or conifers contain oils or irritants that are offensive to certain pests? After all, most of these substances are produced by plants to protect themselves; mightn't the less well-armed neighbours benefit by the proximity to those powerful natural defences?

PART V
Using the Harvest

As self-reliant gardeners we are the ultimate consumers of what we grow. We have control, and thus responsibility, over the entire system from beginning to end, if indeed there can be said to be any beginning or any end. While that burdens us with more responsibility it also gives us many more options. A specialty fruit that is unknown or unmarketable in its raw form may be delectable on our own tables. Food that is damaged or otherwise waste-prone may be salvaged as a new product, such as a puree. Home processing may lead to the creation of wholly new foods.

Having the right kitchen tools or appliances is essential for certain food preparation processes, and I've found that having such equipment in turn opens up numerous opportunities for experimentation with new ways to process and use the wide variety of food I grow.

As with my gardening, my food processing and preserving techniques don't all fall into neat categories. For example, I have a grain mill, which I also use to make oilseed meals and other stuff. So in this part of the book I'll tell you how I process and preserve various foods, with a bunch of my favourite recipes thrown in.

CHAPTER SIXTEEN

Milling, Baking, and Sprouting

I use grains in many ways, as cereals and flour and meal and even to make a kind of "meat loaf." To work with grains it's essential to have a good mill, and perhaps the most indispensable piece of equipment in my kitchen is the Corona hand mill. There are numerous small mills on the market; some are motorized for speed and ease of use, while others have large flywheels with long handles, which grind materials very finely in a single pass. Each has its advantages and drawbacks, but we greatly prefer the Molino Corona for its affordability and versatility. They were manufactured in Colombia, but now in China, and were originally sold for the Third World market, until the 1960s back-to-the-land movement created a burgeoning demand among American whole-food nuts like me. I believe they were originally intended for one purpose only: making soaked hulled corn into masa for tortillas. Ironically, therefore, they are the ideal tool for making a wide variety of food products, wet and dry.

One adaptation for the whole-foods market was the added option of stone plate burrs instead of the steel burrs. Supposed advantages are that stone is more natural and causes less heating of the flour and thus less rancidity. In fact the material used is not natural at all, but rather a synthetic carbide. As for the slight heating (an increase of about 5°F, or 2.8°C), that would be a greater concern for me if I were grinding large quantities of flour to store. However I rarely grind more flour than I need at the moment, and since I will put it into a 375°F (190.6°C) oven within the hour, I can't fret too much about a 5°F pre-warming of the flour. On the other hand the steel burrs allow me to do a lot of things that are impossible with the stone plates.

I have long contemplated getting one of those larger longer-handled grain mills for easier, quicker grinding of whole wheat bread flour, but with such a panoply of other uses the Corona will always occupy a secure (sic!) place at the end of the kitchen counter.

I should emphasize that *any* milled grain product is far tastier and more wholesome when freshly cracked or ground on demand, as the oil in the germ starts to go rancid very quickly. Granted that processing small amounts of grain right when we want to cook it can be a nuisance, the difference in quality may offset the inconvenience. After all, it was the removal of the nutrient-rich germ that made industrial-scale milling, shipping, and storage of flour feasible.

Milling Flour and Meal

The Corona does not yield a satisfactory (to me) flour in a single pass of grain through the plates. It is easier on the arms, while heating the flour less, to mill it in two or more passes. The first time I adjust the grinder to make a coarse meal, as fine as I can make it without undue exertion. Usually I then tighten the

The Amazing Corona

Grinding flour is but one of the many uses for my Corona. I can grind oilseeds (sunflower, sesame, flax, pepita, peanut, and more) into a fine oily meal (or butter in some cases), which has further uses that I describe later in this chapter. I can fresh-grind spice seeds like coriander, caraway, and anise into a vastly more flavourful product than the comparatively bland stuff that's been sitting for months on the store shelf. I can make my own mustard powder, as well as paprika and cayenne from dried homegrown peppers, not to mention garlic powder from either the dried cloves or bulbils. I can grind dried herbs, especially celery leaves, green leek tops, and similar garden "wastes" to use in soup stock or infused with oil and tamari for an herby popcorn sprinkle. I can even make apple flour from dried apples (for example, as a dry ingredient in applesauce cake), along with potato flour, chufa meal, and terrasol meal for adding to various creamy soups and porridges.

The trick to working with the Corona is to adjust the plates right. Just right makes a fine powder, but a little too tight and they will heat up and varnish the burr plates. Low temperature is also a big help; when possible I prefer to set up the grinder out in the shed on a below-zero day. The product is dusty fine, but will often cake up hard when stored in a jar in a warm pantry (like carob powder?). I may need to break it up when I use it, like old-fashioned sugar.

the flour anyway, or I may discard it. I don't wish to reject any of the bran (which is a source of several trace elements not found in the rest of the kernel), but the final bit of residue may be from persistent hulls and not the bran.

When grinding any flour I may save effort by sifting after the first coarse grinding, then regrinding both components separately before reuniting them. That's because the finer stuff may tend to clog up the whole grind, making it harder work and generating unnecessary heat.

Flour for Bread Baking

Part of producing high-quality baking flour is grinding it finely enough, especially the bran. While the Corona makes a fine grind (I have little experience with other types) after repeated passes, it would be simpler if you had a slightly finer sifter than what is available in your kitchen supply store. Fortunately you can buy perforated metal screening in a wide range of sizes and improvise your own sifter. Fine grind also depends on the sharpness of your burr plates, and so you may need to replace those occasionally, even though the old ones may still be fine for general use.

Milling Corn

Because the type of field corn we grow for our own use is a flint corn, there is a component—the flints—that resists grinding finely. It is brighter yellow and has a more gummy proteinaceous texture, since the softer starch is ground fine and separated by sifting. This has a very coarse texture and is much like the grits associated with southern cuisine, but not quite; unlike the bland and pasty starch, it has richer flavour and texture. I may use this as a cereal itself, much like cracked wheat or steel-cut oats, or I may use it as you would use rice or, again, grits.

I probably use coarsely cracked grains as much as or more than flour and meal. Of course the macrobiotics approve of this but that's not where I'm coming from. I just like a variety of foods, and as much as I love bread, it's not the only way to enjoy grain.

By the way one way I do *not* enjoy grain is malted and converted to alcohol; been there, done that, no

plate-clamps for a finer regrind. This renders most of the flour fine enough for baking, but with some remaining shorts and bran. I remove these by sifting, but I don't discard them; I simply regrind them yet again on a finer setting, perhaps twice, sifting after each pass, before adding it back in with the other flour. If a tiny bit still won't pass the mesh even after repeated grinding and sifting, I may add it to thanks. Too much work for a highly refined product that tastes like medicine. Being a nonspiritual person I'm not interested in "spirits" or "essences" of a real thing; just give me the real thing, thanks.

Milling Buckwheat, Rice, and Sunflower Seeds

For me the most challenging part of growing and using buckwheat, rice, and sunflower seeds has been processing them—in other words, removing their hulls. With wheat and rye the hulls can be removed by the usual threshing and winnowing process; with oats and barley you can simply grow the loose-hulled or naked varieties. Rice, however—although I have successfully grown it—is a challenge to prepare for the kitchen.

The Colour of Flour

The finer flour is ground, the lighter its colour. When old folktales tell of the king's bread baked from the whitest flour, they do not mean white in the sense we mean it today. Before the development of roller milling in the 1800s, separating the germ and the bran was nearly impossible. Their "white flour" was largely whole wheat flour, ground again and again and sifted or "bolted" through ever-finer cloths.

Figure 16.1. It's easy to use a hand sheller to remove flint corn kernels for grinding.

And what about sunflower seeds, not the oil type but the larger confectionery type? For years I used a friend's Japanese rice-hulling machine to de-hull my sunflowers (never tried it with anything else, including rice, which I have not yet grown in sufficient quantity); however it is inconvenient for me to travel to the machine, and anyway I have always looked for solutions that are available to anyone anywhere. In an old (1979) issue of *Organic Gardening* magazine, Jeff Cox described how he had some success first grading the sunflower seeds carefully for size and then putting them through the Corona at a carefully adjusted tightness. For me this was a complete failure (he mentioned using stone burrs, rather than steel, which probably makes all the difference). Cox further described having very good success using a C. S. Bell #60 hand mill and a Marathon Uni-Mill, both presumably with stone burrs. I haven't tried these, but 35 years later both companies are still in business selling those models.

Japanese buckwheat (*Fagopyron esculentum*) is very easy to process into *flour* with the Corona mill. Moreover it is a truly whole-grain product: The germ is still included, only the hull is removed. I accomplish this by setting the burr plates on the Corona slightly looser than for fine grind. The ground kernels are almost all fine enough to pass through a sieve, whereas most of the hull particles are not. The tiny flakes of hull that do pass are not offensive in taste or texture. I get a ratio of 60 percent flour to 40 percent hulls.

With the Tartary-type buckwheat (*Fagopyron tataricum*) this method was a failure; even a tiny amount of hull residue gives the product a nasty bitter flavour. I've visited the buckwheat mills of the St. John's Valley between Maine and New Brunswick; they avoid this by roller milling, which instead of actually grinding grain, squeezes it between increasingly close-spaced fluted steel rollers, with sifting between each stage. The hulls remain largely intact—broken but empty shells. In fact, looking at a heap of this residue, you would insist that it looks like whole buckwheat. Curiously, those hulls are not wasted, but are shipped to Japan where they are

Figure 16.2. First the pasta maker rolls out a flat, thin sheet of dough, in this case 70 percent buckwheat.

valued for pillow stuffing, and of course it is also sold for landscape mulch.

A problem with the roller process is that it also removes the germ, which is called *gru* in Acadian French and fed to livestock, a net waste of nutrients. Roller milling is the same process that brings us white flour, and over-refined stuff is not what I'm all about.

So at least the Japanese buckwheat is easily made into nice flour, but what to do with this whole flour? Well, of course it will make superb pancakes, but it does *not* give you kasha. Another wonderful use of buckwheat is for making soba noodles (*soba* means "buckwheat" in Japanese), but I had always

Figure 16.3. The next step is to cut the sheet of soba dough into narrow strips.

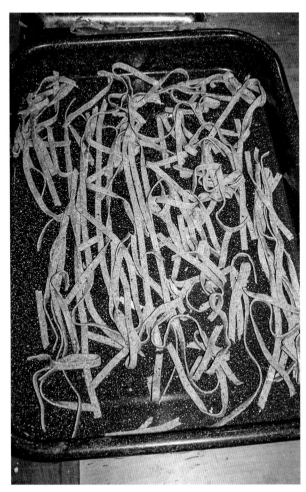

Figure 16.4. These dried, broad soba noodles are rougher than the commercial product, but no less delectable.

wondered how one could possibly make an all-buckwheat noodle with no wheat for gluten. Well, it turns out one can, though I am not necessarily that "one," but it also turns out that some soba noodles are made with as much as 20 percent wheat flour. I tried it with about 30 percent wheat (that was just a guess, as I had not yet heard about the 20 percent) and got a quite acceptable product. I didn't even use durum flour, the high-gluten type used for pasta; I just used what I had at hand, bread flour. The soba noodles I made in my pasta maker were a bit rough-textured, but then my all-wheat pasta never comes out as smooth as store-bought, either, though the flavour is generally better. I believe part of the trick is in adjusting the moisture—mine are probably too dry. In any case I am certainly inspired to persevere.

Now anyone who doesn't love kasha with blueberry raisins for a breakfast cereal needs to seek professional help, but I've found nothing to substitute, certainly not buckwheat noodles. On the other hand blueberry sauce on buckwheat pancakes is none too shabby. And what right-minded person wouldn't also love kasha with savoury dishes? The soba noodles are an excellent substitute for that. As an entrée it goes so well with fried sauerkraut (if you don't love sauerkraut, I see little hope for redemption), rocambole garlic, and shovelfuls of paprika (my ancestors were mostly British and French, but my soul was born in Hungary).

Making Waffles

Grinding grains into flour is the simplest way to process them, but I don't wish for my grain consumption to be limited to bread, especially since some grains don't make especially good bread. For example, oats are not very suitable for baking, but I do love them in waffles. A batter of 5 cups oat flour, 4 cups sunflower meal, 3 cups wheat flour, and ½ cup of amaranth flour makes waffles that taste exquisite. Conventional waffles rely on lecithin-rich eggs and/or baking powder to fluff up the batter. Vegans can substitute a little soy flour for the eggs, or omit it altogether, but something is needed to make them light. Another alternative is the original leavening: yeast.

The main problem with using yeast in my oat waffles is the oats. In fact anything except wheat does not have the gluten necessary to trap air bubbles (usually carbon dioxide, or CO_2), and the amount of wheat flour used in oat waffles isn't enough. Another method we learned years ago is to incorporate air mechanically, as with a blender, in which case you don't need to use any wheat at all (especially if gluten sensitivity is a problem). It is rather inefficient to blenderize only enough batter for two or three waffles at a time, but you can't let it sit around at all or you lose the lightness (you can always reblend it). By the way if you are leavening with either baking powder or a blender, then any wheat you use should be pastry flour (soft white wheat) as opposed to hard red wheat, which is most suited for yeasted products. If you use at least one-third sunflower meal in the batter, you don't have to brush oil onto the waffle iron, as you do with conventional recipes.

Now, amaranth seed is about as hard as a miser's heart, in addition to being very tiny, and so I have to pass it through the Corona mill four times, each time as tight as I can crank it. And then what? No matter how I prepare amaranth alone—as a porridge, roast, whatever—my kids want nothing to do with it; it is gooey, and Daddy grows it out in the dirt. For that matter my homegrown oats and sunflower seeds can

be a hard sell in most forms. Waffles are perhaps the most elegant way of using my homegrown amaranth and other oilseeds. Every Saturday morning we all wolf down stacks of waffles with all that stuff in them, and topped with homemade (gasp!) applesauce or other fruit purees no one recognizes. I can tweak the recipe to my heart's content and no one ever notices; I'll thank you to keep this just between ourselves.

Baking

Bread recipes are so easy to come by, I won't bother to include any here. I will point out that success in bread baking, especially yeast bread, depends a lot on the quality of the flour used. That requires a proper balance of fine bran, hard and soft starches, and proteins, especially gluten, all finely ground and reground with nothing removed. Modern wheat varieties and the way they are grown (in chemical-laden, humus-less soil) and harvested (by combine, when they are in the flint stage) and processed (roller-milled to remove all the bran and germ) are intended for making white bread. Making a light whole wheat bread from such wheats usually requires the addition of more gluten. This gluten is refined from other wheat of which the remaining components were used for other purposes, mainly animal fodder. Therefore it is no more "whole wheat" than the residual flour from which it was extracted, or the white flour minus its bran and germ, and yet it is sold as such. To make truly whole wheat loaves that are not heavy or doughy, you should raise older varieties (Early Red Fife is a good example, but only one) in mineral-rich humusy soil, reap it at the hard dough stage, stook-cure it (as described in chapter 9), and fresh-mill it by grinding, not rolling and sifting. Since those qualities are difficult or impossible to find in the marketplace, you probably need to do it all yourself, but it gives you control over every stage and ensures you a tasty and nutritious loaf.

Baking with Leavenings

The very same gluten that makes wheaten dough into a dense, indigestible mass also enables it to be lightened by the introduction of carbon dioxide (CO_2) bubbles, created by yeast fermentation, by chemical reaction with sodium bicarbonate, or by direct injection. The sodium bicarbonate found in baking soda and baking powder comes from mined deposits: baking powders also include various other chemicals such as tartrate, phosphate, and aluminum compounds designed to acidify the dough for the alkaline sodium bicarbonate to react, producing bubbles.

None of these chemicals is very wholesome, so it would appear that the less of these chemical leavenings we use in our food, the better. A distant kinswoman of mine once explained how she sometimes used new-fallen snow as a leavening in biscuits (which otherwise called for cream of tartar, a by-product of winemaking). It has been suggested that the leavening quality of new-fallen snow is due to trace amounts of ammonia, but I think it's more than that. The snow itself could do the job, if it is fresh and fluffy, and assuming you mix it into the batter quickly just before baking. As the snowflakes melt and turn into steam, they expand and leave countless airholes, just like the carbon dioxide from yeast. I've never tried it myself, but Cousin Phyllis said it worked well.

Yeast, on the other hand, not only lacks any of the harsh chemicals of mineral leavenings, but is itself a source of riboflavin, an important B vitamin (which I have never actually seen, you understand). Baking yeast is derived from the brewing of beer; the residue from brewing is called "barm," and when someone is described as acting a bit "barmy," well, you know where they've been. In the process of converting starches to sugars to alcohol, yeast produces lots of carbon dioxide (CO_2) bubbles, which expand upon heating to produce a deliciously light loaf, absent any harsh chemicals. A complication with using yeast is maintaining it. I mean nowadays you can buy active yeast in either cakes or dried granules, but that kind of convenience is relatively modern. Prior to that, bakers relied on homegrown cultures like farmers yeast—basically the same culture but maintained in a medium of mashed potatoes, sugar, and so on. We've tried it before, but found it difficult to keep it going for long without the culture becoming contaminated.

Saleratus and Sourdough

In the late 19th century both baking soda (called saleratus) and sourdough became important leavenings for westering migrants, gold rush prospectors, and others whose transient lifestyle made obtaining and maintaining common yeast cultures impractical. However, saleratus had the disadvantage of souring the digestion with constant use. In stories, diaries, jokes, and lore of the Klondike and Yukon gold rushes, saleratus is frequently mentioned and always with reference to indigestion and flatulence.

By contrast, sourdough was easily maintained and shared and its bread product was much more wholesome. Its use became so widespread that the very word *sourdough* became synonymous with grizzled Alaskan prospectors.

Sourdough, on the other hand, is a symbiotic mixture of yeasts and bacteria that is easier to propagate long-term, as it promotes the conditions for its own maintenance in an acidic environment where competing and contaminating microbes are held in check.

Because sourdough is not a single strain of yeast, but typically a composite, perhaps from different sources, there are invariably differing populations. The flour itself contains an assortment of spores, which is why kosher matzoh must be shaped and baked within a very short time after being wet: Its own inherent yeast, or chometz, will self-propagate even in the complete absence of added yeast.

My friend Shana Hanson has made sourdough starters from a variety of natural sources. Certain mushrooms have suitable yeasts living on their gill surfaces; the skins of fruits (like apples and grapes) that show a "bloom" on their surface are also good

Bread and Beer

Just by the way, if you disapprove of civilization—and who doesn't nowadays?—blame it all on bread and beer. Some archaeologists have concluded that the sedentary life was mainly a response to the need to stay put long enough for dough to rise and beer to brew. Otherwise, we'd all still be out there stalking aurochs and other carefree activities.

sources (Shana has had good results with juniper berries in particular). Even the loose shaggy bark of old sugar maples is impregnated with sourdough culture. In all these cases you can simply soak the material in tepid water for several hours and then use that water to make the initial starter dough. Henceforth the culture is maintained by saving a small portion of each batch and incorporating it into the new batch of dough. Over time and with proper care, the sourdough yeasts will establish themselves ever more securely, crowding out any off-types among them and creating a community of reliably consistent and vigourous cultures.

One reason for discussing sourdough in such depth is that in addition to being an easily homegrown leavening, it also is a suitable leavening for certain grains, notably rye, which does not have enough gluten to make a toothsome loaf by itself. Many modern rye breads have a considerable proportion of wheat flour added for the sake of gluten, but in Northern Europe, rye breads are often made with rye alone, which a sourdough culture makes possible.

Pocket Breads

In chapter 17 I describe how I make a sort of guacamole salad stuffing for flat bread sandwiches, so I probably should share some tips on how I make the flat bread. I'm talking about pita and chapati, both whole wheat, one leavened, the other not. Pita, often called Bible bread, is basically the same as regular yeasted whole wheat bread but for the flat, thin shape. Well, except that the way it is kneaded is particularly important. The important thing (for both pita and chapati) is to flatten out the dough repeatedly, folding it in half each time. This develops the gluten in horizontal planes, which tends to entrap the water in those layers, so that the expanding steam forms the pocket—I'm not sure why it doesn't just make oodles of little pockets as in regular bread. It does that, too, but it mostly separates into two layers, creating a nifty cavity for stuffing.

Chapatis are pretty much the same story, except hold the yeast. Also, you shape chapatis somewhat thinner by patting them between your hands. Actually, I use a rolling pin, since I never got the hang of the spanking back and forth. One of my sisters-in-law from Uttar Pradesh showed me the correct way of cooking them, so that you get a pocket instead of a brittle cracker. I'll share it with you: After patting or rolling out a 5- or 6-inch (12.7–15.2 cm) disc about 3/16 inch (0.5 cm) thick, you drop it onto a preheated cast-iron pan—but only for a few seconds, enough to sear the surface. Then you flip it over quickly and let it brown lightly on that side. In prepping for this you should have a thick cloth folded in quarters; I use a clean dish towel. Keep it handy. The moment the second side is done enough (*not* scorched), you flip it back onto the first side, the one you seared. Immediately begin patting gently and broadly with the cloth over the whole surface, starting in the centre and working outward. Bubbles should begin to form, and as you continue to press them around they grow and merge until you have a huge hollow ball. (Press it too hard and you'll just squash it back together into a sink stopper.) Then set it aside quickly before it scorches, and start another one. If the bubble bursts sometimes a gentle poke will seal it up again and it will continue to rebound with every gentle pressing until the bubble, or pocket, goes all the way to the edges. I have had them puff up until they practically rolled off the pan. If the pocket forms completely, you can stuff it with all sorts of things; otherwise you can just roll it up or fold it around the filling like a taco. Chapatis should be eaten fresh; once they cool and dry out they turn into Frisbees.

As for pitas they should be made from bread flour—in other words hard red wheat. I would assume the same for chapatis, since the gluten is what makes them stretch; it's what I use, though I have heard that pastry flour will work.

I will repeat the concern expressed earlier about quick-cooked breads, especially chapatis. The phytic acid in undercooked grain products ties up certain vital minerals like calcium and zinc. Indeed the main populations that suffer from zinc deficiency are those that rely on such quick-cooked breads.

Snowflakes

Another family favourite is Snowflakes, a traditional Winter Solstice (Esperian New Year) treat. I use the following ingredients (I can't tell you exact measurements, but I've listed them in order of quantity): oat flour, sunflower meal, maple syrup, water, fresh-ground aniseed, salt.

This makes a stiff dough, which I roll out to roughly ⅛ inch (0.3 cm) thickness on a smooth pine board dusted with oat flour to minimize sticking. I shape them with an improvised cookie cutter—a medium-sized tin can—and carefully lift them onto an unoiled baking sheet with a cake turner. Then I lightly make an imprint of a snowflake with a carved wooden stamp (that's the tricky part, unless you happen to be a woodcarver or know one). I made mine from a ½-inch (1.3 cm) basswood disc, with the snowflake pattern raised about ⅛ inch. If you make the carving too intricate, it is wasted effort; after all, it's only dough. The snowflake barely shows until you bake the wafer in a hot (400°F, or 204.4°C) oven for a very few minutes—watch it like a hawk! When done the surface will be a light golden colour, and the depressed lines of the snowflake figure will still be off-white, an elegant contrast. The mildly sweet biscuits are tender and crisp, and I always make lots and lots of them, as they are extremely popular.

Crackers

You can make delightful crackers from any assortment of homegrown grains; however they work best if you use at least one-third wheat flour. Other low-gluten flours may be just as good if they will hold together, but there's the rub. Also, up to one-third of oilseed meal, usually sunflower, will keep them somewhat tender, else they'll be more like raw lasagna noodles—okay for boiling maybe, but not much for dipping.

By the way crackers are a great way of using lots of homegrown poppy seed, although they are best left whole to avoid scorching. I use just enough water to make a stiff dough. I find the main trick is rolling out the dough wafer-thin without having it stick to the board or rolling pin. Frequent light dusting with flour is the remedy, although the oil meal in the dough also helps. Before baking I prick them with a fork and sprinkle salt on them. They should be baked in a quick hot oven (400°F, or 204.4°C), which means you must check them often as they bake; I don't rely on a timer.

Sprouting and Malting

A little-used form of whole grains is sprouts. In particular wheat and barley, when slightly sprouted, or malted, have a completely different taste, texture, and digestibility. The grain is transformed in two important ways. First the phytate is broken down. Phytate is a phosphorous-based acid that inhibits the absorption of some essential minerals, notably calcium, iron, and zinc. It is broken down or neutralized by three things: sprouting, fermentation, and heat. It is significant that chronic zinc deficiency is most common in countries such as India and Egypt, where flat breads are quickly cooked (or half cooked?), often over cow-dung fires. On the other extreme there is a particular type of sourdough bread made in Sweden that is first sprouted, then fermented (with sourdough culture), then baked very long and slow. The phytate level in that type of bread is extremely low.

Second malting causes the starch to break down by enzymatic action to form simple carbohydrates (no, I never actually saw an enzyme or a carbohydrate; this is all hearsay). The enzyme is maltase and the products are simple sugars; in fact that's how they produce the malt syrup used in brewing beer.

Apparently the same chemistry that increases the sugars reduces the gluten, because in fact an excess amount of malt in the dough completely ruins the ability of the dough to rise, and the result is leaden. The sourdough bread mentioned earlier is baked very long and slow and is still a relatively dense loaf. A small amount (1 tablespoon per loaf) of fine malt flour is ample for starting the yeast.

One of the uses for hard red wheat is the production of gluten for various meat analogues. I want nothing to do with such highly refined stuff, any more than TVP (texturized vegetable protein)—another perverted food product masquerading as something it is not. I enjoy the tough, chewy texture it imparts to entrées, but not at the expense of all that stuff that's been stripped out. It is no more a whole or natural food than is white flour.

Wheat Berry "Meat Loaf"

We have a rather simple way of enjoying some of that glutenous effect without sacrificing the nutrient balance of whole wheat. Basically, it consists of soaking the whole wheat berries (kernels) overnight and then pouring off (but saving!) the liquid—hold that thought. Then I grind the wet berries in a meat grinder, the kind with a ridged plate, not holes. What comes out is not just a coarse dough; it is a highly texturized lump in which the gluten has been developed by the grinding process, yet still contains all of the starch, B vitamins, et cetera. Sometimes a second pass through the grinder can make it even better. Unlike seitan, it is a whole food, not a perverted protein concentrate. But then what to do with it? Well, here's my favourite: I put a lump in a bowl with various dried crumpled herbs (especially savoury and sage), what we call "baloney seasoning" (ground coriander, paprika, garlic powder), mustard powder, and if I wish either black pepper or ground shepherd's purse seed. By the way the black pepper is the only ingredient I can't grow myself; and anyway, it is considered by some to be bad for you, although shepherd's purse has its detractors as

well. If you have difficulty making paprika (I find it possible but quite tricky), you can just take ripe bell peppers (raw or steamed) and some crushed dried dill flowers and blend them (yes, they have blenders in Esperia; they are pedal-powered). If you're the one eating this then you add whatever you want; those are just my preferences.

Now, depending on how dry the lump is, I may need to add lots of tomato juice (preferably from the watery but high-flavoured Gardener's Delights). If the lump is soft enough to knead in the other stuff (like firm bread dough), then I hold back most or all of the tomato juice at first. I place this loaf into a large casserole, with plenty of space all around it, and I fill the "moat" with the tomato juice—yes, I drown the loaf, cover it in juice. I often stick big pieces of carrot, onion, turnip, or whatever, into the moat as well. The loaf may look ridiculous drowning in a lake of juice, but after an hour in the oven the liquid mostly will have disappeared, absorbed into the wheat loaf, leaving a coating of thick tomato sauce. I might mention the obvious fact that you could add any favourite barbecue seasonings to that juice before roasting. I keep the lid on the casserole most of the time, so that the juice will cook into the loaf, not just evaporate.

Anyway, it makes a some-good substitute for meat loaf kinds of things, but that's only the beginning. Let's go back to that leftover soak-water we saved. While soaking overnight those wheat berries have imparted a lot of stuff to the water, like vitamins and enzymes, including maltase. Some folks call it rejuvelac and credit it with all sorts of miraculous health benefits: regrows hair and missing teeth, cures collywobbles and housemaid's knee, and so on. Well, I don't know about any of that, but this I do know: If I add to it enough sunflower meal to make a very thick cream, and let it set overnight in a warm spot, it undergoes a curious transformation, either exquisite or gross, depending on how you feel about things like blue cheese and sauerkraut. In the morning it has separated into a tart watery bit on the bottom and a frothy curd on top. The surface is usually turned a light brown with a greyish layer beneath, from oxidation. I stir this all back together into a cream-coloured dairylike substance with a flavour reminiscent of sour cream (at least to someone who hasn't tasted sour cream in over 30 years). In fact one favourite way to use it is to plunk a dollop into a bowl of Esperian borscht for our traditional Winter Solstice feast.

Another way I use this "sun cheese" is to spread it thickly on top of that wheat roast after the tomato juice has all been absorbed. If I put the uncovered casserole back in the oven for a few hot minutes, the "cheese" will brown slightly and mellow in flavour, somewhat like a sharp cheddar.

Still another way to use sun cheese is in dairylike spreads. For example, whip it together with a bit of miso (especially chickpea miso), and add blended ripe bell peppers, some mustard powder or sauce, and some crushed dried dill flowers. Dipped with carrot sticks or slathered on fresh-baked sourdough bread, it's none too shabby. If you make too much sun cheese for immediate use, no problem: In a covered crockery bowl it can keep in the root cellar or fridge for several weeks, preserved by its own lactic acid. (More ideas on spreads in chapter 17, but remember this ingredient.)

Instead of making the soaked wheat kernels into wheat roast, you could continue the sprouting process to make malt meal instead. For this you just rinse the soaked berries (after pouring off and saving the rejuvelac) and put them in a sprouting jar. You can make this out of any size of glass jar with a two-piece canning lid. You remove the metal disc from the threaded ring and replace it with a circular piece of nylon window screening. Every day, after you rinse the wheat berries, you place the jar upside down leaning against something (I use a dish strainer) so it can continue to drain thoroughly. Watch the wheat carefully until the sprouts have grown at least as long as the kernels, then spread them out on a nonmetal drying rack for a few days until they are brittle-dry. (Since the kernel is still alive, it will continue to grow slightly, but as it dries out it will quickly stop sprouting.) When they're completely dry you can either store the sprouts for later use or grind them in a Corona mill, either into coarsely cracked meal or into a fine flour depending on how you plan to use it. I prefer the cracked meal for use by itself as a hot cereal, but if I am mixing it with something like cornmeal, I grind to match the texture of the corn.

CHAPTER SEVENTEEN

Freezing, Fermenting, and More

Grains are easy to store in their whole form. All you need are cool, dry conditions and rodent-proof containers. But when it comes to fruits and vegetables, some form of processing is often needed both to preserve the food and to make it more widely acceptable. Here are some ideas for doing that.

Fruits

Although fruits are generally rich sources of sugars, they also tend to be rather acidic. Indeed some of the sweetest fruit is also the sourest: A lemon has far more sugar than a peach or a banana. Much of the sugar in fruits is a result of the breakdown of acids during ripening, persimmons being an extreme case. A paradox about jelly making is that pectin requires both acid and sugar to make the juice gel. There is more pectin in unripe fruit, as well as acid, but often not enough sugar. Jelly recipes typically call for the addition of lemon juice and pectin (often a product of lemon peels), plus enough cane sugar to offset the other two. That always struck me as a bit absurd: The basic fruit becomes merely a flavouring agent, a juice extract from which the pulp has been separated. I mean I could make a really lovely jelly by adding lemon juice, sugar, and pectin to windshield wiper fluid, but I prefer to use fruits as whole as possible. If that means a raw apple or pear, so much the better, but for fruits that have objectionable seeds or skins, we needn't reject all but the juice. By pureeing and straining we can still have most of the fibre, vitamins, and minerals. I find that many "minor" or neglected fruits that are incredibly rich in good stuff (bioflavonoids, antioxidants, and more), but whose overwhelming flavour hinders their wider acceptance, can become exquisite food items by blending with a milder-tasting puree like apple or pear. The same applies to juices: A bit of elderberry juice added to apple cider makes a drink that is both nutrient-packed and beautiful, yet doesn't need any justification ("but it's good for you!").

Modern fruit varieties may produce a rather insipid product, but older varieties of apples, for example, will give a rich, thick sauce that requires no sugar or cinnamon to interest the palate. My favourites include Baldwin, Black Oxford, and Followwater; Stark is particularly rich in pectin, although by itself it yields an ugly-looking cider.

Both oliveberry and hardy kiwi make an elegant puree, but *only if raw*. Cooking them at any stage releases tannin from oliveberry and calcium oxalate from kiwi, which are as delightful as sucking on leaky batteries. I mash the raw fruit and grind it through a sieve before freezing it (uncooked). Oliveberry puree is so rich in pectin that it soon sets up into a hard gel, and when thawed some of it separates out into a tart yellowish liquid. I reblend the mixture so it is a somewhat softer gel, then mix it with three or four times

About Sweeteners

Most of my sweetening needs (if indeed they can properly be called "needs") can be met by maple syrup. I have plenty of maples and plenty of firewood; what I lack is plenty of time. For many years I tapped and boiled sap to make what others consider a gourmet luxury—we just couldn't afford sugar. In recent years, my seed preservation project has competed for my time, the need to fill and mail requests coinciding with the rising of the sap. The Scatterseed Project is a "mission"; it is not lucrative. When I can afford to hire some labour, the sugaring will be a very prudent investment of time. At this writing we use lots of Sucanat, an unrefined cane sugar product that is much inferior to maple syrup.

There are other sweeteners that can be homegrown, but I'm still investigating their practicality. Sugar beets are pretty easy to grow, but processing them into a usable "molasses" has eluded me so far. I haven't given up; I just need to keep refining the technique, which involves coagulating with pickling lime (CaOH) and filtering before boiling down the final syrup.

Malt syrup is a very tasty product resembling sorghum molasses. (I love sorghum molasses, but am too far north to successfully grow it.) Malt syrup can be made from homegrown barley. Not the hullless barley I described in chapter 9, but common barley of the two-rowed type. The barley malt syrup is considered by most as a stage in the brewing of beer, but for me it is a final product ready for table use. The directions for making it sound easy enough, but I have not tried it. To date I haven't grown more barley than I could use as whole grain (the naked type), and I'm not sure how practical it would be. Nevertheless, I plan to explore that sometime.

Although honey is technically not a vegan food, that is not my main reason for eschewing it. After all, it doesn't require heavy fuel use to boil it down, it doesn't occupy space that could be used by other crops—in fact the presence of hives enhances the other crops. I do not refuse gifts of honey, but I don't purchase it or keep bees myself. I can't spare the time to do it properly (careful management is required to harvest only the surplus; anything else, including replacement with sugar, I consider inhumane). Furthermore, honeybees worldwide are under stress from human intervention, and I prefer not to meddle with them. I encourage the wild populations, and I benefit from them in every way but the honey; I'm satisfied.

To some extent you can use either perry (pear cider) or pear puree as sweeteners, especially if cooked down to a concentrated form. Due to their bland sweetness they work especially well blended with strong-flavoured fruit purees and juices such as oliveberry and elderberry.

the amount of applesauce to tame the over-robust flavour: a combination of raspberry, strawberry, and rhubarb. It is not unpleasant to eat straight, but I can enjoy a lot more of it cut with applesauce.

When I first planted grapevines my main concern was hardiness. I had no interest in winemaking, and I assumed that my need for grape juice would be much less than my production, while raisins would be the greater demand. I now question that assumption, given how much I adore grape juice, but nevertheless raisins are a biggie, and since we are in the habit of using seedless raisins, my priority shifted to those grape varieties that are both hardy *and* seedless. That still left several choices, and I have collected those. Subsequent experience has convinced us that raisin making is not a realistic proposition for us here: Climate doesn't give enough sugar in the grapes, plus the weather is not right for drying them without mould. Fortunately we discovered a way of keeping grapes, seedless or not, for fresh eating. Our attic is dry and airy and apparently an ideal temperature for keeping grapes in excellent condition for several weeks after they're harvested. We spread the clusters out in a single layer on long drying racks made of framed nylon screening. We support the racks on crosspieces between the rafters, being careful not to break the skins.

When drying grapes any that do get broken or bruised will likely mould within a few days and need to be culled out, but the whole undamaged grapes will remain intact for several weeks, withering very slowly and growing sweeter all the while. Since fresh eating and grape juice and puree are now my main objectives, seedlessness has become less crucial—I'm happy to chew on seeds in my fresh grapes; I only find them a nuisance in raisins, especially in my oatmeal.

We can also make "raisins" from things other than grapes. Well, technically not, because the word *raisin* comes from the French word meaning "grape"; what we call raisins, they call *raisins secs*, or dried grapes (likewise in Spanish: *prunas* are plums, prunes are *prunes secas*). However, if you're not quibblesome over such linguistic fine points, dried (lowbush) blueberries and elderberries are suitable alternatives in some recipes, especially muffins. I describe how to make these later in the chapter.

Apple Products

We have always been blessed with such an abundance of apples, whether our own or foraged, that the dilemma was how to use them all. It became obvious early on that apples are not generic; they differ hugely in flavours and textures, storability and uses.

Take Wolf Rivers, a very old apple that originated in Wisconsin. It is a huge and beautiful apple, superb for certain purposes, useless for others. It is so soft that high-hanging fruits are often damaged by blue jays pecking. They bruise easily and store poorly. They are rather light, not particularly juicy; the pressman hates to see them coming, as they tend to clog up the feed-chain with their great bulk. Yet for stuffing and baking, they are without peer. We core them, fill them with sunflower meal and maple syrup

(dried blueberries do no harm, either), and fill up a deep baking dish with them. If they are individually cupped in aluminum foil, their syrupy juice stays with them better. In the oven they practically explode into fluffy white mounds of deliciousness. This recipe sort of works with other apple varieties, but none like Wolf Rivers. They also dry phenomenally well.

For stewing in compote we prefer to dry firmer apples like Baldwins or Black Oxfords. However, many people use dried apples mainly as a raw snack, and for that you cannot beat Wolf Rivers. They slice into huge wedges, 16 or 20 per apple, with stay-white flesh and a tender chewiness like candy. We used to gather and dry 20 to 30 bushels of Wolf Rivers for the snack-food market, reserving the harder stewing types for our own use.

There is no variety that I consider unsuitable for applesauce, nor is there any variety I consider all-sufficient for sauce. The same applies to cider: The larger and more divers the assortment, the better the juice, the better the sauce. Bland sweetings with tart crabs with aromatic pippins all work together to make a nuanced and delightful blend such as you will not find in any supermarket or farm stand. Those establishments use the rejects from their fresh dessert apples, Macs and Red Delicious and the like, as their cider stock. Shoppers who have never tasted anything other than blah commercial cider have no experience of real cider flavour such as our great-grandparents knew.

The same goes for applesauce: Our applesauce does not require the addition of sugar or cinnamon to make it exciting. The most that we could ask of it is that it be more apple-y, and we do that the same way the Shakers did. While putting cut-up apples into a kettle to cook for sauce, we add enough water to ensure they will not stick to the bottom. The Shakers replaced the water with cider to intensify the sweetness and bouquet of the sauce; we copy their method whenever fresh cider is available for a truly exquisite applesauce.

Another way we use cider is boiled down to a syrup or soft jelly. It is very sweet but also very sour, and so it's a bit strong to spread on toast. However, it

Figure 17.1. We spread out apple wedges on screen racks to dry.

Figure 17.2. The Saladmaster shreds apples coarsely for dried chips instead of wedges.

Figure 17.3. Mike Bouchard suspends a rack of apple wedges below the ceiling over the woodstove to dehydrate, which will take a few days.

Figure 17.4. These dehydrated apple wedges and chips are ready for storage.

Figure 17.5. This one-variety applesauce made from Followwater apples is delicious—but ugly.

is one of my favourite things to use on sweet-and-sour tempeh. Combined with sunflower or pumpkin seed meal and spooned over the diced tempeh in a hot cast-iron pan, the mixture quickly evaporates to leave a tasty toasty glaze on the tempeh, like barbecue sauce. Needless to say the addition of paprika and garlic powder and prepared mustard to the mixture, along with ground cumin or coriander, only enhances the barbecue effect. Basically the cider syrup just replaces the vinegar and molasses typically used in barbecue, only it's a lot nicer.

Making Cider Vinegar

Speaking of vinegar, that's an obvious use for surplus cider. We make it by simply dumping fresh raw cider into barrels such as olive oil comes in. The problem

with making it in our cellar is that when we press the cider, usually in November, the cellar is already cooling down rapidly. That is good for apples and carrots and other foods being stored there, because it preserves them; however, we don't *want* to preserve the cider, at least not as cider. We *want* it to spoil (in a good way, that is).

Understand that when cider "spoils" under certain circumstances, such as in the absence of air, it turns to wine, which is not our objective. Given air, wine continues to form vinegar. In the deepening chill of our cellar, the cider hardly begins frothing and fermenting before it grinds to a near halt, and alcohol forms faster than it can turn into acetic acid, or vinegar. Therefore, come spring, our "vinegar" still has a slight booziness to it, though it certainly is not wine. With the return of warmer weather the process continues apace, and by late summer we have good pure vinegar.

One winter I made a barrel in my upstairs workshop, which is semi-heated, at least compared with the cellar. The vinegar "made" in a few months, though that part of the house had a noticeable pickle scent. The quality seems no better or worse. Since I usually cannot spare that space in my shop, we continue to make it in the cellar, knowing it will require several months longer to be usable. Since we always have a partial barrel from the previous year, the delay is no problem.

In the process of becoming vinegar a gelatinous layer forms on the surface, often an inch (2.5 cm) thick. That is called mother of vinegar or just mother because, like a sourdough starter, it can be used to start a fresh new batch. That is probably a good idea, although I rarely use it, relying instead on the natural yeasts on the apple skins. A second layer, the nether mother, forms on the bottom of the barrel. It, too, is gelatinous, but it is more fibrous and opaque, containing much of the pulp sediment that precipitated out of the juice like a fine snow.

The barrel has been unlidded all this time, to permit aeration. It is only covered by a taut-stretched multilayer cheesecloth to keep out fruit flies and dust. I remove the covering and carefully peel off that top mother, putting as much as I can grab into a bucket. It is kind of like groping for jellyfish, except these jellyfish tend to fall to pieces, leaving slimy strands floating about. I put all the mother I can get into a sieve and drain off the vinegar. A lot of vinegar remains within the mother, though, so I nudge it this way and that to release all I can through the screen before discarding the mother. Then I continue to bail out all the vinegar, straining it into a new barrel.

When I near the bottom I am careful not to rile the sediment unnecessarily. At that point it is a rusty red, basically the colour of cider; however stirring it up reoxygenates it, and the sediment continues to darken to a less attractive coffee colour, so I am happy to leave behind as much of that as possible to give a purer, lighter vinegar. I also strain out the bottom vinegar rather than waste it; it may not be pretty, but it makes an excellent window cleaner. Before the vinegar is completely stable, some weeks or months later, I strain it again, removing any newly coagulated mother or sediment. After that the vinegar will stay good for years. There's nothing at all bad about the mother—some folks actually believe in eating it as a health enhancer, although I believe it is mainly pectin and cellulose and dead yeast—it's just not aesthetically pleasing to have slimy little blobs and strands in your dressings or pickles.

Apple Compote

Few things are as nice to start off a cold winter morning as an apple compote with yeast dumplings. The compote is as simple or as complex as you care to make it: I've used dried apples, blueberries, or elderberries with ground coriander or dried rose petals. Whenever I've made this during sapping season I've used fresh maple sap instead of the water. In autumn I may replace at least some of the water with cider, although I add that toward the end so as not to interfere with the "plumping up" of the dried fruits.

While these fruits are stewing I start a yeast dough rising. When it has risen I bring the compote to a full boil before dropping in serving spoonfuls of the soft dough. I make a nearly solid layer of dumplings, but the boiling liquid will keep welling up around them, which is good, as it cooks them more thoroughly. A traditional Pennsylvania Dutch way of determining when they are ready is to wrap a thread around one and pull it tight, cutting the dumpling in half. If the thread comes through clean, they are ready. I just break one open with a fork and see that it isn't doughy.

A nice variation is to add some oil meal like sunflower or hazel meal to the liquid just before dropping in the dough. It makes the liquid creamier, as if it weren't rich enough already.

Fruit Leather

Just about any fruit can be pureed and dried to make a fruit leather. Apple and pear are the all-around best for this, due to their thick pulp, but other juicy berries can be mixed with those to give them more body and make them go farther.

We use the same principle as for drying sweet corn: We cover a baking sheet with a thick (¼ to ⅜ inch, or 0.6–1.0 cm) layer of sauce and put it in a low oven with the door ajar. As soon as the sauce reaches a dryness that allows you to slide or flip it onto a nylon screen drying rack (no metal!), then continue to air-dry it right over the woodstove.

The trick, insofar as there is any, is to use more heat than you would for veggies, yet watch closely so as to remove the leathers before the edges scorch, finishing them off on drying racks. If you tried to start them on the screening, they would drip through and mould.

My friend Liz Lauer makes a to-die-for blackberry leather by spreading the puree on glass windows placed on shelves in her greenhouse. She must spread it very thickly indeed, for her finished product is thicker than harness strapping, yet never moulds.

Cousin Tom makes all sorts of fruit leathers, as described in chapter 12.

Dehydrating Food

Long before we had a freezer we preserved a great deal of food by dehydration. There were a lot of cute little dehydrators on the market in those days, but they were hopelessly inadequate for the volumes of food we had to handle. Also their claim to fully dry food in 12 hours or less was lost on us, because we felt the high heat required to do that was more harmful than beneficial; we didn't want to lose half the nutrition before even cooking them. In hindsight those probably worked fine for drying fruits, but we felt the green and yellow veggies were too delicate for such temperatures. We preferred to dry most things on overhead nylon screen trays suspended from frames built into the ceiling. Each tray could hold several pounds of chopped food, and most things would be largely dry within 48 hours, though it might take another day or so to ensure that they were perfectly dry before putting them away.

We avoided sunlight, because exposure to UV destroys chlorophyll and carotene. In fact we wanted as little heat as possible, preferring a flow of warm, dry air to carry away the humidity arising from the chopped food. The natural airflow under our ceiling was ideal most of the time, often with no supplemental heat at all. During very sultry weather we might keep a low fire in the woodstove for the sake of dryness and air movement rather than the heat itself, which was certainly not desired. A small electric fan would have helped a lot, but at that time we had no electricity.

For quick dehydration we cut most foods to no more than ¼ inch (0.6 cm) thick, especially the cross sections of leaf stems, like celery, chard, and kale. Root crops like carrots we slice into ripple chips in our Saladmaster, which creates more surface than a flat cut would. We also use the coarse-shred attachment to make something comparable to a shoestring french fry. We use that for carrots, too, but especially for apples. Our main way of apple drying is as "schnitz," the wedge-shaped segments (typically 12 per apple) that are so perfect for compote; for using like raisins with cooked cereal, however, the small chips produced by the coarse-shred attachment are more like raisins in size and more appropriate for spoon eating.

Most of the literature on dehydration insisted on blanching: precooking all food, even greens, to halt the enzymatic action within it. The problem we had with that was that food became a shrunken, sodden mass that was nigh impossible to dry before it spoiled. I got to try some of those dehydrated blanched greens once, a commercial trail food product dried in special vacuum dehydrators. My dad and I took some on a backpacking trip, and although they were tasty enough I could see no improvement in flavour over our own slow air-dried greens.

We greatly prefer greens cool-dried and stored without too much compression. That is, we gently stuff them into glass jars to minimize storage space, but not so tight as to crush them too much or make them too hard to get back out. Moreover a fair amount of air in the storage containers prevents the slight moisture in the greens from concentrating and causing deterioration over time. Of course some food crops can best be fresh-stored in the root cellar, but only to a point. In late winter garlic threatens to sprout, along with leftover carrots, onions, and potatoes. In the years when I have a big crop, my approach is to dry some in the fall and then dry more in late winter rather than lose it altogether. The dried garlic can be ground into powder. Carrots and onions can be reconstituted and used like that, although they are particularly good in soups and stews, where the long, slow cooking brings out the best in them. Some potato varieties will keep until the new crop comes in, but some early types, especially the yellow fingerlings, are quicker to sprout; we coarse-shred them and air-dry them until they're brittle-hard, at which point I can grind them into flour. That potato flour is especially appreciated for instant mashed potatoes for camping trips, or it can be used to thicken "cream soups" and sauces. It's a bit like using cornstarch or arrowroot, but much more flavoursome and nutritionally balanced. When raw potatoes are dried they developed a slight sweetish taste as some of the starches convert to sugars. If we wish to avoid that, we steam the potatoes lightly before shredding and drying.

Table peas definitely must be blanched lightly before drying to halt the ripening enzymes. Even then they taste much like canned peas, which I find acceptable; it's just a completely different food from fresh or frozen peas. We have done green beans blanched and unblanched; both work, but I prefer them unblanched.

Bell peppers are very easy to dehydrate; indeed the flavour is somewhat intensified, even when they are reconstituted. That is partly because the peppers continue to ripen while drying; if they are partly red when sliced, they will be all bright red when dry.

Tomatoes are a particular challenge; only the dry-fleshed paste types can be dried at all, and they must be sliced thin and watched very carefully. If they are ever so slightly unripe when sliced, they will have a much better chance, as they continue to ripen while drying, much like peppers. Also, if the wet seed-gel is scooped out first, they are less likely to mould.

In Italy the variety Principe Borghese is a popular drying tomato. Supposedly they can just be cut in half—well, maybe in Italy. Personally I feel little need to dehydrate tomatoes. Canned tomatoes fill all my needs and are vastly easier, especially for large quantities. The use of dried tomatoes stored in olive oil for gourmet dishes is swell, but I don't consider it very important.

Sweet corn is a special case. It must be blanched first, or it will turn all tough and starchy as it dries. Drying sweet corn-on-the-cob, like field corn, gives a very inferior product. That's the way to handle sweet corn for seed, but not for eating. We process it as for fresh eating: steamed on the cob and chopped off when it is cool.

So far the process is the same as for freezing, except now we spread it out on baking sheets in a slow oven with the door ajar, stirring occasionally to let moisture escape from the interior. We used to completely dehydrate sweet corn in the oven, but we had difficulty controlling the process toward the end. The thinner areas at the edges of the pan would often brown or even scorch; moreover it tied up the oven when we had lots more corn waiting to be processed. We knew that if we spread fresh-chopped corn on the drying racks, it would make a drippy mess and the corn would invariably sour. We discovered that if we oven-dried it only to a minimal level of dryness and then spread it thinly on the nylon screen racks to finish off, we got a superior product, with much more efficient use of the oven space, about twice as fast.

Sweet corn is one food we continue to dehydrate, even though we usually freeze most of it. Dried sweet corn is just different from fresh; even after reconstituting it has an intensified corny flavour that is especially appreciated for some things. For example, ground into a flour, it is a superb addition to flint cornmeal porridge. By itself it would make a poor cereal—too vegetably—but a couple of tablespoons added to the grain form imparts an especially rich corn flavour. Likewise in casseroles and some soups, sweet corn flour can be a huge enhancement.

Blueberries, dried into "raisins," have a more intense blueberry flavour, even when reconstituted.

Figure 17.6. Whether it's being used for drying or freezing, the boiled corn is chopped off the cob. Photograph courtesy of Sylvia Venter

The problem is that high bush blueberries, the big ones you buy in the store, are difficult or impossible to dry before spoiling. What you need are the wild lowbush berries. If you haven't a source near you, better create one (see chapter 12). We dry lowbush blueberries whole; the bruised ones tend to mould before they dry. Whole berries take forever to dehydrate, often two to three weeks, but that's no problem as long as they don't spoil in the meantime, and they won't if handled properly. We spread them on large drying screens (20 × 48 inches, or 50.8 × 121.9 cm), much like regular drying trays but longer. We place those trays outside (sun-drying is good for most fruits, increasing the availability of minerals like iron), on wooden skids on a wooden platform where air circulates freely. To deter birds or other pests we place a second screen upside down over the fruit-laden racks. The main nuisance is that we must bring them into the house every night before dew and during rainy weather. Surely those commercial dehydrators would finish them in a much shorter time but would never handle the necessary quantity—we have often dried 40 to 80 pounds (18.1–36.3 kg) of blueberries. Properly dried blueberry raisins can be kept for several years with no noticeable loss in quality.

Using Dried Veggies

To use dehydrated vegetables, especially leafy greens, it is important to cook them properly for best flavour and nutrition. We steam them as opposed to boiling them. The real difference is that you use just enough water that the dried food will absorb it all by the time it is finished cooking; there should be nothing left to pour off. It doesn't take long to get a sense of the proportions—actual measuring is difficult, as the food may be more or less compressed—but erring on the side of too dry is better *if* you keep a small kettle of boiling water handy to add more if necessary. An excess of water means you need to cook until that steams away, which may mean overcooking the veggies.

By the way, that is our general practice when cooking all vegetables: to use just enough water that it will

have steamed away by the time the food is cooked, with no nutritious liquid going down the drain.

Thanks to dehydrating the Corona mill is useful for oh-so-much more than grinding and cracking grains. For example, I grind powdered herbs or dried things like leek tops or celery leaves to add to vegetable oil for a sprinkle on popcorn or veggies. Fresh-ground coriander, cumin, caraway, and anise have a much richer flavour than preground, whether you grow your own or purchase them.

Of Oil and Oilseeds

As I described in chapter 11 we once decided to grow a small crop of canola (rapeseed). It didn't work for us as an oilseed crop, but we ended up with lots of rapeseed. We noticed that brassica seeds (like cabbage) were sold in the whole-foods store for the purpose of growing your own sprouts, so why not rapeseed? We tried it and were delighted at the very mild turnipy flavour in midwinter. Later, an extension agent told me that there was a problem with raw rapeseed being toxic to some creatures, causing blood thinning. I suppose he was referring to the erucic acid that has been largely bred out of modern canola—which, by the way, is the main distinction between canola and rape, besides the hard-to-market name. In any case we never felt any ill health effects, which may also be due to the sprouting, which completely alters the seed's chemistry. Anyway we acquired a wonderful new food, but so much for vegetable oil production.

I've also grown a few pounds of the seed type of flax for eating, but not for extracting oil. The problem with flax oil or linseed oil is that when heated it oxidizes to form a natural plastic, which is why it is used in oil-based paints (other oils, like soy, require the addition of dryers). That's why you buy linseed oil at the paint store, not the grocer's. Yet flax is supposed to contain lots of omega-3 oils (not that I doubt it; I've just never actually seen them), so I'm all for eating lots of flaxseed, only in its complete form. I've used a bit sprinkled on cooked cereal, but the main way we found to consume it in quantity (and even get the kids to eat it) is in French toast. Of course "real" French toast is dipped in a batter of eggs and milk and nutmeg. How to make a clingy batter from plant foods alone? As it turns out flaxseed meal with a bit of fine oat flour and soaked in water for a few minutes makes a gelatinous batter almost as gross as real eggs. Like I said, my kids love it. Of course if you really want to imitate the taste, as well as the texture, of real eggs, you could always add some homemade soy milk, and the high level of lecithin will further mimic the leavening effect of eggs, as in a soufflé.

We still had not discovered how to produce our own vegetable oil, for sautéing veggies, frying pancakes, dressing salads, popping popcorn, and so forth. We looked into sunflower seeds, reasonably easy to grow, but again, what about the hulls, which we assumed had to be removed to press oil? About then I began to get discouraged about the whole idea, especially when I discovered how much space it took to grow a pound of sunny seeds, and how many pounds of seed it took to press a gallon of oil. One thing that led me to rethink my whole approach to the business was a comment by some Seventh-day Adventist friends. They offered a quote from their prophet, Ellen White, to the effect that oil should be used "as in the olive." A radical fringe within the denomination interprets this to mean whole unextracted oil, in its natural context, a truly whole food. I am not an Adventist, but this concept rang very true for me, especially when I realized that the oil in oilseeds is perhaps the least nutritionally critical part of the seed. After all, our own bodies can and do synthesize lipids (fats or oils) from starches and sugars, but the oilcake, or residue from pressing oil, is full of protein, B vitamins, calcium, phosphorus, magnesium, and other goodies, which are all discarded (or fed to livestock, which is almost the same as discarding it).

Why, sez I to meself, should I go to the work of growing and pressing all that seed, then throw away the most nutritious part? Even if the seedcake is fed to livestock, most of the nutritional value is lost to us by the inefficiency of the food pyramid. In any case, if my vegan diet is predicated on someone else raising animals to consume the waste, then how vegan is it, really? (A similar logic applies to tofu, though less so.)

What About Deep-Fried?

For deep-fried stuff like french fries, onion rings, and doughnuts, we have no real substitute, except merely baking in a hot oven. For example, we make "french bakes" by slicing raw potatoes into strips and putting them on cookie sheets in a very hot oven, turning once or twice until they are crisp, but not greasy-crisp.

Therefore, we reasoned, it behooved us to develop a wholly new cuisine independent of this refined food ("unrefined" oils are only unrefined in that they are unfiltered and unbleached and unsolvented, a very good beginning). We experimented and found that all these seed meals are very easily ground in our Corona hand mill. From there it was not difficult to come up with many ways to use meal instead of oil.

Frying foods with oilmeal. For frying foods like pancakes, burgers, pacoras, and croquettes ("little nibbles" in French), we use a preheated cast-iron griddle, sprinkling a bit of fine-ground oilseed meal (usually sunflower), then quickly spooning on the batter or burger mix before the meal begins to smoke and scorch. This effectively avoids the use of extracting oils or Teflon, but it has an additional advantage. Though I am using a much smaller amount of oil (contained in the meal) to get the same results, the cooked surface is not merely greasy; it also has a crisp, nutty-flavoured coating that adds to the taste and texture of the cake. And of course, *all* the nutrients are there, not just the oil.

My favourite way of using leftover oatmeal is to either slice it (if firm enough) or shape it into patties and dredge them in sunflower meal before dry-frying them in a hot cast-iron pan; no greasing necessary.

Making mayo with oilmeal. For mayo and other dressings, we again use the complete ground oil meal. I make a brine of salt, water, and vinegar, bring it to a boil, then add enough oil meal to make a thick cream. I simmer that for another minute or two before removing and letting it cool. Sitting overnight in a cool place only improves it. This product is not like real mayo in that it doesn't form those silky peaks, as it lacks free oil *and* eggs, and but its creamy texture is an excellent replacement. Moreover, I often add in further ingredients at the same time as the oil meal, things like paprika, ground mustard seed, dried dill weed, leek powder, et cetera. Although sunflower is my standard oil meal, for dressing I love to substitute pumpkin seed (pepita) meal for part of that. The sulphur compounds in the latter impart a distinct flavour reminiscent of eggs or cheese (that from one who has tasted neither in over 40 years).

Using meal in baked goods. For foods that call for oil as an intrinsic ingredient, like cakes and cookies, we simply add sunflower meal, using twice as much meal as the oil it replaces, and reducing the flour portion slightly to allow for the oil meal solids. Again, this adds not only the oil but also the distinctive flavour of the seed meal, whose nuttiness can be enhanced by very lightly toasting the seeds before grinding.

Meal with steamed veggies. Another way I use oilseed meal (especially sunflower or pumpkinseed) is when steaming veggies, especially greens. When there is just a last little bit of liquid left in the bottom of the pan, I stir in a few tablespoons of oil meal. This immediately turns into a rich creamy coating on the veggies. Pepita (pumpkinseed) meal gives a particularly rich flavour. A late-summer favourite of mine is kale and chard and rocambole steamed together with pepita meal added at the end, and rolled up in a chapati or stuffed into a pita. I find I consume more whole greens in this form than any other way; they are the heart of the meal. Lightly toasted poppy seed meal is also exquisite used this way, although the blue-grey varieties make everything look ghastly; if you're into appearances use the white-seeded varieties.

I am truly glad we invented all of those culinary strategies before we learned how to press our own oil, or we might have denied ourselves a creative avenue that has led us to many new foods. For better or worse we have since conquered that initial hurdle— better because there are other uses for vegetable oils, like soap and lubricants, and worse because it is now too easy for us to generate a waste product that used to be food. Fortunately, given the effort to grow the hull-less seeds, we have an inherent incentive to continue with the complete seed. Yet we also have the option of getting oil from other sources that would otherwise be useless, as I will explain anon. By the way here is another case where the marketplace enables us to do things that are less wholesome and less eco-efficient, by obscuring the inputs.

An Oil Press

I learned of a small home-scale oil press that can be clamped onto a heavy bench or counter (see appendix A); it costs less than $200 and comes in a padded mailer. It includes a tiny kerosene lamp that goes under the feed barrel to preheat the meal slightly for more thorough expelling. That's fine, but I don't fancy having kerosene around my food (or around my farm, if possible), so I replaced that with a small candle stub. By the way most if not all oils are heated somewhat for extraction, even so-called cold-pressed oils. But more on that anon.

This oil press (Piteba) is manufactured by a fellow in Holland whom I found very willing to help with problems, and I had plenty of problems, at least at first. Again and again the thing clogged up and I had to clean out the hard-packed plug from which I had extracted very little oil. At the maker's suggestion I added a couple of tablespoons of water, mixed it thoroughly, and let it sit around for a day or two. I was using old birdseed for the experiment, and it was just too dried out to release its oil. Sure enough the water soaking made the whole a bit more pliable so the seeds could be extruded from the little exhaust ports. I think a small candle would have done as well or better; perhaps a birthday candle if it would not be used up so quickly. It persisted in plugging up, but

I finally realized I was simply underestimating the strength of the tiny press. When I exerted myself a bit more (still less than I would for grinding grain), it packed tighter and tighter until suddenly the residue came squeezing out the exhaust ports like toothpaste and oil began to flow, sometimes in rapid drips into a little jar and sometimes as a steady thin stream. I was impressed and delighted; I believe I would have gotten still better results by tweaking the adjuster nut at the end and perhaps reducing the candle flame.

This is a very small, funky-but-rugged device, certainly useless for large-scale production, but ideal for the subsistence gardener/farmer. There is a pirated version on the market made in China. The alloys in some parts of the Chinese model have tested positive for lead, and I am sure I would not have gotten the kind of free technical backup service I got from the inventor, Edwin Blaak, who continues to refine the design and technique. Having succeeded with the sunflower oilseed, I already had other schemes for the device. I knew I could press oil from naked-seeded pumpkins—though why should I wish to, when I can use them entire with no waste? But what about pumpkinseeds *with* hulls? I assumed that they probably wouldn't work well, else why would the Austrians be growing the hull-less type? Quite probably because the yield ratio is greater without the hulls, but what if I had hulled pumpkin- or squash seeds, which are only a waste product from growing the squash to eat the flesh? *Any* yield would be an asset. I tried a pound of outdated zucchini seed, again adding a couple of tablespoons of water and letting it soften up for a day or two before pressing. The yield was slightly disappointing compared with the sunflowers, but only slightly, considering that it was a waste product. I also tried some Guatemalan Blue squash seed, a fat-seeded maxima squash likewise headed for the Dumpster. The yield was no better, maybe worse, then the zucchini, a pepo—a baby food jar from a pound and a half of seeds; but again I grew the squash to eat the flesh, and the seed was a wasted by-product. No more, and by the way the flavour was lovely; indeed I have never tasted such flavoursome oil as from these fresh-pressed (though old) seeds.

Another Extraction Method

I was interested to read F. H. King's description in *Farmers of Forty Centuries* of how Chinese peasants extracted oil from cottonseed meal: First they boiled the meal, then pressed the cake (in cloth "cheeses") in a primitive press, using wooden wedges instead of a screw (which is basically a spiral wedge), and then skimmed off the oil that floated to the top.

Other Uses for Oilseeds

Even if we adapt our cuisine for using only whole oil-seeds, there are still other uses for extracted oil. Lots of nonfood products such as soaps, salves, ointments, and lubricants are or can be based on vegetable oil. If we aspire to be self-sufficient for those items as well (not that we are), the question of how to press our own oil remains, and sunflowers continue to be the most attractive option for that. We know that we need to grow the oilseed type as opposed to the confectionery type. I remember reading somewhere that Native Americans made oil by boiling ground sunflower seeds and skimming the oil off the top. Whenever I make sunflower cream (from hulled and ground seeds of the confectionery type), there is always a film of oil bobbles on top. Presumably with the oilseed type there would be twice as much oil, but even that is pretty meagre for the quantity of seeds used—maybe enough to whip up a batch of war paint, but forget about deep-frying doughnuts (I know no other alternative for the latter, nor have I tried very hard to find one).

To digress a bit: There are a number of alternative technologies for machines, including cars and tractors, that rely on used vegetable oil for fuel. I say that's fine, as far as it goes: The justification is that the old fryolator residue is just going to be wasted anyway, and insofar as that is true it probably makes sense to use it instead of fossil fuel. But I have to ask how this all stacks up in eco-efficiency; if it takes the produce of an acre of sunflowers to fuel the tractor to plough that acre, where is the sustainability? I'm not saying it does; just that we need to ask such questions. How do they do that in Esperia? We'll discuss this issue in greater depth in the conclusion. With pumpkins, as mentioned earlier, we can avoid the whole problem of removing hulls by growing hull-less varieties. Then it is as simple as scooping out the seeds, drying them, and rubbing off any debris that clung to them before drying. Then they are ready to eat, in various ways, as I will discuss anon. You see, we don't *need* to extract the oil for any of my uses. In fact it seemed strange to me that the hull-less pumpkins widely grown in Eastern Europe are not intended for whole eating, but for pressing oil for salad dressing. A little bit is exported to the United States as a prostate health remedy, even though you could get the same benefits more cheaply by just eating the whole seeds; but then of course it wouldn't seem like medicine, so it probably wouldn't work.

In *Farmers of Forty Centuries,* King describes street vendors in China peddling snacks like roasted peanuts, popcorn, sunflower and pumpkinseeds. He also mentions *melon* seeds as a roasted snack, although he doesn't mention the species. This inspired me to try pressing oil from melon seeds and from ripe cucumber seeds (surplus and outdated). It worked, and the flavour? I haven't tasted butter in 40-plus years, but I don't remember it being quite as tasty and rich. And what about chufa, or nutsedge? It supposedly has a high-quality oil that can be used for food or industrial purposes. Clearly I have more experimenting to do.

You might suppose that even a greedy fellow like me would be satisfied to have salvaged that bit of lovely oil from the compost pile, but not so. At least with the pumpkinseed residue, or seedcake, there was still all that valuable nut meal, minus most but not all of its oil, and still laden with protein and all those other goodies. I added enough water to make a thin slurry and, after simmering for a while, sieved out a rich cream with no detectable hull in it. Delicious. Don't try this by grinding the seeds finely in the Corona. The fine hull particles will pass through the strainer and ruin it. The reason it works with a press is that the hulls are only coarsely crushed.

So far I use this "pumpkin cream" mainly in bisque-like soups, such as kale with leeks and potatoes. Not too shabby at all. Though I haven't tried it, I doubt that would work with the sunflower seedcake: The hulls contain a lot of dark pigment and bitterness, which would likely leach into the "cream."

Dips and Spreads

Hummus-bi-tahini is one of my favourite ways of eating chickpeas. Moreover I see hummus as a general concept that can be applied to most or all legumes: mashed and mixed with something oily and something sour and salty and spread on something. In the case of soybeans and field peas, I first split them in the Corona mill (set very loosely), then winnow the hulls away to save cooking time and improve the texture of the "hummus." (With real hummus made from chickpeas there is no need for that; just pre-soak them several hours before cooking.) Any of the other legumes, including kidney beans, runner beans, field peas, and lentils, can be inserted into this pattern with happy results.

Another classic hummus ingredient, sesame (tahini), is difficult or impossible to grow in the Far North. Again, there are cool-climate substitutes, like sunflowers (best if lightly roasted before grinding). To my taste there is nothing like pepita meal (hull-less pumpkinseeds) for a nuanced creaminess with a hint of cheese or eggs (it's partly the sulphur).

But what about the lemon juice? Lemons surely won't grow in my gardens. Vinegar is a decent substitute, but not quite as sprightly as the lemon. I've thought about using quince, which has a fruity sourness comparable to lemon. I know that juice can be extracted from quince, but I don't know how to store it. I have dried grated quince flesh and reconstituted it. That could work in hummus if you don't object to a bit of quince pulp. This is wild conjecture; in reality I mostly use lemon.

Figure 17.7. Edamame for guacamole: What next?

Toasted Cheese, Anyone?

By the way we make another kind of cheesy spread or sandwich filling with pepita meal mixed with miso and a little liquid (water or vegetable stock) and things like paprika, ground caraway, and garlic powder. It makes something vaguely reminiscent of a toasted cheese sandwich. Stirred into a white sauce it works as a fondue dip.

Cousin Tom has come up with an idea that puts this all in a different light. He puts steamed edamame (green soy) in his blender with oil, lemon, salt, and garlic, and makes a very rich tasty spread/dip that he views as a substitute for guacamole. I make it often, and it certainly has the appearance and texture of guacamole, but there the analogy breaks down somewhat, as the edamame has much more protein than avocado, yet less fat (I don't add any oil to mine). Yet if guacamole is not the model for this food item, neither is it quite like my hummus analogy, which is starchier due to the use of *mature* legumes, like kidney beans and field peas. Indeed I see Tom's invention as something altogether different, something that inspires still further variation. The big difference here is that the edamame is an *immature* legume, quite different from mature soybeans, more quickly prepared and easily digested. Chickpeas and lentils are, to my knowledge, not practical to eat in the green shell stage, but kidney beans and runner beans certainly are, giving a delightful variation on taste and texture.

Of course we all know immature peas; they're what most folks think of as just peas, or table peas. While those can certainly be used for this "mock guacamole," frankly it's just a bit *too* vegetably. A more substantial product can be made by using field pea varieties at the immature stage, when they are more creamy than succulent.

This general formula lends itself to sandwich spreads and dips, but it also serves to make a salad into a complete cold entrée, especially useful for picnics and party occasions.

Any number of condiments can be added to this ersatz guacamole to perk it up even more. Prepared mustard (presumably from your own homegrown ground mustard seed) is a no-brainer, especially with *any* form of peas. It goes without saying that consuming peas without mustard is just barbaric; less well known, however, is that all the other legumes are enhanced by some form of the pungent crucifer, whether as ground seed or as greens. That's not just my opinion; "studies have shown" that individuals who do not like peas and mustard together are often prone to other antisocial behaviours. In parallel double-blind controlled tests those same subjects were found not to appreciate savoury and cumin with their legumes, an inexplicable perversion. Paprika or some form of pepper is pretty much de rigueur, as well as any sort of pickle relish. My favourite is Pennsylvania Dutch chowchow (heavy on the corn, please), but shredded or fine-chopped terrasols are none too shabby, either. Oh, and that sun cheese we discussed earlier? This would be an excellent occasion to bring some of that out of the root cellar.

As I mentioned this sham guacamole can be used to enrich salads, but an even more synergistic recipe (meaning, in this case, greater than the sum of its parts) calls for all this to be stuffed into a pocket bread or rolled up in a chapati. The salad part could include almost anything, but my own favourites include shredded cabbage, lettuce, grated carrot/celeriac/terrasol, and minced onion/ripe bell pepper. And of course we mustn't omit powdered dill flowers. Oh, we haven't discussed that yet; let's do so now.

Condiments for Spreads and More

I grow lots of dill. I love the taste, and it is easy to grow, plus it attracts many beneficial pest-eating insects. I use some ripe seed, mostly in cabbage and beet recipes, but by far my greatest use is for dried dill *weed*, or anet (not a coincidence that it is

Figure 17.8. Dill is at the height of its aromatic glory at the flowering and green seed stage, not after it ripens.

a close cousin to anise). Dill weed (as opposed to seed) can be stuck in the garden almost anywhere: a few empty feet of carrot row, a section of lettuce that got harvested early, an area under a pole bean trellis that will soon be shaded by vines, wherever. I rarely reserve a space for dill weed; like scallions, it just fits in here and there. Sometimes I've planted dill in between cabbage plants, knowing it may not have time to mature *seed* due to late planting, but may just make it to the flowering stage before freezing weather shuts it down. The first year that we did that, I was hauling the gangly aromatic plants to the compost pile when it occurred to me that those tiny umbel blossoms might be suitable for drying, just like the gone-by anet (which I cut before it bolts, often at 8 to 10 inches high—20.3 to 25.4 cm—and usually get a second cutting from the regrowth). Indeed it dried very nicely, and a sweet resinous flavour from the aromatic blossoms was even more intense than the anet. I crushed and sifted the dried flowers, and now I rarely make my sunflower mayonnaise without it, among other uses. Steamed fresh peas and carrots just cry out for mustard and dill flowers (oh, and savoury, of course).

I have not yet mentioned another condiment, which belongs in this or any salad that aspires to greatness, especially potato salad: pickled nasturtium seeds, harvested in the immature stage, when they are green and succulent and not very fibrous. The pickled *buds* have been likened to capers, a rather strained comparison I think, since the flavour is much more suggestive of radish. Either of those is, to my taste, a shoddy substitute for nasturtium pickles.

Rotten Foods That Are Supposed to Be That Way

I am certainly not able to tell you anything very authoritative about fermenting foods. I am no Sandor Katz, and my knowledge of the subject is extremely limited. I will briefly share the basic ideas that enable me to enjoy eating out-of-season cabbage family crops out of season.

I always thought sauerkraut making required the hard late Dutch heads, which take lots of fertility, water, and proper weather to grow. I've learned otherwise. While those are no doubt the easiest to shred into a delicate white "floss," especially using a krauting board with multiple blades, I have had fine luck with any cabbage, white, green, or purple, even Early Summer Wakefield cabbage. I am careful how I hold the heads to chop the slaw, rotating this way and that to keep cutting across the grain; there is a knack to it, which you will figure out quickly enough. Also, I'm not quite so picky about the fineness of the shred, as long as it is consistent for proper fermentation and air exclusion. To every 4 pounds (1.8 kg) of shredded cabbage, I add 2 tablespoons of sea salt, mix it, and let it sit in the bowl while I shred the next batch. By the time I have another 4 pounds ready, the first batch is wilted and sweltering in its own juices. I dump that into an appropriately large crock (I have 15-, 20-, and 25-gallon models; 56.8, 75.7, and 94.6 l) whose bottom I've carpeted with scraps of the outer wrapper leaves, split so they lie flat.

I pack in that first layer very tightly with a wooden mallet, adding the second and successive layers as they are ready and tightly packing each in turn, to exclude as much air as possible. Within a few inches of the top I cap it with several more layers of wrapper leaves (some use a slit plastic bag), then place a large enamel plate on top, one only slightly smaller than the inside of the crock. Since I often use an enamel-coated steel plate, I'm very careful to check for any chips or cracks, as exposed metal would react very badly. China is better. Onto the plate I heap several fist-sized jasper rocks, with smooth hard surfaces that can be scrubbed thoroughly between uses, and secure cheesecloth (several layers) over the top to keep out fruit flies, while the crock bubbles away happily in a warm corner for two to three weeks. The smell is not unpleasant and reminds us to check on it every so often.

The salt and the weighted plate are intended to draw out the cabbage juice, replacing the air pockets with the brine. Fall cabbage, after a chilling frost, has lots of natural sugar, which comes out in the brine and ferments to form lactic acid, the same natural preservative that keeps cheese. The sweeter the cabbage, the sourer the brine, the better it keeps. Since we are counting on abundant cabbage juice to exclude any air, the cabbage should not be allowed to sit around too long and dry out or commence to spoil, but should be krauted within a few weeks after harvest. It may keep fresh enough for eating way into the winter, but will make an inferior kraut.

That's the main problem with kraut from early summer cabbage: There is less sugar in the juice, so it will have less preservative just when the weather is warmest. Of course you can pasteurize the kraut in glass jars, but half the point of making kraut is so that it will keep all by itself in the cold root cellar. Come spring, the warming cellar will no longer prevent the kraut from spoiling, so *if* there is any left, *then* it should be canned.

Of course, early cabbage (like Wakefield) can be planted later to mature in October, and then it will make as fine-keeping a kraut as any.

Figure 17.9. The fresher the cabbage, the better the sauerkraut.

Figure 17.10. Shredded and salted cabbage is ready for packing in the crock.

We have also made kraut from true turnips (not rutabagas) and from daikons (Japanese winter radishes); these krauts are interesting garnishes, not belly-filling staples.

One of my favourite sauerkraut dishes, modified from a Transylvanian recipe, is to steam or fry it with lots of onions, garlic, whole green beans, hard winter apples (like Followwaters or Ben Davis), and Sweet Dumpling winter squash (subbing for sweet potatoes). This recipe would traditionally be accompanied by sausages. I sometimes use soy burgers with lots of baloney seasoning (ground coriander, paprika, and garlic powder). Any kind of apple would be okay, except most would turn to mush before the dish is cooked. Large noodles or rye sourdough are both quite at home with this dish.

Another favourite sauerkraut recipe is Runaground Stew, made with canned stewed tomatoes, fresh or dried shell beans, onions, garlic, and baloney seasoning. Great over mashed potatoes. Again, you decide the proportions.

Soy Foods

I can easily grow my own soybeans, even here in Maine, but how to eat them in like amount? Tofu is a highly refined, high-fat, high-protein, low-fibre wimp food that has no discernible flavour unless you add enough other stuff to it. And what about all the okara, the residue from its production? Much like the spent seedcake from oil pressing, it is fed to livestock; hogs especially love it, turning it into pork, yum. Heads up, fellow vegans! Someone else's carnivorism is subsidizing your "compassionate diet." Even if we compost all that rich stuff, the wastefulness is overwhelming.

A friend who used to run a commercial tofu business told me how he used the okara (pulp) to make a very tasty canned mock meat, which he also sold. Basically he added some flour, salt, and spices; packed it very tightly (his emphasis) into tin cans closed at both ends; and boiled it in a steam bath for a considerable time, sort of like brown bread. I tried some and agreed that his salvage product was far superior to the tofu. Finding a creative use for his by-product made Peter's business more holistic, but the two food items were not. Eaten by different people at different times, neither was a whole food.

Soy milk beverage has most of the same issues; it is after all the starting point for tofu. There is a patented commercial process for soy milk that supposedly uses the whole bean. I don't know whether it is amenable to home-scale methods.

Aside from the wastefulness homemade soy milk (as I make it) has a chicken-y flavour that's not so bad of itself (after all, most people like to eat chickens; most of us who don't are not particularly faulting the taste), but it is a bit off-putting in a beverage. Adding sugar and vanilla doesn't make it taste like eggnog; it just makes it taste like chicken soup with sugar and vanilla. If we quit trying to make soy into some dairy analogue that it isn't, we might find that homemade soy milk is quite acceptable for some uses. As the "milk" in chowder, for example, that chicken-y flavour is a fine complement to the flavours of corn, leeks, and sea veggies while it is also a perfect nutritional match for the potatoes. Mind you, I'm talking about fine-ground soybeans boiled in the broth, not some extract; everything is there.

Of course, burgers or patties are the ultimate meat analogue, whether made from pinto beans, chickpeas, soy, or any other legume. The fact that most of these need some sort of grain flour to bind them together is itself a plus: The complementary proteins of pulses and grains make them as complete as meat, without the cholesterol or other undesirable features. But whereas ground-up muscle meat tends to hold itself together, a mashed legume tends to fall apart, and the flour that gives cohesion often makes it pasty. I mentioned earlier how we soak and wet-grind wheat berries and develop the gluten without rejecting the rest. That works pretty well, but if you then wish to incorporate mashed legumes into the "dough," it's not so easy. Kneading and regrinding works, but it is tedious and rather messy.

Here is what *not* to do: For years our method was to boil soybean meal and wheat meal together with lots of herbs and spices until it was very thick, then add a lot of raw oilseed meal (sunflower or pepita) to make it even thicker and drier. Shaped into burgers and dry-fried on a cast-iron pan, they always came out like what they were: delicious but essentially . . . fried mush. The problem here is not wholly with boiling the grain; it is about boiling the soy meal.

More recently we soak and boil the split soybeans and put them through a meat grinder with burr plates. This produces a finer and drier burger mix, regardless of what we do with the grain. Indeed sometimes instead of using flour I grind wheat berries that have been soaked overnight (see the section in chapter 16 on "meat loaf") and then mix them together. The soy depends more on itself and less on the grain for its coherence and thus doesn't need to be so doughy. It is a big improvement.

I obsess a bit over burger technology because I think it is important. They are the vegan's burgers and franks and cold cuts, the convenient protein-rich entrée for cooks on the go. You can prepare the mix way ahead of time and use only what you need at the moment. Nor is my only thought for the vegan. However much the carnivore might prefer a grilled steak (not sure I ever had one), when the job market dries up and the kids are whining, it's handy to know some main-course dishes that can come completely from your own garden and your hands with so little expense.

My all-time favourite way of eating soybeans is as tempeh. Now, *that* is a whole food, fit for real men and real women who don't fancy anyone taking anything away from their beans. That's not to say that it has no culture; in fact it owes its existence to a rice mould culture, *Rhizopus oligosporus*. It is quite easy to make, once you have the starter spores. As I understand it, in Indonesia where it originated, you only need to press the cooked beans with certain plants to expose them to the naturally occurring inoculum. Here in cooler climates, you need to purchase the initial starter culture (see appendix B), but henceforth it is not rocket science to propagate and maintain your own, as we have done. You also need to maintain the right conditions (very warm and humid) for over 26 to 30 hours. Most home tempeh making is done in improvised incubators using things like Styrofoam coolers and lightbulbs. We made do with open-air drying racks hung over the cookstove; it took some attention to maintain a rather steady temperature, but it worked fine all but once. With a bit of care you can avoid the once.

We split soybeans in the Corona mill and winnow away the hulls and any fine bits. Then we wash them well, floating off any remaining hulls. We boil them long and slow until they are tender all through, then drain off any liquid and pat them down thoroughly dry on a bath towel. Surface wetness will tend to support urea-forming bacteria, which interfere with the mould growth. When cooled to room temperature, we sprinkle the recommended amount of inoculum over the split beans in a large bowl; in our experience, the quantity of starter is less important than its uniform distribution, so we mix very thoroughly. Then we distribute the beans into several deep enamel pie plates. In hindsight I should have made some square 2 × 6 × 6-inch (5.1 × 15.2 × 15.2 cm) wooden trays, which would have allowed denser packing. Most tempeh makers put the trays inside plastic bags with holes pierced in the top for air exchange. We cover the pie plates with pierced aluminum foil, which never touches the beans.

If you can keep these trays at a fairly constant 85°F (29.4°C) without wide fluctuations, the tempeh will be finished within about 26 hours. Some hours earlier you will see the white mycelia spreading over and through the mass of soybeans, binding them in a mat of soft white strands. When the mat completely covers them, you can call it done. One way to be absolutely sure is to watch for grey spots that will begin to form. These indicate that the mould is beginning to sporulate, or "go to seed." Even if it becomes grey all over, the tempeh is no less edible; in fact some folks prefer it that way. The thing is, if you plan to keep this for a while, you should immediately cut the tempeh into slabs and steam it briefly to stop the mould before you put it in the freezer, tightly wrapped. It will store like that for many months in perfect condition. Lacking a freezer you must eat it very quickly, like meat. Of course, if you live where I do, you can make it in the winter, and all outdoors is your freezer. Watch for animals, though; they know a good thing.

Propagating our own culture for later use turned out to be very easy. We just left one section of the tempeh on the rack to continue sporulating. When it became dark grey all over we sliced it thinly and spread it on those same racks to dry in the open air—no more bags or foil covering. It continued to sporulate as it dried, and finally we had several pieces of brick-hard tempeh. We ground it finely in the Corona and stored the powder in a glass jar in a cool, dry place until we wanted it, several months later. With some nervousness we sprinkled that powder on a new batch of cooked split beans. It seemed to take slightly longer to get established—perhaps a few hours—but it did "take" and made as fine a product as the commercial culture.

Of the many batches we made only one spoiled and smelled like old urine—a sorry waste. I think we never discovered what went wrong: Did we not pat the beans dry enough or regulate the temperature carefully, or had our own homegrown inoculum finally gotten too old? All are possibilities, given the funkiness of our setup, but I found the process forgiving enough that I feel sure with a modicum of diligence, you will never have anything but success.

Any discussion of tempeh making is apt to lead to the subject of vitamin B_{12}, or cyanocobalamin, a vital nutrient supposedly found only in foods of animal origin. Actually all the cyanocobalamin on the planet is the result of bacterial synthesis. It has been rumoured that minuscule amounts occur on the surfaces of at least some leaves (from bacterial action on plant exudates), or that some people have the bacteria in their mouths and digestive tracts. None of those reports is unquestioned, and they may be without base.

What is known for certain is that B_{12} is involved in the digestion and metabolism of proteins. Therefore most vegans, while lacking significant amounts of B_{12} in their diet, have demonstrated a greatly reduced *need* for the vitamin. A doctor friend who is himself a vegan points out that what is a normal healthy B_{12} blood level in a vegan would cause alarm in a meat-eating patient, but also the other way around: A carnivore's high B_{12} numbers would cause concern if found in a vegan's blood, as abnormally *high* levels are associated with certain diseases, including some forms of cancer.

I always test very low for B_{12} in the blood, yet have never had any symptoms of pernicious anemia. I not only haven't eaten any animal products in decades, but I rarely eat any commercial baked goods, some of which have been enriched with synthesized B_{12}. It's more remarkable that my B_{12} level isn't even lower. Perhaps it actually helps that I am not a very hygienic fellow: I nibble fresh stuff out of the garden without bothering to wash it, I gnaw carrots while barely knocking the dirt off, I don't use mouthwash or alcohol, I don't take antibiotics, generally my protein level is modest and easily digestible. Does any of that help? I don't know, but most people consider me singularly healthy and I feel like it. For the last couple of years I've taken a supplement (about thrice yearly) just to keep everyone happy, but I don't feel a need for it and seem no better or worse than before. None of this is meant as suggestion for anyone else, but it brings us back to tempeh. It is claimed that tempeh made with authentic Indonesian cultures may contain significant amounts of B_{12}, due to the natural "contamination" of *R. oligosporus* by certain bacteria.

Sprouts for Fresh Use

In chapter 16 I described sprouting grains for cooking and baking, but of course fresh sprouts are a great way to enrich a salad or lighten a sandwich (which really should be called a shrewsbury, since it wasn't the Earl of Sandwich who invented it, nor did he discover the Islands; he was just a better poker player). Asians thought up bean sprouts long before we hippies thought up anything. Mung beans are not a reliable proposition where I am—possible but not reliable. Some early varieties of aduki beans are fine, but what I prefer are pea sprouts. Any field pea will work, provided they are not split, but since the part we want most is the sprout and not the seed itself, my favourites are some of the tiny-seeded varieties that are usually grown for a forage crop. They're not on the general market, at least in the United States, but I am trying to identify a number of varieties that are superior for that particular purpose and introduce them into the seed trade. Meanwhile any field pea will do.

Clover and alfalfa sprouts are extremely popular and for good reason, though I haven't found the seed to be easily produced on a homestead scale. Rapeseed, on the other hand, is very easy to grow, process, and sprout, and it makes a lovely substitute for cabbage in late-winter salads. For that matter any brassica will serve nicely, especially if you end up with 10 times as much, say, kale seed as you can possibly use for planting (not hard to do). You could even include radish seed sprouts if you really must. Sunflower sprouts are also very tasty and delightfully crunchy, once you get past the de-hulling problem.

Sprouting seeds is easy enough; even I can tell you how to do it. Cut nylon screening into 6-inch (15.2 cm) squares (or circles, if you want to be fussy), each with a heavy-duty rubber band. Put those over the tops of 1- or 2-quart (0.9–1.9 l) glass jars. Size depends on how big a batch you're making, but err well on the side of roominess. For sunflower sprouts use a quantity of seeds equal to a third to half the volume of sprouts you mean to end up with, as they will not expand all that much compared with pea or bean sprouts. Likewise rape, although they will easily quadruple in size. Some of that depends on just how far you wish to sprout them. Cover the dry seed with a lot more cool, clean water than they can possibly absorb; let them soak in it overnight, and in the morning pour it all off through the screening, rinse them in clear, cold water, drain that thoroughly, and place the jars at an angle in some supporting frame (we just use a dish strainer).

Rinse the sprouts at least once, better twice, a day until they are done, which is however big you like them but before they get too strong-tasting. They should never be allowed to dry out too much, nor should they ever be allowed to sit in the rinse water, hence the screen and the angle position. For best quality you should keep them covered against light until the last few hours or day. That way you have mild and tender sprouts, yet the light will form chlorophyll and boost the nutrient level. Sunflower sprouts react to light by becoming somewhat bitter and resinous-tasting, so maybe best forgo the light altogether for those.

Some of the smaller seeds like alfalfa and rape will have a lot of tiny hulls mixed in, which some may disprefer, though they're not very unpleasant. Most of them can be easily removed by floating in a surplus of cool water and skimming off the hulls that come to the top. You cannot really preserve sprouts once they're ready, as they just keep growing, but you can bring them to a pretty complete stop by refrigerating them for several days, until they're used up. There are very many other seeds, including grains, that can be sprouted for fresh eating. These are just the ones I've had the most experience with.

CONCLUSION

Throughout this book I have been portraying the marketplace as a problem, as indeed I believe it is; however there is no denying that the marketplace also holds solutions to many problems. My ideal world is *not* a place where everyone is restricted to using only foods, tools, ideas, or other resources that are homegrown, homemade, or of local origin. What a dismal, restrictive, backward place that would be. The marketplace of products—goods and services—is inextricably linked to the marketplace of ideas; it is no coincidence that the agora and the forum are near to and part of each other.

"Free market" has always been an oxymoron, never more than in our present times. I am more interested in the freedom of the market*goers*, and I see clearly that an unfettered, unregulated marketplace makes slaves of all the players, *even those* who appear to profit greatly from it. And now the marketplace appears ready to not only enslave the world but destroy it. When we neglect to confront global warming because we believe the economy cannot stand it, we are allowing the marketplace to ring the death knell of civilization.

Though we hear much about the tyranny of government and the desirability of the individual being free from it, we seem nearly oblivious to the even more pervasive tyranny of the marketplace. Even the most detached philosopher has to go there. My friend Denis Culley was once told by a Zen master: "It seems we all have to buy and sell." And of course being a homemaker or a homesteader does not count as a "job"; that's just something you do—some people collect stamps or play boccie.

Yet many of us who try to impact the world by influencing government neglect to wield an even greater power than the ballot. Once a year I get to cast my vote for our collective political life, yet every day at the checkout counter I get to express my opinions in a forum where I am clearly heard. All those product surveys you're pestered with tell you that someone cares very much what you think. Whenever you manage to avoid the marketplace altogether, they get your message: It is not between product A and product B; maybe they are in the wrong line altogether. How sad that we don't exercise that vote more often. How naive to meekly accept big business while opposing big government, the only defence we have against it. I might be much more interested in such libertarian limits on government if there were similar restrictions preventing the corporations (legally citizens, mind you) from gobbling me up.

You must realize that Congress and the corporate board are sitting right in the middle of your garden, and you cannot throw them out; in a garden-without-borders there is no "out."

As you may have noticed I'm not a great believer in panaceas, or all-or-nothing victories. I am all for the half measures, incremental changes, compromises. Neither the marketplace nor complete self-sufficiency can solve all our problems. In fact we have a spectrum of problems that call for a spectrum of solutions. The marketplace has become such an all-pervasive tyrant that I am well disposed to rebel against it, overthrow its complete power, but *not* destroy it (as if that were possible). The marketplace is older than civilization: Hunter-gatherers would meet in a forest clearing to exchange blocks of obsidian, salt, amber beads, ochre, but never meat or grain or things they could provide for themselves. Trade items were largely limited to mineral products that were irregularly distributed on the earth and thus disproportionately valuable. Eventually two minerals, gold and silver, came to epitomize all trade, though their usefulness for other purposes was very minor. Paradoxically self-reliance for staple commodities made the marketplace a *more* useful and equitable place: The less you are dependent on buying and selling, the more you can profit from both.

Energy and Alternative Technology

Today virtually everyone is completely dependent upon the marketplace; hardly anyone is self-reliant for anything. When I was a lad my mother cut our hair and sewed many of our clothes, and that was considered pretty radical. Now even home cooking, especially from scratch, has become a specialty for gourmets and foodies. Clearly we would have a very long way to go to regain the merest vestige of personal and household autonomy. We can't do everything—I would know that—and so we have to set priorities: What is doable? What is worth the effort? Compared with what else? Those are decisions you must make for yourself. In this book I have tried to share some ideas that may help you assume more control and responsibility over your life, but far more than that I hope I've helped you see that your garden is indeed without borders.

Near the commencement of this book I described what I call my Esperian premise—a set of assumptions that underlie my methods and reasons. Applied to agriculture alone they are organic (far beyond OMRI standards) and profoundly sustainable. And what about the machines and their fuel? We must return to the question: How do they do that in Esperia? Mowing can of course be done with a scythe, which has almost everything going for it—except that it's relatively slow, especially if you have a large area to mow and a lot of other things to do. But that begs the question: Compared with what? Again, if we rule out gas power, as ultimately we must, then how else are we to do it? My friend Peter Vido is committed to hand-scything and teaching others how (when the *Titanic*'s stern lifts up is *not* the time to start learning). Trying to mow with an improper style of scythe, poorly adjusted and sharpened and with bad form is a sure path to exhaustion; doing it right can be fairly effortless and pleasurable (I'm somewhere between

the two poles myself). I am not a Luddite or traditionalist or romantic; I'll embrace any newfangled technology, as long as it is sustainable, but so far that has not shown up, to my knowledge, so the scythe looks pretty good. I use my Gravely-powered walking rotary mower for most of my fields, especially areas where woody brush has grown up and needs to be not only cut but shredded as well. For green manures and the more improved hay land, I use the scythe whenever I can.

My rototiller sees minimal use, especially since my close spacing renders it useless for cultivation. Green manures and grain stubble can be chopped in by hand (grape hoe), which does a better job, but takes longer. I use both according to circumstances.

Chipping brush and shredding leaves are two very important operations on my farm for which I can hardly imagine an alternative. However, that is *not* to say that such machines must be fossil-fuel-powered. While some equipment like mowers and chain saws must be portable, always moving to where the work is, stationary operations like chipping/shredding, pumping water, and cutting wood allow more alternatives, if only we can deliver some kind of power to a fixed power-takeoff shaft. That power might come from a farm-scale windmill or waterwheel or wood-fired steam engine, and I do *not* assume it must first be turned into electric current, then back again. In Esperia I imagine (since I've never been there) that pulley-driven cables transmit the force from the source to the work site. I suppose the main limit to distance would be the number of pulleys required to hold the cable off the ground, especially if they're not in a straight line, which would add friction at each bearing. Various gears, clutches, and governors might give control over the amount, speed, and frequency of such delivery.

Steam engines have real potential if fired by wood blocks or farm-produced charcoal. Apparently they are not limited to stationary applications; I read an account of forestry in communist China that described wood-fired steam tractors being used in logging operations. I always wondered how a wood or charcoal fire stands the constant shaking of a tractor bouncing over rough ground and how sparks are confined. The Stanley Steamer auto and the Lombard Tractor (both invented in Maine) are not relevant examples, as they burned kerosene, which I assume is easier to control.

I've long dreamed and designed some such appliances (even having the audacity to describe such untried technology in detail in my novel), but never yet had the spare time to build anything, and so I've been bogged down with the primitive technology of the 21st century. Abandoning the internal combustion engine (and perhaps even the electric dynamo) need not return us to the Neolithic age; rather we need a new vision of what we mean by *high-tech*. We should not embrace drudgery; we must move forward, not backward, but to know which direction is truly forward, we must constantly consult our compass of sustainability.

Obligations of a Garden-Without-Borders

I have never paid any income taxes in my life, or at least very little, and the IRS has always insisted on giving that all back to me. In their eyes I am a pauper, though if they were to come visit my house and share my table, they might have second thoughts. I do nothing to deceive them; they just don't want my money, which to them looks paltry. Indeed it is, but

Figure C.1. It's always fun to check out the nearby brook, but let's face it: The real gold is over in the gardens.
Photograph courtesy of Arica Bready

if they had a line on the 1040 form asking the "value of goods produced and consumed on the premises," it would be a *very* different matter; I would have to pay through the nose, but in what? It would be grossly unjust to require payment in specie I never possessed. (Actually, *nobody* pays taxes in specie anymore, you're not allowed to, despite the charming myth that it is acceptable "for all debts public and private"; plastic only, please.) So how about 15 percent of my carrots, 15 percent of my wheat, 15 percent of my pumpkinseeds, 15 percent of my firewood, et cetera ("Hey, it's in the mail")? Even in Esperia that creates endless complications. Among the obvious problems is the question of value. As long as you're eating those carrots the government doesn't need to care two farts what *you* think they're worth, but if you are offering them as part of your contribution to the welfare and defence of our nation . . . no, no, that just won't work, you understand. We need something extremely standardized, something the marketplace recognizes and appreciates, something so highly symbolic that it can be transferred instantly and stored in astronomical amounts in data storage devices, not in Fort Knox. Zucchini just doesn't fill that need.

If you *wish* to shirk your civic obligations and dodge responsibility for the society around you, well, self-sufficiency is a great way to do that. Not buying and selling, bartering when necessary, keeping everything under the table, below the radar: Yep, that works very well and is to my knowledge perfectly legal; nobody knows or cares that you exist.

But what if you *want* to contribute your full share to our collective life? What if you are eager to support those things we all value: defence, education, science, arts, health, and welfare? What if you want to *maximize*, not minimize, your contribution to the common good? It can be done, but you must be more creative about it than just sending the IRS a cheque. You can volunteer time or talent to things that benefit others. Of course extreme self-reliance tends to leave little free time, but like everyone else everywhere, we must make swap-offs and compromises and just do the best we can. In fact I am aware of a whole underground economy that exists among the very poor. It consists of things like shared child care, garden veggies, car repair favours, rides to town, poached game, and various other currencies that the government never created and never knew existed.

How much sharing is enough? With money the government tells you how much, but when it is some other currency (food, service, whatever), you must decide. However maybe there's a better way of looking at it than how much is enough. Perhaps instead of restricting our contributions to a certain percentage, we should rather vie with one another in supporting the vision we share for a sustainable society. Rather than resent those who contribute less than their share, maybe we should pity them because their gardens have such confining borders.

Tools

I don't profess to be an expert on tools; in general I have tried what comes my way, typically from auctions and barn sales, often by the boxful at less than a dollar each. I get a mixture of treasures and junk that way; I mainly discover the difference by trying them. Because of my acquisitions approach many of my favourite tools are not easily found today (unless in museums). I can only describe or show them to you and let you seek them online or wherever. When I know of a present-day source I'll mention it, which doesn't necessarily mean it's the only or best source.

Garden Tools

Earthway Precision Seeder

This small and affordable push-tool slices a furrow, drops the seed, and covers it in one pass. It comes with six feed plates for different sizes of seed. Although there is no plate specifically designed for grain, I have good results with the "beet/chard" plate. Moreover, the company also offers a blank plate from which you can fabricate your own, though I have not tried that. For vegetable seeds this tool is

Figure A1.1. Earthway seeder.

probably best suited for market gardeners—I find it easier and more precise to hand-sow most vegetable seeds. However the Earthway is just the thing for garden-scale sowing of grains.

Broadhoe

This tool, which is also called an Italian grape hoe, is sometimes confused with a grub hoe or mattock, but it is not the same thing. The grape hoe has a relatively light forged-steel blade with an eye for the handle attachment. The blade is slightly curved, as the handle is recurved, for an effective swing. I think of this as the poor man's plough, although it is superior to the mouldboard in so many ways, except for speed. It doesn't open the soil deeply (that's a task for the broadfork), but it easily breaks up the top 6 inches (15.2 cm) of soil, if reasonably stone-free. Larger stones can usually be chopped loose and scooped out; both the blade and the handle are very stoutly made. If it's kept sharp and used at shifting angles, a grape hoe can break up old sod that a rototiller would bounce over, and it chops in lush green manures that would hopelessly tangle in the tines of a tiller. It doesn't obliterate the crumb structure or annihilate the soil life. Instead it reduces and incorporates the residue and lots of air, so that bacterial action can hasten the decay and release nutrients. Ploughing and rototilling are more efficient than using a grape hoe only if you equate efficiency with speed. Available (with handle) from Lehman's, among others.

Wheel Hoe

An incredibly effective and efficient tool for cultivating spaces wider than 14 inches (35.6 cm). Mine was a relic, a wedding gift from an antique tool aficionado. I ignored it for a year or two until I realized what I was missing. Especially for young weeds (under 3 inches, or 7.6 cm) you can get thorough control in a very short time and with comparatively little brawn. If a single pass doesn't exterminate them a second pass a few sunny hours later will finish off the stragglers and take half as long as the first. Mine came with a set of five tines on a toolbar, which I can flip over to present a horizontal stirrup knife for shearing

Figure A1.2. Broadhoe.

off taproots. Eventually I came to prefer the latter; indeed I have lost most of the tines and never bothered to replace them, whereas the originally stirrup blade wore out from constant and hard use, so I cold-forged another one from slightly heavier bar stock. Unlike a rototiller, the wheel hoe does not "osterize" the soil, but retains much of the crumb structure and does little damage to the useful soil biota. I've seen several modern designs on the market, including pneumatic rubber tires and other attachments, but I'll not swap.

Garden Hoes

These simple tools are incredibly underutilized, being replaced by space-hogging, exhaust-belching

Figure A1.3. Wheel hoe.

Figure A1.4. Warren hoe.

rototillers. Those who do use them often use them improperly. For one thing I constantly see folks take a little chop, then back up for the next chop, so that they are burying the very ground they want to work next. You should first chop a little divot, then reach forward a bit so the next chop fills up that divot while digging a new one, and so forth. Yes, you are treading on the new-tilled ground, but at least the ground is broken and the weeds uprooted; you can stand to the side and hold the hoe offset to avoid any compaction.

For making furrows and for cultivating between close-spaced crops (less than 8 inches, or 20.3 cm), I prefer a triangular warren hoe. Aside from making seeding furrows also use its two sides for shallow cultivation, controlling the width of chop by twisting it slightly and using whichever side gives the right angle to the wrist. For a slightly wider cultivation (8 to 14 inches, or 20.3–35.6 cm) I use a type of hoe I never see offered for sale in stores, but apparently it was once widely used, because I find them in used-tool shops and at barn sales. It closely resembles what is today sold as an onion hoe or beet hoe. It is perfect for shallow cultivation where a wheel hoe will not fit, yet it is also deep-bladed enough for hilling up corn and potatoes. One of mine is so very shallow (that is, as opposed to deep, whether from wear or design) that the soil tends to drop off the back of it—much like a collinear hoe, but at a different angle. It's great for tiny young weeds but little use for hilling.

Broadfork

For many years I've used a broadfork for loosening the subsoil, especially before taproot crops and brassicas. This tool isn't meant to turn the soil, but rather to open up the structure of deeply compacted soil without inverting the natural layers. I don't remember where I got my original, which I love, but Johnny's Seeds offers five models, of which I trialed four. Each has its own advantages.

#727. This extra-wide (27 inches, or 68.6 cm) model does a large area quickly, assuming that larger stones have been removed in the past. No broadfork is suitable for removing large stones; that's what spades and pry bars are for. By the way I'm talking about a heavy-duty pointed steel bar such as you rarely find anymore. I've seen some fairly good bars online; mine came with my land and my neighbour Lucian—it was made from the driveshaft of a Model T truck, about 9 feet (2.7 m) long. Essential on my kind of land, unless you want to constantly hire an excavator.

#415. This narrow (15 inches, or 38.1 cm) model is great for tighter areas, including between growing crops, in which case it should not be levered

Figure A1.5. Broadforks.

too much lest you damage adjacent roots. My wife has found this one useful for working in confined landscape beds.

#920. The tines of this model (nine tines in 20 inches, or 50.8 cm) are actually too tight-spaced for subsoiling, but its forte is lifting root crops like carrots. For that use alone it's not worth it for me, but it could be very valuable for market gardeners.

#520. This model is especially meant for deep, hard subsoil. Its five triangular tines (totaling 20 inches, or 50.8 cm, wide) are more resistant to bending. With wide gaps between tines this fork won't quibble over every small stone.

Johnny's generously allowed me to trial all their models, but if I had to choose a single all-purpose one, it would be the #727, since most of my digging is in soil that has been deeply worked and de-stoned over the years, yet I need to cover a lot of ground quickly and efficiently. All models are rugged, quality-made

Figure A1.6. Austrian scythe.

tools. I had first thought that the handles were a bit undersized, but now I think not; no need to heft that extra wood thousands of times.

Scythes

Properly handled, a scythe is a reasonably efficient way of mowing tall grass and weeds for compost and mulch. The traditional British-American scythe is quite inferior to the European or Austrian scythe. It is heavy, clumsy, and easily broken. There are several sources of European scythes in the United States: Johnny's Seeds in Albion, Maine, and Scythe Supply in Perry, Maine, are the ones I know best personally.

Food Processing Tools

Davebilt Nutcracker

This is a good-value machine for hand-cracking a variety of nuts, although I use it mainly for my American hazelnuts. You adjust the spacing merely by adding or removing one or more of the washers stacked on the shaft. Since hazels are often variable in size, it is most effective if you first grade the crop through a ½-inch (1.3 cm) mesh screen and feed them separately at the appropriate setting. The machine is constantly binding (which makes the demo video look ridiculous)—which is no trouble at all, since the cracker works in either direction. You can simply crank the handle back and forth, reversing direction whenever convenient. Although the machine does a very efficient job of cracking the shells open, they are still together and you must still pick out the kernels, which is a huge improvement and not impractical for home use.

Piteba Oil Press

In my opinion this device is a huge breakthrough in homestead self-reliance. It will express oil from any seed containing over 25 percent oil (not olives). A diminutive device but quality-made in Holland by the inventor, who is very helpful at fielding questions (in English or Dutch, anyway). It comes in a padded mailer, is easily assembled, and clamps onto a sturdy bench or shelf that will stay put when you crank on it. For best yields it requires a tiny burner to preheat the press chamber; a tiny lamp is included, although I used paraffin oil (a highly refined mineral oil) instead of recommended kerosene (a small candle would probably work as well or better, as very little heat is required). I should point out that there is a cheap knockoff of this device made in China, in violation of patent law and made of alloys containing lead. The Piteba is red.

Squeezo Strainer

An indispensable kitchen gadget for processing large quantities of applesauce, tomato, or other purees (for less than a gallon it's hardly worth the cleanup time compared with a regular crank strainer). A separate screen can be purchased for coarser foods like pumpkin, though I rarely use mine. It's wasteful to discard the pulp after going through just once; a second and third pass will get much more yield and of a richer, thicker quality. Made mostly of aluminum, so don't let hot food sit in it long.

Saladmaster

This heavy-duty stainless-steel tool saves me much time and effort but also enables me to create lots of food items that would otherwise be impractical, such as dried apple chips for a raisin substitute or dried terrasol chips as a first step in making meal. By the way Saladmaster also makes a pasta maker, which I've never tried—but it looks nearly identical to mine.

Corona Grain Mill

I describe this mill and how I use it in chapter 16. There are several suppliers, but my experience with other makes has not been as favourable.

How to Make Your Own Flail

Although it was not my intention in this book to give instructions for woodworking, I think that I cannot in good conscience neglect to tell you how to make a flail. Of all the tools essential to self-sufficient living, this is one that, to my knowledge, cannot be purchased. Although you can improvise a handle to

some extent, the dasher or swingle really needs to be fabricated by yourself or someone who is good at such things, and therefore it seems that at least one of the paths leading away from the marketplace needs to run through your workshop. Even if you do not have a woodworking shop (although any garden-without-borders should include one), it is not difficult to make do. The three or four tools necessary for making a flail are basic hand tools that, once gotten, will enable you to make any number of other essential items, from handles to wheelbarrows to buckets and much more. You don't need a lathe or any power tools.

Here is what you do need:

- A bench vise. If you haven't one perhaps you can borrow the use of one from someone.
- A drawknife. I have several that I've picked up at auctions and secondhand shops, but they are available from woodworking suppliers and are unlikely to bust your budget.
- A scraper for smoothing rough-worked surfaces. I've never bought one; an old man once showed me how to use broken shards of window glass to get a superior finish. (Note! If you have to break a perfectly good window pane, be sure to cover it with a cloth when you strike it, to avoid flying chips getting into your eye or elsewhere.)
- A coarse rasp is also very handy to have, especially if your wood has any knots or off-grain that resists smoothing with the glass.
- A handsaw or table saw to cut the original square.

The best woods to use are white ash or oak, as these are strong and dense and easy to rive, or split, due to their straight, coarse grain. Any other hardwood could be used but may be difficult to rive. Softwoods or conifers are usually too weak and not dense enough to have the proper impact.

First of all you can avoid all of this by simply finding a straight knot-free section of limb that is already the right thickness (this isn't critical; in fact it's ideal to have two or more sizes). Finding such a piece you may be able to just peel off the bark, cure it for a few weeks, and use it like that. Failing that, you can fabricate your own from any size piece.

The first step, after sawing the piece to length, is to either saw it (if you have access to a table saw) or rive it. Although a froe is the ideal riving tool, you can probably manage with an axe. The problem is that unless you are very accurate with an axe, you will probably miss splitting it right where you need to. Too small means the piece is ruined; too large means you still have a lot of wood to remove with the drawknife, just to get it to the basic square. You can be much more precise by placing the axe blade right where you want it on the wood, and then striking it with another tool such as a large ball-peen hammer. Now, axes are not meant to be struck by other steel tools, and it is not good for them. It swedges the poll and tends to spread the eye. Lacking a froe, however, you may have no choice. Strike seldom and carefully; you needn't use great force to just start a split and then gradually force it open. Wear safety glasses; steel sparks may fly. This is all assuming you do not have a table saw, but even if you do, riving may be the easiest way to produce a rough slab, slightly oversized, from which you can re-saw the final square. Lacking a table saw you could hand saw the final square, but ripping (cutting along the grain) hardwood is rather slow and tedious; the properly wielded drawknife will save much time and aching. Clamping one end of the rough square in the bench vise, I use the drawknife to smooth one side. Assuming the piece is straight-grained the grain itself will guide you; sighting along it by eye will help. Once that side is smooth and straight, you want to make the opposite side parallel with it. Again the grain will help guide you. Great precision is not required, but you can use a scribe to make the parallel cutting lines, or simply use a pencil and yardstick. Once you have flattened that second side, then flip it 90 degrees in the vise and do the same with the third and fourth sides. You now have the basic square; with a table saw,

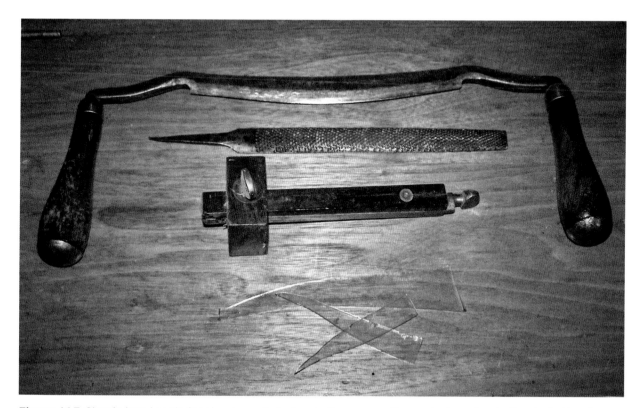

Figure A1.7. Simple hand tools for shaping a flail: drawknife, rasp, scribe, and glass shards (top to bottom).

you could have that part all done by now and start in fresh on the dowel.

After all you don't want a square swingle; you want a dowel, so you've just begun. Without using a lathe (I've never had one), the simplest way to make a cylinder from a square is to first make an octagon. You can compute a formula for determining the width of each flat, but for this purpose your eye is probably adequate. Holding the drawknife at 45 degrees, I shave off the two adjacent corners at the same time, taking care to keep them at the same angle and width. The more you shave, the wider the surfaces become, until their widths are equal to each other and to the remaining surface between them (which has been diminishing as they grow). I rotate the square and do the same on the opposite side, so that I now have an octagon, with pretty much equal sides. If you have misjudged and gouged here or there, there is considerable room for fudging—consider this your practice for more precise work. Now I use the drawknife to further "break" those eight edges to make a 16-sided piece. However, I don't pay too much attention to this; on a piece of this size I can usually work directly toward the cylinder at this point. The rasp may be very useful now, especially if I work it almost lengthwise along the grain, leaving "cat claws" in the wood as I work it toward round. You can do this all with the drawknife if it seems to be going well, though I find the rasp saves time. Soon you have a nice hardwood dowel, finished except for those cat claws or any irregularities left by the drawknife. The final smoothing is best done with glass shards, the greatest technological innovation since knapped flint. Obviously you'll want to hold on with some caution, not pushing too hard against any sharp edge. If a piece snaps it's no big deal; you may have just created a fresh cutting edge, or you can simply discard it and grab a new piece. Broken windowpanes are not precious. Just be careful not to push too forcefully and possibly cut yourself. What is actually a greater risk than cutting yourself (which I believe I never have done) is little chips that may flake off under pressure; be sure to wear eye protection.

Scraping with glass shards will produce a smoother surface than anything, as long as you promptly discard any nicked edges; using sandpaper, however fine, will only scratch it up.

Now that you have your cylinder you need to drill a ⅜-inch (1.0 cm) hole through one end. Taper and smooth both ends of that hole so the connecting string will not abrade. Likewise I round and smooth both ends of the flail somewhat to minimize abrasion of the string and denting of the floor (if in a shed or barn) or tearing the tarp (if outdoors on the ground).

The same kind of tapered hole should be made at the end of the handle, for which I always use a broken shovel handle (these are far too common around my place). I make the connector string from three tightly braided pieces of hay binder twine. Rawhide can work better until it gets dry and brittle. Wire is unsuitable; dog-run chain might work.

I used to be in the tool-handle-making business, and so I can work one from scratch, but if you have a broken spade handle kicking around, that's a great beginning. You just saw it off even and break the raw edge smooth with a rasp and coarse sandpaper. Then I bore a hole through the end and likewise break the rims of the hole at both ends, so that the connecting string that passes through will not fray too quickly.

Now you have a perfectly serviceable tool to add to all your store-bought hardware.

Resources

Recommended Reading

Farmers of Forty Centuries, or Permanent Agriculture in China, Korea, and Japan; republished by Courier Dover, as well as Rodale Press. In 1911 Frank H. King, former chief of the USDA's Bureau of Soil Management, made an extended tour of China, Korea, and Japan. His purpose was not to teach the people of these countries the methods of western "progressive" scientific agriculture, but rather to observe how those nations had managed to maintain such dense populations on limited land area for four millennia without exhausting the land. Among his many conclusions: The reliance on a largely plant-based diet and soil fertility, the huge investment of effort in constructing long-term infrastructure such as terraces and canals, and the application of sustainable technologies like wind and water power have enabled those societies to support large populations at a reasonable level of prosperity. I have derived a massive amount of information and inspiration from his observations, though I was also able to read between his lines and pick up on something that he himself failed to recognize: The imperial structure he reported was tottering on the brink, and within months of his return home the ancient Manchu dynasty would be swept away in the first of a series of revolutions that would radically transform China, Asia, and the world.

Craftsmen of Necessity, by Christopher G. Williams, 1974, Random House. This is something of a coffee-table volume in that it is packed with interesting photos and thought provoking text, designed to inspire you while not intending to teach you how to do any of the particular things shown there. It

does that very well, yet I have also managed to copy many of the specific ideas, such as a gate counter-weight and gatepost all from a single tree, stone house and barn walls with wooden components (like door frames and hinges) incorporated into them, annual crops grown in the shade of taller tree crops, and using the natural curves and crotches in a tree to make a scythe handle or ship's keel stronger than you could assemble. Reading this makes me appreciate that organic is about much more than agriculture.

Facts for Farmers: also for the Family Circle, by Solon Robinson, 1865, Johnson & Ward, New York. Because this twin-volume work is so old it is intrinsically organic; pest remedies include stuff like soap suds and tobacco tea. It is truly exhaustive, treating heavily on corn, cotton, and wheat (soybeans were a novelty then), but also asparagus, "pie-plant" (rhubarb), cane fruits, tree crops, and more. The author was the agricultural editor for the *New-York Tribune*, and thus had a very wide correspondence with farmers of all kinds from around the nation. He begins with livestock and proceeds to farm buildings, crops, tools, fertilization, drainage, and irrigation. Plenty of details; I have gained a lot of insight from it. Some of Robinson's other writings are less endearing: This Connecticut-born Indiana farmer was an unapologetic proponent of slavery.

Seed to Seed, by Suzanne Ashworth, 2002, Seed Savers Exchange. For anyone serious about saving their own seed and getting it right, it's hard to find a better source. This much-researched book goes into considerable detail about a very large number of vegetable crops.

Sources of Seeds and Supplies

A number of seed companies specialize in organic, or at least GMO-free, seeds, and although most offer some F1 hybrids, they all offer many open-pollinated varieties suitable for seed saving. The following are selected largely because of my positive experience with them, and because they offer many of the varieties mentioned in this book that may be difficult to find elsewhere. Also they are all self-owned companies, not properties of larger corporations. This list doesn't pretend to be extensive or complete.

Fedco Seeds/Fedco Trees/Moose Tubers/Garden Supply. A co-op business focusing on non-GMOs. Lots of hybrids, lots of OPs. They source a lot of their seeds and trees from local/regional organic contract growers. Superb selection of varieties, many hard to find elsewhere. P. O. Box 520, Waterville, Maine 04903; (207) 873-7333; www.fedcoseeds.com

Fruition Seeds. A new small self-owned company in the Finger Lakes Region. A modest but excellent assortment of varieties, all non-GMO, all organic, no hybrids; they produce most (70 percent) of their own seeds. 5920 County Road 33, Canandaigua, New York 14424; (585) 300-0699; www.fruitionseeds.com

Grassroots Seed Network. This is a new organization of seed savers who prefer a more democratic structure where decisions are made by the actual seed savers themselves. Anyone offering any seeds in the GSN Source List is automatically a full member entitled to nominate, vote for, or be a candidate for the Board of Directors, plus participate in the online Forum. All of the varieties currently offered by my Scatterseed Project can be found there, along with those of many other members. Membership $15, plus at least one listed variety. 249 Bailey Road Industry, Maine 04938; (207) 491-4259; grassrootsseednetwork.org

High Mowing Seeds. This family-owned seed company has evolved from very modest beginnings, but still produces a lot (30 percent) of their own seed.

Some hybrids, mostly OPs, all GMO-free, all organic. 813 Brook Road, Wolcott, Vermont 05680; (802) 888-1800; www.highmowingseeds.com

Johnny's Seeds. Founded by Rob Johnston, this company is in the process of becoming worker-owned. Too glossy and too many hybrids, but GMO-free, lots of organic and lots of OPs, including lots of good stuff not easily found elsewhere Also a good source for tools. 955 Benton Avenue, Winslow, Maine 04901; (207) 869-3901; www.johnnyseeds.com

National Germplasm System. The ultimate source of genetically diverse propagation material (seeds, tubers, cuttings, etc.) is the USDA's National Germplasm System, a network of regional facilities conserving and distributing hundreds of thousands of accessions of nearly all cultivated crop species. Although their main server group is plant breeders and researchers, they are generally receptive to requests from private individuals, with the understanding that you will further propagate the small samples they send you. They are not a seed company and may not be used like one; do not expect to return to them repeatedly for the same variety. They also greatly appreciate feedback of any sort (trials notes, etc.), as it helps them to better serve others. The simplest way to access their collections is by going online to GRIN (Germplasm Resources Inventory Network) and type in the particular crop species that interests you. They will walk you through the ordering process, which is ridiculously easy. Prepare to be astonished. www.ars-grin.gov/npgs

Nichol's Garden Nursery. 1190 Old Salem Road NE, Albany, Oregon 97321-4580; (800) 422-3985, www.nicholsgardennursery.com

Redwood City Seed Company. A radical small business specializing in non-GMO, organic non-hybrids, lots of native Mexican varieties sourced from native growers. Focus on gardeners growing food for themselves rather than commercial growers. P. O. Box 361, Redwood City, California 94064; (650) 325-7333; www.ecoseeds.com

Seed Savers' Exchange. Since 1975 this organization has been making thousands of food crop varieties (mostly beans and tomatoes, but many others, emphasis on US heirlooms) available to other seed savers. Although they also have a commercial seed catalog, that should not be confused with the general membership, whose listings are found in the annual Winter Yearbook. Also, don't be confused by the word "Exchange," you do not have to offer seeds yourself in order to request seed from Listed Members (there is a sample fee). $40 membership; 3094 North Winn Road, Decorah, Iowa 52101; (563) 382-5990; www.seedsavers.org

Southern Exposure Seed Exchange. Despite its name, it's a co-op-owned and -run seed company that is the principal livelihood of Acorn Community Farm in central Virginia. Their geographic focus is the mid-Atlantic and Southeast; many unusual southern heirlooms (including clonal onions). No GMOs, nearly all (98 percent) non-hybrid. Much of their seed is grown there or by a small group of co-operating growers. P. O. Box 460, Mineral, Virginia 32117; (540) 894-9480; www.southernexposure.com

Turtle Tree Seeds. A non-profit community-run seed business offering exclusively non-GMO, non-hybrid organic seed grown by a team of Bio-Dynamic cooperators (they are the original source of my favourite bulb onion, Clear Dawn). Camphill Village, Copake, New York 12516; (518) 329-3038; www.turtletreeseed.org.

INDEX

Note: Page numbers in *italics* refer to photographs and figures; page numbers followed by *t* refer to tables.

ABOUT THE AUTHOR

WILL BONSALL has worn many hats before and since going "back to the land" including prospector, draftsman, gravedigger, hobo, musician, language tutor, logger, and artist. However he considers subsistence farming his only true career.

In addition to raising food for himself and his family, he is the director of the Scatterseed Project, which he founded in 1981 to help preserve our endangered crop plant diversity. Bonsall is also a founder of the Grassroots Seed Network, a national member-run organization devoted to encouraging the growing, saving, and sharing of all crop seeds. Over the years, he has offered seed and planting stock for more than 5,000 varieties of crops to many thousands of recipients. Varieties emanating from his collections are scattered among commercial seed catalogs throughout the United States and abroad. He is internationally renowned for his large collection of rare and not-so-rare potato varieties, and he also curates over 50 species of other crop plants. In spite of going to great lengths to avoid being a plant breeder, he has developed a few varieties that are offered in the trade, including Gardener's Sweetheart tomato and Mountaineer sweet pepper. Bonsall's first book, *Through the Eyes of a Stranger* (Xlibris, 2012), is an econovel set in a sustainable society of the future. He is currently working on the sequel, *In the Larger House*. He lives and farms in Industry, Maine, with his wife, Molly Thorkildsen, and two sons, in a house that he built largely from materials on his land.

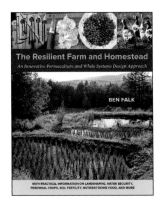
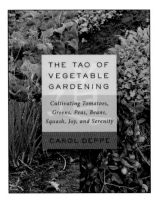
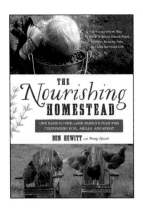
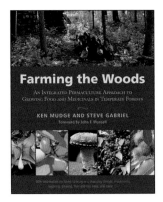